"I have always been impressed by LabVIEW's ease of use and friendly interface. It is exciting to know that LabVIEW is supporting signal processing applications. *LabVIEW Signal Processing* makes your entry into LabVIEW even easier through its simple, illustrative examples and clear descriptions. I highly recommend it for students in both academia and industry at all levels as the ideal introduction to LabVIEW."

—*Prof. Alan C. Bovik,*
Professor of Electrical and Computer Engineering,
University of Texas at Austin

"LabVIEW analysis provides built-in functionality that engineers and scientists have not even considered possible...and *LabVIEW Signal Processing* shows how to do it! The authors communicate a very complicated topic—very effectively. *LabVIEW Signal Processing* is much easier to understand than most books or texts that address these subjects. Value is added for the reader in extensive theory that one can understand as well as in how to do things in LabVIEW."

—*Lisa K. Wells,*
Author of LabVIEW for Everyone

"Between the first and last page of *LabVIEW Signal Processing*, I found a lot of material that improved my knowledge about topics such as signal theory in general, filter design, statistics, and, most importantly, about real-world applications. The examples in the real-world applications are given by LabVIEW users, and they show how the Analysis Library can be used in practical applications."

—*Prof. Jing Zhou, Dean and Professor,*
School of Civil Engineering and Architecture,
Dalian University of Technology, Dalian, P.R. China

"*LabVIEW Signal Processing* is not only a great refresher in signal processing for engineers, but it also serves as an excellent tutorial for nontechnical programmers who want to become more knowledgeable in the fundamentals of signal processing and advanced analysis. It combines well-written text with easy-to-follow examples to clearly explain traditionally difficult concepts while demonstrating how to maximize the power of LabVIEW in developing signal processing applications."

—*Jim Baker,*
Porter Technical Inc., Lexington Park, Maryland

"The theoretical concepts of signal processing are difficult for students and professionals without formal engineering training to understand. Most books on the subject are written in a style which makes it difficult to utilize the concepts in everyday real situations. The authors of *LabVIEW Signal Processing* have presented the material in a simple, straightforward way, thus making it easy to understand and utilize the program VIs associated with LabVIEW/BridgeVIEW.

"The book is well-organized and the background sections provide the reader with the basic concepts and tools necessary for each chapter. The unique aspects of signal processing, starting with the basics and building to the more difficult, are clearly defined in each chapter. Care has been taken to reinforce the reader's knowledge base by providing many program examples. The CD-ROM accompanying the book will be of great help to researchers."

—Robert M. Wise, Research Supervisor,
University of Maryland, School of Medicine

"*LabVIEW Signal Processing* provides a good discussion of digital signal processing with real world applications in LabVIEW that the reader can use as a basis for future experimentation. The authors have carefully chosen a lot of activities which show the reader how to use LabVIEW VIs for solving interesting problems. This book reflects the real power of the graphical programming paradigm for signal processing applications."

—Prof. Norbert Dahmen,
Fachhochschule Niederrhein (University of Applied Sciences),
Krefeld, Germany

LabVIEW Signal Processing

▲ Mahesh L. Chugani
▲ Abhay R. Samant
▲ Michael Cerna

Prentice Hall PTR, Upper Saddle River, NJ 07458
http://www.phptr.com

Library of Congress Cataloging-in-Publication Data
Chugani, Mahesh.
    LabVIEW signal processing / Mahesh Chugani, Abhay R. Samant, and
  Michael Cerna.
      p.   cm. — (Virtual instrumentation series)
    Includes index.
    ISBN 0-13-972449-4 (pbk.)
    1. Signal processing—Digital techniques—Computer programs.
  2. LabVIEW. 3. Computer graphics. I. Samant, Abhay R.   II. Cerna,
  Michael.   III. Title.   IV. Series.
  TK5012.9.C485   1998
  621.382'2—DC21   98-14696
                                                      CIP

Editorial/Production Supervision: James D. Gwyn
Acquisitions Editor: Bernard Goodwin
Editorial Assistant: Diane Spina
Marketing Manager: Kaylie Smith
Manufacturing Manager: Alan Fischer
Cover Design: Talar Agasyon
Cover Design Direction: Jerry Votta
Series Design: Gail Cocker-Bogusz

© 1998 Prentice Hall PTR
Prentice-Hall, Inc.
A Simon & Schuster Company
Upper Saddle River, NJ 07458

Prentice Hall books are widely used by corporations and government agencies for training,
marketing, and resale.

The publisher offers discounts on this book when ordered in bulk quantities.
For more information, contact
     Corporate Sales Department,
     Prentice Hall PTR
     One Lake Street
     Upper Saddle River, NJ 07458
     Phone: 800-382-3419; FAX: 201-236-7141
     E-mail (Internet): corpsales@prenhall.com

Printed in the United States of America

10 9 8 7 6 5 4 3 2

ISBN 0-13-972449-4

Prentice-Hall International (UK) Limited, London
Prentice-Hall of Australia Pty. Limited, Sydney
Prentice-Hall Canada Inc., Toronto
Prentice-Hall Hispanoamericana, S.A., Mexico
Prentice-Hall of India Private Limited, New Delhi
Prentice-Hall of Japan, Inc., Tokyo
Simon & Schuster Asia Pte. Ltd., Singapore
Editora Prentice-Hall do Brasil, Ltda., Rio de Janeiro

To Aai and Bhai, the source of my inspiration
—Abhay

To Clara, the love of my life
—Mahesh

To Traci, with all my love and respect
—Michael

Contents

Background   1

▼ # 2

Signal Generation   35

▼ # 3

Signal Processing   61

▼ 4

Windowing   97

▼ 5

Measurement   133

▼ # 6

Digital Filtering   171

▼ 7

Curve Fitting   227

▼ 8

Linear Algebra   277

▼ **9**

Probability
and Statistics   **349**

■ **Real-World Application**

▼ **10**

Control Systems   391

■ **Real-World Application**

▼ **11**

Digital Filter
Design Toolkit   451

■ **Real-World Application**

▼ **12**

G Math Toolkit   511

■ **Real-World Application**

Preface:
Good Stuff to Know before
You Get Started

What Is LabVIEW?

LabVIEW, or Laboratory Virtual Instrumentation Engineering Workbench, is a graphical programming environment based on the concept of data flow programming. This programming paradigm has been widely adopted by industry, academia, and research laboratories around the world as the standard for data acquisition and instrument control software. LabVIEW was aboard the mid-October 1993 Columbia space shuttle mission as part of a Macintosh-based research project which helps astronauts study space motion sickness. LabVIEW was used to analyze data received from the Mars Pathfinder. Researchers at the University of Maryland used LabVIEW to develop an application software that resident physicians could use to perform cardiothoracic research. These and other success stories (visit our Web site at www.natinst.com for many more user solutions) show how a powerful software like LabVIEW can transform your computer into a virtual instrument which can then be used in any test and measurement application.

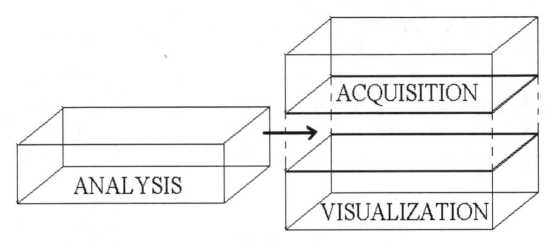

As shown in the above figure, there are three important components involved in test and measurement applications, namely, data acquisition, data analysis, and data visualization. LabVIEW features an easy-to-use graphical programming environment which covers these vital components. The data flow programming paradigm behind LabVIEW is different from the sequential nature of traditional programming languages. It cuts down the time spent in developing an application, thereby allowing scientists and engineers to better utilize their resources.

What Is Data Analysis?

Data is essentially a set of numbers. You can acquire data in a number of ways. For example, the modem connected to your computer may receive a data stream consisting of only 0's and 1's. A basketball fan may create a data set containing the number of points scored by his favorite player. An engineer may acquire signal data from an oscilloscope connected to the instrument that is being tested. In each of these cases, the data by itself does not reveal any important information. For example, the figure below on the left shows a bit stream received by your modem.

Before Analysis

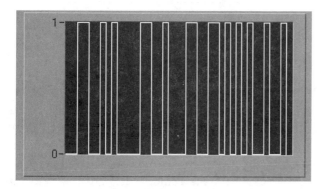

After Analysis

This bit stream by itself looks quite lifeless as it simply consists of 0's and 1's. As shown above on the right, simple analysis on the data reveals that the bit stream actually transforms to the most beautiful word in the English language, "LOVE." As a second example, an engineer may acquire the signal (shown below on the left) from a data acquisition device.

Before Analysis

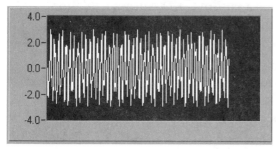

After Analysis

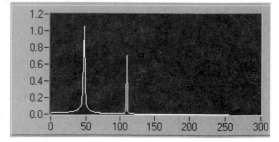

The signal by itself does not reveal any information. However, analysis on such a signal tells the engineer that the signal actually received is the sum of two sine waves of frequencies 48 Hz and 110 Hz. These and other examples show that raw data does not always immediately convey useful information. Data analysis is required to transform the raw data, remove noise disturbances, correct for data corrupted by faulty equipment, or compensate for environmental effects, such as temperature and humidity.

LabVIEW's Analysis Capabilities

The LabVIEW Full Development System features the Analysis Library. The functions in this library are called Virtual Instruments (VIs). This library is also a part of BrigeVIEW, which is the graphical programming language for industrial automation applications. These VIs allow you to use classical digital signal processing and numerical analysis related algorithms without writing a single line of code. The LabVIEW block diagram approach and the extensive set of analysis VIs simplify the development of analysis applications. These VIs are arranged in subgroups, some of which are listed below:

- Signal Generation contains VIs for generating signals such as a sine wave, a square wave, a chirp signal, and white noise.
- Digital Signal Processing contains VIs for computing the Fast Fourier Transform (FFT), auto-correlation, cross correlation, power spectrum, and other functions.
- Windows contains VIs for implementing windows such as the Hanning, Hamming, Blackman, exponential, and the flat top window.
- Curve Fitting contains VIs for performing a linear fit, an exponential fit, a general polynomial fit, and the general least square fit.
- Linear Algebra contains VIs for performing basic matrix operations, matrix inversion, and eigenvalues and eigenvector calculations.

In addition to the Analysis Library, there are a number of specialized toolkits developed by National Instruments. These toolkits are aimed at making new technologies in the area of signal processing, controls, and numerical analysis easy to learn and use. Some of these toolkits are the Digital Filter Design Toolkit, the Wavelet and Filter Bank Design Toolkit, the G Math Toolkit, and the Control and Simulation Software for LabVIEW.

Objectives of This Book

This book is written for students, scientists, researchers, and practicing engineers who are interested in learning how to use the LabVIEW Analysis VIs in practical applications. It is also intended for LabVIEW customers who want to choose intelligently between several similar options available for solving a problem, or for those who may need help interpreting their analysis results. Although the title of the book is *LabVIEW Signal Processing*, we have devoted several chapters to areas outside classical Digital Signal Processing (DSP) theory. One chapter covers the basics of Control System theory and also discusses some of the VIs in the Control and Simulation Software package. The chapter on curve fitting discusses different curve fitting algorithms such as linear fit and exponential fit. The chapter on linear algebra presents detailed material on the solution of general linear equations, eigenvalues and eigenvectors, singular value decomposition, QR decomposition, and Cholesky decomposition. In addition, this book includes an overview of specialized analysis toolkits devoted to specific application areas such as interactive design of digital filters (the Digital Filter Design Toolkit) and solution of mathematical problems (the G Math Toolkit).

In each chapter of the book, we have included numerous activities which are aimed at practical usage of the Analysis VIs in real-world applications. Through these activities, you will learn to

■ Build an arbitrary waveform generator by entering formulas directly on the front panel

■ Avoid aliasing by properly choosing the sampling frequency

■ Separate two signals very close in frequency but widely differing in amplitudes

■ Predict the cost of producing a certain quantity of a product

■ Remove noise from an electrocardiogram

■ Determine stability of systems and design linear state feedback control systems

In addition, this book covers the basic fundamentals necessary for understanding and interpreting your analysis results. General guidelines are provided to help you choose among several options that are available for selecting an appropriate data window, curve fitting algorithm, or filtering technique.

This book is intended for readers who have taken an introductory course in digital signal processing and control systems. Familiarity with simple concepts such as signal representation in the time and frequency domains, Fourier transforms, different forms of system representation, and matrix and vector data types will be useful. However, a brief review of the basic concepts is provided in each chapter as a refresher. It is assumed that the reader is familiar with programming in LabVIEW or at least knows how to read and understand LabVIEW's block diagram approach to graphical programming. It is not the purpose of this book to teach LabVIEW, or discuss the algorithms of each and every VI in full detail. It is application oriented and does not go into the details of the graphical programming language. A very good reference for programming using LabVIEW is the book *LabVIEW For Everyone* by Lisa Wells and Jeffrey Travis, which is also published by Prentice Hall. This book covers in detail all the aspects that you should know to create efficient LabVIEW programs.

Organization

The book is organized into 12 chapters. These chapters have a special structure to facilitate learning:

- *Overview, goals,* and *key terms* describe the main ideas covered in the chapter.
- *Activities* at the end of each section reinforce the information presented in the discussion.
- *Wrap It Up!* summarizes important concepts and skills taught in the chapter.
- *Additional activities* at the end of each chapter give you more practice with the material that you have learned.
- *Real-World Applications* exposes you to solutions of practical problems using the analysis capabilities of LabVIEW.

Here is a brief description of the type of discussion in each chapter:

In Chapter 1, you will learn about sampling an analog signal, aliasing, and the need for antialiasing filters. It will provide the background in basic digital signal processing and discuss some simple concepts.

Chapter 2 covers the theory behind some of the signal generation VIs. These VIs are used to generate signals such as a sine wave, a square wave, and others. It discusses the concept of normalized frequency and shows you how to build a simple function generator.

In Chapter 3, you will learn the basics of transforming a signal from the time domain into the frequency domain. You will learn about the Discrete Fourier Transform (DFT), the Fast Fourier Transform (FFT), and the power spectrum, and how they are different from one another.

Chapter 4 teaches different windowing techniques and how they affect the spectral characteristics of a signal. You will learn about the differences between various types of windows in the Analysis Library and the applications for which they are commonly used.

Chapter 5 shows you how to use the VIs in the Measurement subpalette to perform various signal processing operations such as calculating the amplitude and phase spectrum of a time-domain signal and determining the total harmonic distortion present in a signal.

Chapter 6 introduces you to the world of digital filters. You will learn about the characteristics of different types of filters and how to use them in practical filtering applications.

In Chapter 7, you will learn about the analysis VIs that are used to fit curves to data points. In particular, you will learn about the General Least Squares Linear Fit VI and the nonlinear Levenberg-Marquardt Fit VI.

Chapter 8 will expose to you the basic theory behind the Linear Algebra VIs in the Analysis Library and show you how to use them in different applications. Besides the simple matrix operations, you will learn about eigenvalues and eigenvectors, and the singular value decomposition.

In Chapter 9, we will cover some of the fundamental concepts in probability and statistics such as the mean or average, variance, histogram, and the Normal distribution function.

In Chapter 10, you will familiarize yourself with the theory behind the design and analysis of linear, time-invariant systems. You will learn to use tools for designing linear state feedback systems and graphically analyzing a control system.

In Chapter 11, you will learn how to use the Digital Filter Design (DFD) Toolkit to design filters to meet required specifications. You will also see how to use the DFD Toolkit to analyze your filter design in terms of its frequency response, impulse and step responses, and its pole-zero plot.

Chapter 12 covers the G Math Toolkit which is intended for use by scientists, engineers, and mathematicians to solve mathematical problems in a simple, quick, and efficient manner. You will learn about different types of parser VIs and about solving differential equations using the differential equation VIs.

You will find a glossary, index, and appendices at the end of the book.

In Appendix A, we provide the mathematical background which will be useful while reading the material in some of the chapters. Appendix B gives you a brief description about the error codes that you might encounter while running the VIs. Appendix C answers some of the frequently asked questions. In Appendix D, we list some of the references that might aid you in further understanding the theory discussed in this book. Information about how to contact National Instruments, and the various resources and toolkits available, can be found in Appendix E.

Conventions Used in This Book

The following table describes the conventions used in this book:

bold	**Bold** text denotes VI names, function names, menus, and menu items. In addition, bold text denotes VI input and output parameters, controls, and indicators. For example, "Select the **Sine Wave** VI from the **Analysis** >> **Signal Generation** subpalette."
italic	*Italic* text denotes emphasis, or an introduction to a key term or concept. For example, "A matrix is *orthogonal* if its columns are *orthonormal*."
`Courier`	`Courier` type denotes text or characters that you enter using the keyboard, LabVIEW library names, and folder or directory names. For example, "Save the VI in the library `Activities.llb`."
	Note. This icon marks information to which you should pay special attention.
	Watch out! This icon flags a common pitfall or special information that you should be aware of in order to keep out of trouble.
	Tips and Hints. This icon calls your attention to useful tips and hints on how to do something efficiently.

Your Very Own Software Disk

The CD-ROM included with this book contains an evaluation version of LabVIEW, additional software for the toolkits discussed in this book, activities, and solutions to the activities. The CD-ROM also contains software associated with the real-world applications in Chapters 5 (Measurement), 6 (Digital Filtering), 8 (Linear Algebra) and 10 (Control Systems). If you do not have the full version of LabVIEW already installed on your computer, you can use the evaluation version to work through all the activities in Chapters 1–9. The CD-ROM contains only a portion of the Control and Simulation Software for LabVIEW, the Digital

Filter Design Toolkit, and the G Math Toolkit needed to work through the activities in Chapters 10, 11, and 12 respectively.
The folder structure on the CD-ROM is as follows:

```
Activities and Solutions
     Activities.llb
     Solutions.llb
Additional Software
     Digital Filter Design Toolkit
     Controls.llb
     GMath.llb
LabVIEW
Real-World Applications
```

How to Install the Software:

1. Install the LabVIEW evaluation software on your computer. To install this software, run the `setup.exe` program from the `LabVIEW` folder on the CD-ROM. Follow the instructions on the screen. If you already have the full version of LabVIEW installed, you do not need to install the evaluation version.

2. Create an `Addons` folder in your LabVIEW `Vi.lib` directory, if it does not already exist.

3. Copy `Controls.llb` and `GMath.llb` from the `Additional Software` folder on the CD-ROM to the `Addons` directory.

4. Copy the `Digital Filter Design Toolkit` folder from the `Additional Software` folder on the CD-ROM to your hard drive.

5. Copy the `Real-World Applications` folder from the CD-ROM to your `LabVIEW` folder. This folder contains software corresponding to the real-world applications in Chapters 5, 6, 8 and 10.

6. Copy the contents of the `Activities and Solutions` folder from the CD-ROM to your `LabVIEW` folder. This folder contains the libraries Activi-

`ties.llb` and `Solutions.llb`. While performing the activities in this book, you will open and/or save VIs in `Activities.llb`. The solutions to the activities can be found in `Solutions.llb`. You will have to change the access permissions for both these libraries. To do so, follow the instructions below:

a) Right click on either `Activities.llb` or `Solutions.llb`.
b) Select Properties, then select General.
c) Deselect the Read-only attribute.

7. Repeat Steps 6a, 6b, and 6c for the libraries `Controls.llb` and `GMath.llb` in the `Vi.lib` >> `Addons` folder.

Read the contents of the `readme.txt` *file on the CD-ROM very carefully.*

Restrictions of the LabVIEW Evaluation Version

To run the evaluation version, launch the `Labview.exe` program from the folder in which you installed LabVIEW. Select the `Exit to LabVIEW` button in the lower-right corner. This opens a window which gives you general information about LabVIEW and National Instruments. After reading this information, click on the OK button. This will then open the LabVIEW window. You can access all the features of the full version in the evaluation version with some restrictions. See the `readme.txt` file on the CD-ROM for information about these restrictions and minimum hardware and memory requirements.

Performing the Activities

While performing the activities in this book, you will open and/or save VIs the library Activities.llb. The solutions to the activities can be found in the library Solutions.llb. After completing the installation instructions, both these libraries can be found in your LabVIEW folder.

As shown in the example below,

 Sine Wave VI (**Analysis >> Signal Generation** subpalette)

in an activity, when you come across a VI icon followed by the name of the VI (**Sine Wave VI**) and its location (**Analysis >> Signal Generation** subpalette) in the **Functions** palette, choose the particular VI from the specified subpalette and drop it on the block diagram. Then follow the instructions in the activity to complete the block diagram.

Purchasing LABVIEW

LabVIEW runs on the following platforms: Macintosh, Power Macintosh, Sun SPARCstations, HP9000/700-series workstations running HP-UX, and PCs running Windows 3.1, Windows NT, or Windows 95. If you would like information on how to purchase LabVIEW, contact National Instruments.

National Instruments
6504 Bridge Point Parkway
Austin, Texas 78730

Telephone: (512) 794-0100
Fax: (512) 794-8411
E-mail: info@natinst.com
Worldwide Web: http://www.natinst.com

Or contact the local National Instruments branch office for your country. See *Appendix E* for branch contact information.

About the Authors

Mahesh Chugani received a diploma in Electronics and Radio Engineering from the Cusrow Wadia Institute of Technology, Pune, India, in 1984, and a Bachelor of Engineering (Electronics and Telecommunications) from the College of Engineering, Pune, India, in 1987. He received his M.S. and Ph.D. degrees in Electrical Engineering in 1990 and 1996, respectively, from Rensselaer Polytechnic Institute (RPI). His Doctoral Thesis was titled "Feature Analysis of Doppler Ultrasound Signals obtained from Stenotic Arteries." Mahesh is a member of the IEEE and Eta Kappa Nu. He has published several papers and won numerous awards. Some of these include the Charles M. Close Doctoral Research Prize at RPI in 1996, the Whitaker Award at the 20th Annual Bioengineering Conference in March 1994, and the RPI Presidents International Service Award in 1993. His interests include learning languages, traveling, ballroom dancing, and playing the piano.

Mahesh has been a DSP software engineer at National Instruments since May 1996. He serves as a reviewer for the IEEE *Potentials* magazine. His areas of interest include filter design, biomedical signal processing, dynamic signal analysis, and educational activities.

Abhay Samant joined National Instruments in August 1996 as a Software Engineer in the DSP Software (DSPSW) Group. He received his Bachelor of Engineeering (Electronics) from the University of Bombay, India, in 1992, Master of Science in Electrical Engineering from the University of Kentucky in 1994, and Master of Science in Computer Science from the University of Illinois at Urbana Champaign in 1996. His theses dealt with the development of fast and numerically stable computational techniques for solving electromagnetic scattering and radiation problems.

Abhay's interests are in the areas of numerical algorithms, digital signal processing and communication systems. His work has been published in a number of reputable journals and has been presented at a number of conferences. At National Instruments, he is responsible for the Analysis Library in LabWindows CVI and ComponentWorks. He has also developed part of the Control and Simulation Software for LabVIEW and is working on software targeted toward applications in the area of telecommunications. He is a member of the IEEE and Phi Kappa Phi. He has served as a reviewer for a number of journals. When he is not at work, Abhay likes playing his guitar, jogging, or cooking.

Michael Cerna has been a software engineer at National Instruments since January 1989 and is currently developing DSP software for the company. Michael earned his B.S.E.E. degree in 1987 and his M.S.E.E degree in 1989, both from the University of Texas at Austin. His master's thesis in the areas of acoustics and digital signal processing was titled "A Digital Implementation of a Constant Beamwidth Acoustic Array."

At National Instruments, Michael is responsible for the Analysis Library for LabVIEW, digital signal processing functions for LabVIEW, LabWindows/CVI, and ComponentWorks, and dynamic signal analysis software. He is a member of the IEEE and the Acoustical Society of America, and his technical areas of interest are digital signal processing, dynamic signal analysis, acoustics, vibration, and sound analysis.

Acknowledgments

This book is the result of continuous support and contributions from a number of people. Without their support, this book could not have been completed.

First and foremost, we would like to thank Dr. Dapang Chen, manager of the DSP Software Group of the Research and Development Division at National Instruments. Dr. Chen allowed us to invest a significant amount of time to write this book. We would also like to thank Shie Qian, Dinesh Nair and Quihong

Yang for developing the Analysis Library, reading parts of the manuscript and providing valuable feedback.

We specially acknowledge the efforts of James Baker, Crystal Drumheller, Archana Shrotriya, James Humphrey, Jim Nagle, Robert Steimle, Vijay Malhotra, Siming Lin, Lothar Wenzel and Shy Aviram for carefully reviewing the book and meticulously checking the activities and the solutions. Our special thanks go to Murali Parthasarathy and Greg Richardson for their help in building the CD-ROM. We would also like to thank Don Clinchy for formatting the original manuscript, Gary Herrera for drawing some excellent figures, and Carman Juarez for his administrative support. Our special thanks to Ash Razdan, Ray Almgren, John Hanks, and Lisa Wells for their support. We would also like to acknowledge the efforts of the authors who have contributed the real-world application articles.

Finally, we would like to thank Bernard Goodwin, Diane Spina, and James Gwyn from Prentice Hall for their hard work, flexibility, and devotion to producing a timely publication.

Summary

There have been a lot of books written on signal processing. Unfortunately, most of these books are written without intuitive explanations and illustrations. These classical texts tend to leave concepts at a theoretical level, often alienating people who want to use signal processing in a more practical, real-world setting. For this reason, we decided to write this book in a simple and straightforward manner with numerous illustrations and activities. The Real-World Applications section at the end of each chapter serves to show how these concepts can be used in a practical setting. We hope that we have been successful in our efforts to explain difficult signal processing concepts in simple language and that you find our book a rewarding and challenging experience.

Mahesh L. Chugani
Abhay R. Samant
Michael Cerna

Introduction

An Introduction to Digital Signal Processing for Instrumentation

By Ajay Divakaran
Mitsubishi Electric Information Technology Center America
Advanced Television Laboratory
New Providence, NJ

We introduce Digital Signal Processing for instrumentation to the beginner. The purpose of instruments, from the first rudimentary measuring devices to today's sophisticated "data-acquisition" systems, is to measure the attributes of phenomena so as to analyze and interpret them and thus gain further insight. Availability of ever faster and cheaper digital computation has spurred development of sophisticated algorithms that take full advantage of available computational power, and vice versa. Such synergy between digital signal processing theory and practice has enabled modern devices to go beyond measurement to analysis and interpretation. A deeper understanding of digital signal processing theory is therefore essential to make the best use of today's sophisticated instrumentation systems. Modern design tools such as LabVIEW help the user understand, as well as effectively use, theory by making the task of building and testing complex systems easy.

Measuring Devices Through the Ages

Measurement of a quantity is finding its magnitude on a numerical scale. Numerical representation enables use of analysis and interpretation, which in turn leads to precise understanding and further insight. No wonder then that the progress of science has been predicated on constantly deepening and broadening the ability to measure.

While it was relatively easy to directly measure quantities such as length and mass, new measuring devices had to be developed to measure quantities that were equally palpable yet not directly measurable. One of the first such measuring devices was the first seismoscope, invented in 132 A.D. by Zhang Heng, a Chinese philosopher. It consisted of a copper vessel with eight dragon heads attached to it with a frog below each of the dragon heads. Each dragon head held a ball in its mouth which would drop into the frog's mouth if the seismoscope were shaken by an earthquake. By noting which frogs had balls in their mouths, it was possible to tell how the earth had moved in response to the earthquake. In other words, the dragon–frog combination enabled a numerical characterization of a difficult-to-measure yet uncomfortably palpable phenomenon. This instrument was sensitive enough to detect tremors that could not be felt. Once, it was said to have detected an earthquake 600 km away. However, it evidently did not measure the intensity of the earthquake and did not produce a permament and precise record of the motion of the ground. By the end of the nineteenth century, several electromagnetic seismographs had been invented that were operated by suspending a magnetic mass, or pendulum, within an electric coil. Since the moving of a magnet inside a coil induces an electric current in the coil, the shaking of the ground could now be converted into an electrical signal. The electrical signal could then be used, for instance, to move a needle across paper to record the wiggles of the earthquake. Modern day seismometers use digital signal processing to record ground shaking over a large range of frequency and amplitude. These seismometers are able to sense ground motion over frequencies from thousandths of a second to less than a hundredth of a second. Instead of measuring the deflection of the magnet within the coil, which limited the range of nineteenth century seismographs, they measure the energy required to keep the magnet centered in the presence of ground shaking.

Other instruments also followed the pattern we see above. First, better sensors, or transducers, were developed that recorded the phenomenon with better precision and accuracy. The sensors converted the observed effect into an electrical signal, which was then used to deflect the needle of a meter or move a pen on paper, so as to provide a more precise measurement of the original effect. This, of course, opened up the possibility of improving the sensor sensitivity and accuracy. The next significant step, however, was the improvement of the processing of the electrical signal produced by the transducer. This was achieved by converting analog signals into discrete-time sequences of numbers, that is, sampling, thus enabling the use of digital computation in the processing of the signal—in other words, digital signal processing. This required not only the development of the technology of digital computation but also the development of the theory of digital signal processing.

The Evolution of Signal Processing Theory

Until the 1960's, almost all of the signal processing was analog. The tremendous advances in both digital computing technology and signal processing theory led to a changeover to digital signal processing. The main advantage of digital signal processing is that it is more flexible since it is done in software with a digital computer as opposed to working with a fixed analog circuit à la the nineteenth century seismographs or with a mechanical system like the dragon–frog seismoscope. The disadvantage initially was the slow computation, large size, and high expense of digital computers. The flexibility, however, provided a testbed for new and more powerful algorithms whose only drawback was computational complexity.

The size, speed, and cost of digital computers began to improve dramatically as a result of breakthroughs in microelectronics. The invention of the microprocessor and the high device densities achieved by inexpensive integrated circuits reduced the size and cost and dramatically increased the speed of computers. This trend continues even today, as both speed and memory per dollar double every year. Interestingly, Gordon Moore, a cofounder of Intel, predicted in 1965 that the transistor density of computer chips would double every 18 months. This prediction has come true so far (see figure on next page).

The most significant theoretical breakthrough at this time was the Fast Fourier Transform (FFT) disclosed by Cooley and Tukey in 1965. The FFT sped up the Discrete Fourier Transform computation by orders of magnitude. Such a speedup made a large set of sophisticated algorithms realizable. The FFT also lent itself to custom hardware, which further motivated investigation of FFT-based algorithms. Furthermore, it caused discrete-time analysis to be studied in its own right as a distinct area of investigation, since it provided a foundation for a rich set of powerful operations in the discrete-time domain.

This last point is crucial, because the realization that digital signal processing can do much more than merely duplicate analog signal processing techniques has led to development of a constantly growing and rich set of techniques that make fuller use of the possibilities offered by digital computation. Modern-day speech recognition and synthesis, for example, would be impossible without digital signal processing.

Some common applications of signal processing in instrumentation are:

1. Signal Manipulation—Signal manipulation involves processing of the input signal to get an output signal. A common example of such processing is the elimination of a chosen set of frequencies from the signal, termed filtering. The purpose could be any or all of

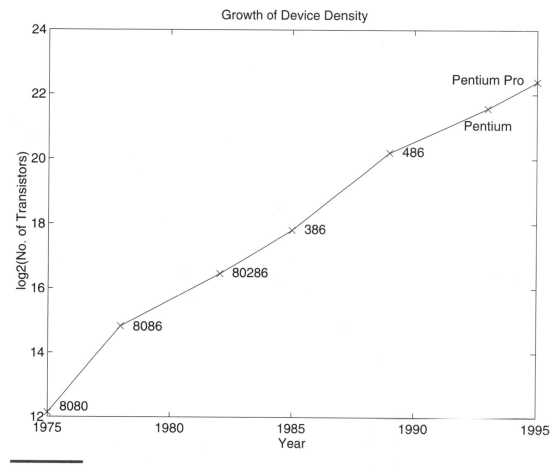

Growth of device density from 1975 to 1995

the following: noise alleviation, signal analysis, signal enhancement for perception, or reduction of data complexity before analysis. Since digital signal processing began by imitating analog signal manipulation, this is the most developed aspect of digital signal processing.

2. Signal Analysis and Interpretation—Signal analysis and interpretation involves attempting to extract information or meaning from signals. For instance, the aim of a seismograph is not to shape or alter the

earthquake but to help characterize it and thus perhaps help predict the next one. Another example is speech recognition, in which the aim is to extract the semantic content and possibly render it in text or to otherwise respond to it. Such systems make use of signal manipulation followed by pattern recognition. One of the oldest approaches to using signal manipulation to aid pattern recognition is speaking loudly in the hope of being better understood. This approach lives on in modern systems in the form of "preamplification" of the signal before it is processed for content.

3. Signal Generation for Automatic Control—Automatic control of a system requires automatic application of appropriate input(s) in response to automatically acquired data. Digital signal processing has enhanced both the data acquisition process (the analysis) and the signal application process (the response). Control systems span the gamut in magnitude and complexity from large chemical reactors and satellite attitude control to simple temperature control systems for homes.

Possible Future Trends

Since digital computer technology continues to advance rapidly, there is immense scope for further advancement of digital signal processing. We can divide the future trends in the following broad categories:

1. Impact of faster and cheaper computation on existing techniques.

   a) The techniques that are implemented in a simplified or diluted fashion because of computational constraints will become more amenable to full-blown implementation.

b) More computationally intensive combinations and/ or iterations of techniques will become feasible.

Computationally intensive techniques such as image restoration and recovery, image analysis, speech recognition, and so forth, will benefit immensely from both of the above. A common example of a computationally demanding application is real-time video processing, in which there always seems to be too much to do in too little time. The volume of data is high and, since the processing is real-time, the available time to process is low. Modern signal processing devices have already made real-time processing of digital video a reality in the form of video-conferencing and high definition television systems. However, such systems are expensive and limited in functionality. Future video systems are expected to provide powerful processing, such as content creation at consumer prices.

2. More elegant and effective solutions—an outstanding example from the past of elegant solutions is the aforementioned Fast Fourier Transform, which makes use of a divide and conquer strategy to cut down the computation while preserving the exactness. Incidentally, historical research has shown that Gauss discovered the fundamental principle of the Fast Fourier Transform. This goes to show that not only development of new theory but also innovative application and extension of existing theory can lead to advancement of the field. More recent examples of elegant solutions include Discrete Cosine Transform (DCT) based signal compression, wavelet signal decomposition based processing, and so forth.

3. Solution of open problems—new techniques will be discovered that may or may not make use of the additional computing power. For example, in attempting to mimic human perception with image or speech recognition systems, the problem is not necessarily so much the lack of computing resources as the lack of a

clear understanding of the underlying phenomenon. Even so, the processing is expected to still be digital and will certainly benefit from advances in digital computing.

Seismic signal processing provides a striking illustration of solution of open problems by synergy between theory and practice. The first wavelet functions were developed to help analyze seismic signals. They were then applied to image processing, which led to advancements in image processing as well as refinement and extension of wavelet theory. Today wavelets find application in a wide variety of fields such as image compression, pattern recognition, speech recognition, biomedical signal processing, and so forth. Signal analysis has indeed come a long way from the days of counting the balls in the frog's mouths.

The broad trend in signal processing systems is toward making the details of the processing increasingly transparent to the user, thus enabling him to put together complex systems very quickly. The biggest thrust is expected to be in the area of signal interpretation, which relies today on substantial human involvement. An increasing number of sophisticated algorithms will become feasible in real-time, thus expanding the power of real-time systems.

Using Design Tools Such As LabVIEW

Design of effective instrumentation requires a good understanding of the physics of the observed phenomenon as well as a good understanding of the principles of digital signal processing. As we mentioned earlier, until the invention of digital signal processing, building of instrumentation was a time-consuming task that required high skill. Even with Digital Signal Processing, it takes some skill and experience to build instrumentation because good programming skills and some mechanical skills are required.

Tools such as LabVIEW enable the user to build instrumentation using software objects called "virtual instruments" on a computer using a graphical interface (see figures below). The

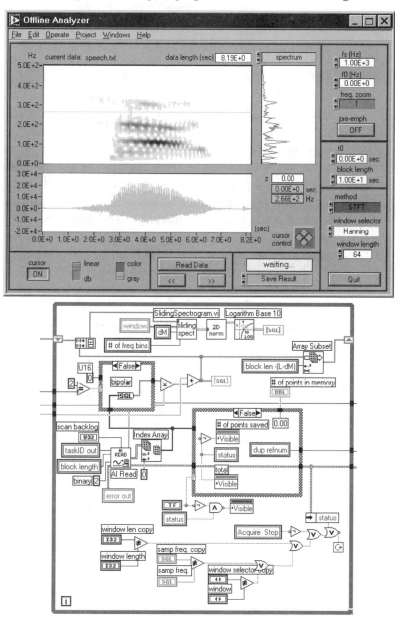

Graphical interface of a virtual instrument

graphical interface allows the user to build the system in software using a block diagram. The source code of each block is built through the graphical interface. Furthermore, it is easy to build a graphical interface to the virtual instrumentation system itself. The graphical interface allows easy interactive experimentation, thus facilitating both learning and execution.

Such tools cannot, however, help defy the "garbage in garbage out" principle. It is the underlying physics of the problem that will dictate the essential instrumentation. No amount of LabVIEW prestidigitation can make up for an incorrect understanding of the experiment. Furthermore, since the range of functionalities and the combinations thereof is large, it is even more important to first thoroughly understand the basic building blocks of signal processing. The more advanced library routines will then be easier to grasp and, thus, building an effective system will be easier. This will also help eliminate redundancies from the design. To misquote Einstein, "The instrumentation should be as simple as it can be and no simpler."

The above concerns are valid only if one seeks to use such tools as an alternative to understanding the principles of Digital Signal Processing. Such tools are, in fact, excellent aids to understanding the principles. Since they take away the effort of building the components, they afford the opportunity to try numerous experiments with the individual components as well as with combinations of components. In short, they provide a practical and illustrative complement to the theory.

References

Oppenheim, Alan V., and Schafer, Ronald W. (1989). *Discrete-Time Signal Processing*, Prentice Hall, Englewood Cliffs, NJ.

Overview

A digital *signal is one that can assume only a finite set of values in both the* dependent *and* independent *variables. The independent variable is usually time or space, and the dependent variable is usually amplitude.*

Digital signals are everywhere in the world around us. Telephone companies use digital signals to represent the human voice. Radio, TV, and hi-fi sound systems are all gradually converting to the digital domain because of its superior fidelity, noise reduction, and its signal processing flexibility. Transmission of data from satellites to earth ground stations is in digital form. NASA's pictures of distant planets and outer space are often processed digitally to remove noise and to extract useful information. Economic data, census results, and stock market prices are all available in digital form. Because of the many advantages of digital signal processing, analog signals are also converted to digital form before they are processed with a computer. This chapter provides a background in basic digital signal processing and an introduction to the LabVIEW/BridgeVIEW Analysis Library, which consists of hundreds of VIs for signal processing and analysis.

GOALS

- Learn about the digital (sampled) representation of a continuous-time and continuous-amplitude (analog) signal.
- Become aware of quantization and its effects.
- Grasp the concept of aliasing and how to prevent it.
- Understand the need for antialiasing filters.
- Understand the need for the decibel scale to display amplitudes.
- Become familiar with the contents of the LabVIEW/BridgeVIEW Analysis Library.

KEY TERMS

- digital signal
- sampling
- Nyquist frequency
- aliasing
- antialiasing filters
- decibels

Background

Sampling Signals

To use digital signal processing techniques, you must first convert an analog signal into its digital representation. In practice, this is implemented by using an Analog-to-Digital (A/D) converter. Consider an analog signal $x(t)$ that is sampled every Δt seconds. The time interval Δt is known as the *sampling interval* or *sampling period*. Its reciprocal, $1/\Delta t$, is known as the *sampling frequency*, with units of samples/second. Each of the discrete values of $x(t)$ at $t = 0$, Δt, $2\Delta t$, $3\Delta t$, and so forth, is known as a *sample*. Thus, $x(0)$, $x(\Delta t)$, $x(2\Delta t)$, ..., are all samples. The signal $x(t)$ can thus be represented by the discrete set of samples

$$\{x(0), x(\Delta t), x(2\Delta t), x(3\Delta t), \ldots, x(k\Delta t), \ldots \}$$

Figure 1–1 shows an analog signal and its corresponding sampled version. The sampling interval is Δt. Observe that the samples are defined at discrete points in time.

In this book, the following notation represents the individual samples:

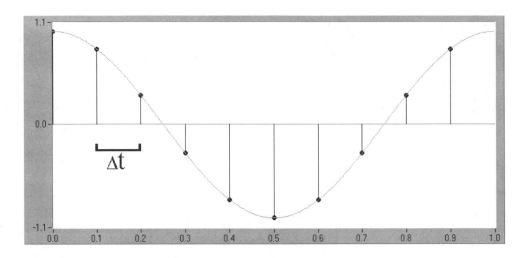

Figure 1–1
Analog signal and corresponding sampled version

$$x[i] = x(i\Delta t), \qquad \text{for } i = 0, 1, 2, \ldots$$

If N samples are obtained from the signal $x(t)$, then $x(t)$ can be represented by the sequence

$$\mathbf{x} = \{x[0], x[1], x[2], x[3], \ldots, x[N\text{-}1] \}$$

This is known as the *digital representation* or the *sampled version* of $x(t)$. Note that the sequence $\mathbf{x} = \{x[i]\}$ is indexed on the integer variable i and does not contain any information about the sampling rate. So by knowing just the values of the samples contained in $\mathbf{x}$, you will have no idea of what the sampling rate is.

Quantization

Information (data, code, etc.) in computers (either in memory or in a file on the hard drive) is stored digitally in the form of a finite number of bits. Every sample of an analog signal thus needs to be converted into a finite number of bits before it can be processed by the computer. This conversion is achieved by an Analog-to-Digital Converter (ADC) at the front end of a Data Acquisition (DAQ) board.

One of the important parameters of a DAQ board is its *resolution*—the number of bits that the ADC uses to represent the analog signal. Since the number of bits (and thus the number of combinations) is finite, and the continuous analog signal can have an infinite number of possible values, a range of analog values is mapped onto one particular combination of bits. This is illustrated in Figure 1–2 where an analog signal (a sine wave), with a peak-to-peak amplitude of 1.00 volt, is converted into its digital representation by 3 bits. The 3-bit representation gives us $2^3 = 8$ possible combinations. In this particular example, analog values between 0.0–0.125 Volts are represented by the bit combination 000, values between 0.125–0.250 Volts are represented by 001, and so on till the values between 0.750–1.00 Volts are represented by 111.

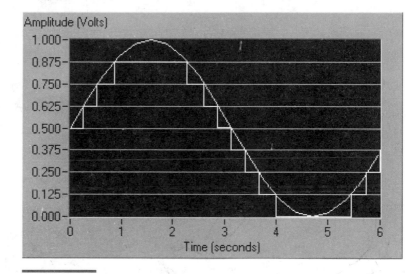

Figure 1–2
Quantization of an analog signal

The higher the resolution, the more the number of divisions into which the range is broken, and therefore, the smaller the detectable voltage change. We can see that the 3-bit digital (binary) signal is not a good representation of the original analog signal because information has been lost in the conversion. By increasing the resolution to 16 bits, the ADC's number of codes increases from 8 to 65,536 (2^{16}), and it can therefore obtain an extremely accurate representation of the analog signal.

Sampling Considerations

Another important parameter of an analog input system is the rate at which the DAQ board samples an incoming signal. The sampling rate determines how often an Analog-to-Digital (A/D) conversion takes place. A fast sampling rate acquires more points in a given time and can form a better representation of the original signal than a slow sampling rate. Sampling too slowly may result in a poor representation of your analog signal. Figure 1–3 shows an adequately sampled signal, as well as a signal that has been undersampled. The effect of undersampling is that the signal appears as if it has a different frequency than it truly does. This misrepresentation of a signal is called an *alias*.

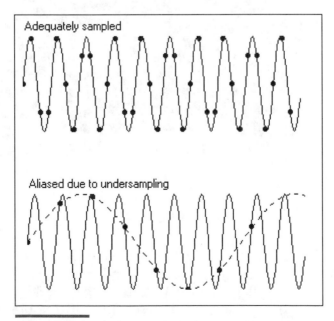

Figure 1–3
Aliasing effects of an improper sampling rate

According to the *Nyquist theorem*, to avoid aliasing you must sample at a rate of at least twice the maximum frequency component in the signal you are acquiring. For a given sampling rate, the maximum frequency that can be represented accurately, without aliasing, is known as the *Nyquist frequency*. The Nyquist frequency is one-half the sampling frequency. Signal frequency components above the Nyquist frequency will appear aliased between DC and the Nyquist frequency alter sampling. The alias frequency is the absolute value of the difference between the frequency of the input signal and the closest integer multiple of the sampling rate. Figures 1–4 and 1–5 illustrate this phenomenon. For example, assume f_s, the sampling frequency, is 100 Hz. Also, assume the original continuous signal contains the following frequencies—25 Hz, 70 Hz, 160 Hz, and 510 Hz. The signal spectrum before sampling is shown in Figure 1–4.

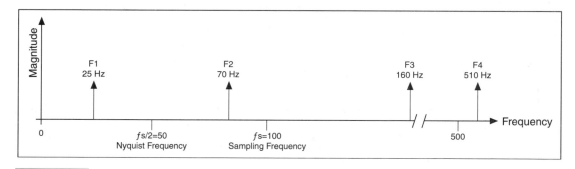

Figure 1–4
Actual signal frequency components

In Figure 1–4, frequencies below the Nyquist frequency ($f_s/2$=50 Hz) are sampled correctly. Frequencies above the Nyquist frequency appear as aliases. For example, f_1 (25 Hz) appears at the correct frequency, but f_2 (70 Hz), f_3 (160 Hz), and f_4 (510 Hz) have aliases at 30 Hz, 40 Hz, and 10 Hz, respectively. The original and aliased components are shown in the spectrum in Figure 1-5. To calculate the alias frequency, use the following equation:

Alias Frequency = ABS (Closest Integer Multiple of Sampling Frequency – Input Frequency)
= | fs • round (f/fs) – f |

where ABS means "the absolute value." For example,

$$\text{Alias } f_2 = |100 - 70| = 30 \text{ Hz}$$
$$\text{Alias } f_3 = |(2)100 - 160| = 40 \text{ Hz}$$
$$\text{Alias } f_4 = |(5)100 - 510| = 10 \text{ Hz}$$

Now let's look at what happens when you increase the sampling rate. Nonaliased components should remain fixed, and since aliased components are a function of the sampling rate, they should change their apparent frequency. For example, when f_s is increased to 150 Hz, f_2 no longer aliases, f_3 changes its alias frequency to 10 Hz, and f_4 changes its aliased frequency to 60 Hz. Now, if we further increase f_s to 400 Hz, f_2 and f_3 no longer alias, but f_4 changes its alias to 110 Hz.

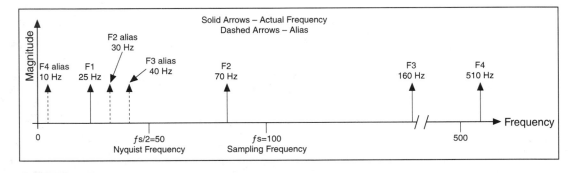

Figure 1–5
Signal frequency components and aliases

A question often asked is, "How fast should I sample?" Your first thought may be to sample at the maximum rate available on your DAQ board. However, if you sample very fast over long periods of time, you may not have enough memory or hard disk space to hold the data. Figure 1–6 shows the effects of various sampling rates. In case a, the sine wave of frequency f is sampled at the same frequency f_s (samples/sec) = f (cycles/sec), or at 1 sample per cycle. The reconstructed waveform appears as an alias at DC. As you increase the sampling to 7 samples per 4 cycles (case b), the waveform increases in frequency but continues to alias to a frequency less than the original signal. Case b is

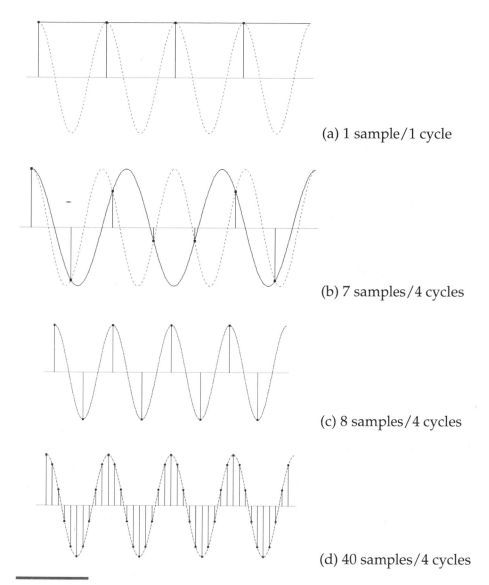

(a) 1 sample/1 cycle

(b) 7 samples/4 cycles

(c) 8 samples/4 cycles

(d) 40 samples/4 cycles

Figure 1–6
Effects of sampling at different rates. In cases a and b, the solid lines represent the reconstructed signal.

sampled at $f_s = (7/4) f$. However, if you increase the sampling rate to $f_s = 2f$, as shown in case c, the digitized waveform has the correct frequency (same number of cycles) and can be reconstructed as the original sinusoidal wave. By increasing the sampling rate to well above *f*, for example 10*f*, as shown in case d you can more accurately reproduce the waveform.

The Nyquist theorem gives you a starting point for an adequate sampling rate—at least two times the highest frequency component in the signal. Unfortunately, this rate is often inadequate for practical purposes. Real-world signals often contain frequency components which lie above the Nyquist frequency. These frequencies are erroneously aliased and added to the components of the signal that are sampled accurately, producing distorted sampled data. Therefore, for practical purposes, sampling is usually done at several times the maximum frequency—five to ten times is typical in industry.

Sampling should be done at least at the Nyquist frequency, but usually much higher.

Why Do You Need Antialiasing Filters?

You have seen that the sampling rate should be at least twice the maximum frequency of the signal that you are sampling. In other words, the maximum frequency of the input signal should be less than or equal to half of the sampling rate. But how do you ensure that this is definitely the case in practice? Even if you are sure that the signal being measured has an upper limit on its frequency, pickup from stray signals (such as the powerline frequency or from local radio stations) could contain frequencies higher than the Nyquist frequency. These frequencies may then alias into the desired frequency range and thus give us erroneous results.

To be completely sure that the frequency content of the input signal is limited, a lowpass filter (a filter that passes low frequencies but attenuates the high frequencies) is added before the sampler and the ADC. This filter is called an *antialias* filter because by attenuating the higher frequencies (greater than Nyquist), it prevents the aliasing components. Because at this

stage (before the sampler and the ADC) you are still in the analog world, the antialiasing filter is an analog filter.

An ideal antialias filter is as shown in Figure 1–7(a).

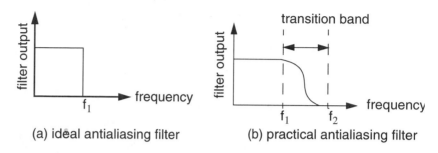

(a) ideal antialiasing filter   **(b) practical antialiasing filter**

Figure 1–7
An ideal and a practical antialiasing filter

An ideal antialiasing filter passes all the desired input frequencies (below f_1) and cuts off all the undesired frequencies (above f_1).
However, such a filter is not physically realizable. In practice, filters look as shown in figure (b) above. They pass all frequencies <
f_1, and cutoff all frequencies > f_2. The region between f_1 and f_2 is
known as the *transition band*, which contains a gradual attenuation of the input frequencies. Although you want to pass only signals with frequencies < f_1, those signals in the transition band
could still cause aliasing. Therefore, in practice, the sampling frequency should be greater than two times the highest frequency in
the transition band, so this turns out to be more than two times
the maximum input frequency (f_1). That is the main reason why
you may see that the sampling rate is more than twice the maximum input frequency. We will see in a later chapter how the transition band of the filter depends on the filter type being designed.

Why Use Decibels?

On some instruments, you will see the option of displaying the
amplitude in either a linear or a decibel (dB) scale. The linear
scale shows the amplitudes as they are, whereas the decibel
scale is a transformation of the linear scale into a logarithmic
scale. You will now see why this transformation is necessary.

Suppose you want to display a signal with very large as well as very small amplitudes. Assume you have a display of height 10 cm and will use the entire height of the display for the largest amplitude. So, if the largest amplitude in the signal is 100 V, a height of 1 cm of the display corresponds to 10 V. If the smallest amplitude of the signal is 0.1 V, this corresponds to a height of only 0.1 mm. This will barely be visible on the display!

To see all the amplitudes, from the largest to the smallest, you need to change the amplitude scale. Alexander Graham Bell invented a unit, the Bell, which is logarithmic, compressing large amplitudes and expanding the small amplitudes. However, the Bell was too large of a unit, so commonly the decibel (1/10th of a Bell) is used. The decibel (dB) is defined as

$$1 \text{ dB} = 10 \log_{10} (\text{Power Ratio}) = 20 \log_{10} (\text{Voltage Ratio})$$

Table 1–1 shows the relationship between the decibel and the power and voltage ratios.

Thus, you see that the dB scale is useful in compressing a wide range of amplitudes into a small set of numbers. The decibel scale is often used in sound and vibration measurements and in displaying frequency domain information. Indeed, human perception is roughly based on a logarithmic scale. You will now do an exercise that shows a signal in linear and logarithmic scales.

Table 1–1   *Power and voltage ratios corresponding to decibel levels*

dB	Power Ratio	Voltage Ratio
+40	10000	100
+20	100	10
+6	4	2
+3	2	1.4
0	1	1
-3	1/2	1/1.4
-6	1/4	1/2
-20	1/100	1/10
-40	1/10000	1/100

Activity 1-1

Objective: To build a VI that displays the signal amplitude in both linear and dB scales

This VI will display the squared values of 100 data points on a waveform graph. The fifth data point will create a spike. You will observe that the spike is visible on the dB scale.

■ Front Panel

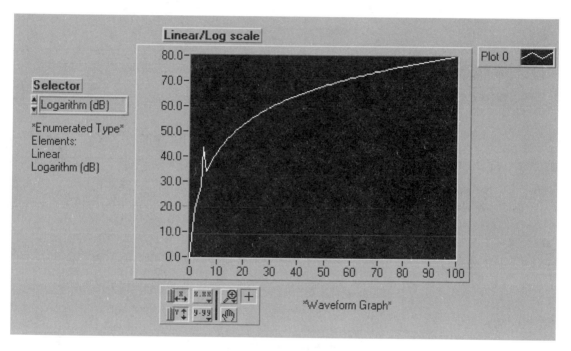

1.   Build a VI with the front panel shown above.

     The **Selector** control (**Controls » List and Ring » Enumerated Type**) has two options, *Linear* scale and *Logarithm* (*dB*) scale.

■ Block Diagram

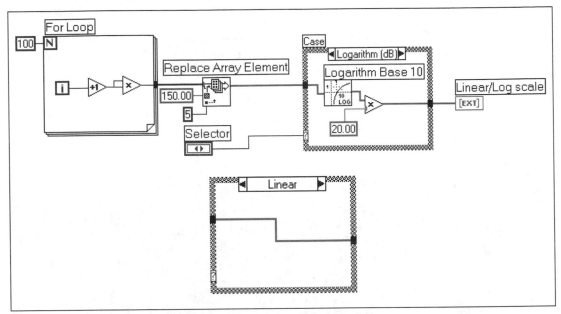

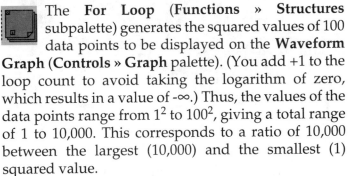

2.   Build the block diagram as shown above.

The **For Loop** (**Functions » Structures** subpalette) generates the squared values of 100 data points to be displayed on the **Waveform Graph** (**Controls » Graph** palette). (You add +1 to the loop count to avoid taking the logarithm of zero, which results in a value of -∞.) Thus, the values of the data points range from 1^2 to 100^2, giving a total range of 1 to 10,000. This corresponds to a ratio of 10,000 between the largest (10,000) and the smallest (1) squared value.

The **Replace Array Element** function (**Functions » Array** subpalette) replaces the fifth data point, which has a value of $5^2 = 25$, by 150, to create a spike at the fifth element. You will see how the spike is barely noticeable on the linear scale but is easily distinguishable on the dB scale.

Depending on the **Selector** control, the **Case** structure (**Functions » Structures** subpalette) either passes the data directly (*Linear* scale) to the **Waveform Graph** or calculates 20 times the logarithm to the base 10 (*Logarithm (dB)* scale) of the data points and sends the result to the **Waveform Graph**.

The **Logarithm Base 10** function is found in the **Functions » Numeric » Logarithmic** subpalette.

3. Select the *Linear* option from the **Selector** control and run the VI. Note that the spike at element 5 is barely visible.

4. Select the *Logarithm (dB)* option from the **Selector** control and run the VI. Note that the spike at element 5 is very easily noticeable.

> *Observe the change in the y-axis scale as you switch between the "Linear" and "Logarithm (dB)" options.*

5. After you have finished, save the VI as **dB_linear.vi** in the library `Activities.llb`.

■ End of Activity 1-1

Overview of the Analysis Library

Once the analog signal has been converted to digital form by the ADC and is available in your computer as a digital signal (a set of samples), you will usually want to process these samples in some way. The processing could be to determine the characteristics of the system from which the samples were obtained, to measure certain features of the signal, or to convert them into a form suitable for human understanding, to name a few.

The LabVIEW/BridgeVIEW **Analysis Library** contains VIs to perform extensive numerical analysis, signal generation and signal processing, curve fitting, measurement, and other analysis functions. The **Analysis Library,** included in the LabVIEW/BridgeVIEW full development system, is a key component in building a virtual instrumentation system. Besides containing the analysis functionality found in many math packages, it also features many unique signal processing and measurement functions that are designed exclusively for the instrumentation industry.

The analysis VIs are available in the **Analysis** subpalette of the **Functions** palette in LabVIEW or BridgeVIEW, as shown in Figure 1–8.

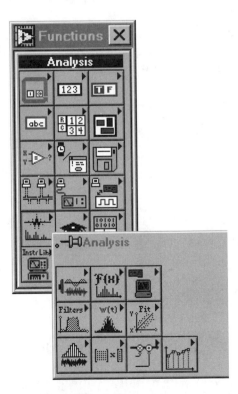

Figure 1–8
Analysis subpalette

There are 10 analysis VI libraries. The main categories are

 Signal Generation: VIs that generate digital patterns and waveforms.

Digital Signal Processing: VIs that perform frequency domain transformations, frequency domain analysis, time domain analysis, and other transforms such as the Hartley and Hilbert transforms.

 Measurement: VIs that perform measurement-oriented functions such as single-sided spectrums, scaled windowing, harmonic distortion, and AC/DC estimation.

 Filters: VIs that perform IIR, FIR, and nonlinear digital filtering functions.

 Windows: VIs that perform windowing for spectral analysis such as the Hanning, Flat Top, and Blackman-Harris windows.

 Curve Fitting: VIs that perform curve fitting functions and interpolations.

 Probability and Statistics: VIs that perform descriptive statistics functions, such as identifying the mean or the standard deviation of a set of data, as well as inferential statistics functions for probability and analysis of variance (ANOVA).

 Linear Algebra: VIs that perform algebraic functions for real and complex vectors and matrices.

 Array Operations: VIs that perform common, one- and two-dimensional numerical array operations, such as linear evaluation and scaling.

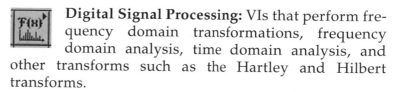 **Additional Numerical Methods:** VIs that use numerical methods to perform root finding, numerical integration, and peak detection.

In this book, you will learn how to design and use the VIs from the **Analysis Library** to build a function generator and a simple, yet practical, spectrum analyzer. You will also learn how to design and use digital filters, the purpose of windowing, the advantages of different types of windows, how to perform simple curve-fitting tasks, and much more. The activities in this book require the LabVIEW or BridgeVIEW full development system. These activities can also be performed using the LabVIEW Evaluation version which can be found on the CD-ROM at the back of this book. For the more adventurous, an extensive set of examples that demonstrate how to use the analysis VIs can be found in the **labview » examples » analysis** folder.

In addition to the **Analysis Library,** National Instruments also offers many analysis add-ons that make LabVIEW or Bridge-VIEW one of the most powerful analysis software packages available. These add-ons include the *Joint Time-Frequency Analysis Toolkit,* which includes the National Instruments award-winning and patented Gabor Spectrogram algorithm that analyzes time-frequency features not easily obtained by conventional Fourier analysis; the *G Math Toolkit,* which offers extended math functionality like a formula parser, routines for optimization and solving differential equations, numerous types of 2D and 3D plots, and more; the *Digital Filter Design Toolkit* for designing and analyzing digital filters; and many others. Some of these specialized add-ons will also be discussed later in this book.

Wrap It Up!

- This chapter introduced the digital (sampled) representation of a signal.

- To convert an analog signal into a digital signal, the sampling frequency (f_s) should be at least twice the highest frequency contained in the signal. If this is not the case, the frequencies in the signal that are greater than the *Nyquist frequency* ($f_s/2$) appear as undesirable aliases.

- You can use a lowpass filter before sampling the analog signal to limit its frequency content to less than

$f_s/2$. Such a filter used to prevent the effect of aliasing is known as an *antialias filter*.

- You saw how to use a logarithmic scale (the decibel) to display a large range of values. It does this by compressing large values and expanding small ones.

- This chapter also gave an overview of the LabVIEW/BridgeVIEW **Analysis Library** and its contents.

■ Review Questions

1. Give some examples of digital signals in everyday life.

2. Given a set of sample values $x = \{x[i]\}$ where i is an integer variable, what is the sampling rate?

3. What is aliasing? How can it be avoided?

4. Given that the sampling frequency is 100 Hz, what is the alias frequency (if any) for the following: 13 Hz, 25 Hz, 40 Hz, 75 Hz, 99 Hz, 101 Hz, 200 Hz, and 350 Hz?

5. Why do we use the decibel scale? In what applications is it normally used?

6. Which of the following is possible using the Analysis VIs?

    a) Finding the mean or standard deviation of census data.

    b) Designing a filter to remove noise from an electrocardiogram.

    c) Detecting peaks in a blood pressure waveform to measure the heart rate.

    d) Interpolating between data points to plot the trajectory of an object (for example, a comet or a cannonball).

■ Additional Activities

1. Modify Activity 1-1 so that the values of the data points range from 1 to 1,000,000 (instead of 1 to 10,000 which corresponds to a range of 80 dB). Run the VI.

    a) What is the corresponding range in dB?

    b) What can you do to make the spike more clearly visible?

Data Acquisition System Virtual Instrument (LabVIEW) developed for Particle Impact Noise Detection (PIND) Testing of Pressure Sensors used in the Space Shuttle Main Engine

Sonia Balcer
Rocketdyne Division
Boeing, Canoga Park, CA

The Challenge Nondestructive Particle Impairment Noise Detection (PIND) testing of approximately 650 Space Shuttle Main Engine (SSME) pressure sensors required the acquisition of numerous runs of data each as well as the archival of the resulting data files. With the added requirement of providing operator control of various parameters, automation of the screening process was a crucial need.

The Solution LabVIEW was used to create a versatile, user-friendly PIND Data Acquisition System (DAS) virtual instrument that automated data acquisition and storage. Over a period of 3½ years, it has been used for non-destructive contamination screening of over 650 flight SSME units.

Figure 1–9
SSME's during ascent of Space Shuttle into orbit

Background

In 1993, several incidences of anomalous electrical behavior were observed in the pressure transducers used on the Space Shuttle Main Engine (SSME) built by Rocketdyne Division of Boeing in Canoga Park, California. Analysis revealed the cause to be tiny (some less than a mm in size) fragments of conductive contamination left from sensor manufacture which, during the vibration encountered during flight use, could produce temporary electrical shorting between conductive surfaces of the internal circuitry. Such contamination could easily be introduced during fabrication of the sub-assemblies in which solder slivers, solder balls, wire fragments, pin fragments, machining chips,

Figure 1–10
Pressure sensor used in the SSME

and other particles could fall undetected into blind holes or be shed as wires are pulled through support structures.

The long-term solution was to evolve the sensor design for simplicity and ease of fabrication into a unit of comparable performance. However, as any such improvements for the Space Shuttle must be established by extensive test experience, it was necessary to develop a nondestructive means of ascertaining which of the approximately 650 units in present flight use contained contaminants potentially large enough to affect performance.

Therefore, a non-destructive screening effort was initiated, utilizing Particle Impact Noise Detection (PIND) testing and real-time microfocus x-ray examination of all pressure sensors in stock, quantifying the loose conductive material within each unit and allocating them accordingly. As the x-ray technology advanced during the time-frame of the effort, the PIND testing became less important and regarded largely as an indicator of total loose material (conductive and non-conductive) within the sensor. It does, however, tend to compensate for the limitations of the x-ray method in detecting extremely thin contaminants (e.g. observed foil-shaped objects) and has therefore been retained for the duration of the screening effort.

Particle Impact Noise Detection (PIND)

PIND testing is a nondestructive means of "listening" to and quantifying the vibrational energy imparted to housing of a sensor by mobile pieces of metallic contamination (e.g. specks of solder or fragments of wires), providing a rough indication of the amount and location of contamination. The PIND tester applies vibrational excitation during sixty runs performed in groups of ten runs done in various orientations in which the sensor is mounted vertically (V1), horizontally in two successive variations (H1, H2), again vertically (V2), upside-down (U1), then vertically once again (V3). The PIND noise contributions of the particles in any given sensor will depend upon their mass (and related material properties), number, cavity location, and resultant set of travel trajectories within the sensor cavity during vibrational excitation. (For example, consistently loud noise in the vertical orientation is indicative of numerous particles in cavity 1, and erratically loud noise in the upside down and horizontal orientations is indicative of large or numerous particles in cavity 2.) Figure 1–11 illustrates the three internal cavities of the sensor (in a cutaway view) and Figure 1–12 shows a block diagram of the test system.

From the 6 × 10 PIND test runs in different orientations, if V,H,U are the maximum outputs in the vertical, horizontal, and upside-down orientations respectively, the Noise is taken to be N = sqrt (sum of squares; V,H,U). This, along with other test results and design configuration information, is factored into the allocation of units for engine use.

Forty sensors built according to the new-design have been nondestructively screened using this method, with dramatically reduced contamination levels being demonstrated, such that non-destructive screening for internal contamination will not be needed in the future.

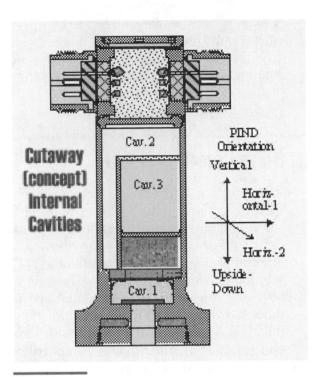

Figure 1–11
Sensor cross-section (representation)

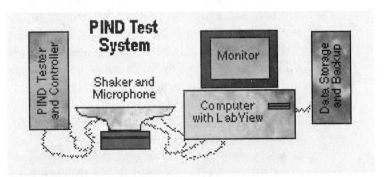

Figure 1–12
PIND-DAS setup

Description of PIND-DAS Virtual Instrument

The Particle Impact Noise Detection (PIND) Data Acquisition System (DAS) virtual instrument was written in the National Instruments LabVIEW graphical programming language in early 1994 and run on a Dell 466/M computer, (486 with 16 MB RAM, and Windows 3) for the next 3½ years on over 650 flight units for the Space Shuttle Main Engine. The PIND VI consists of the top level program and a hierarchy of subordinate virtual instrument modules developed to coordinate the PIND data acquisition process. Additional functions and modules are utilized from the LabVIEW function and VI libraries.

Description and Function of PIND-DAS System

The PIND virtual instrument handles all of the data acquisition, scaling, plotting, and storing to magnetic media for the PIND testing of SSME flight pressure transducers. Analog waveforms generated by the B&W Engineering Particle Impact Noise Detection (PIND) system are detected using a National Instruments AT-A2150C analog data input card, with analog input capable of 11025 acquisition scans per second, equivalent to 16 bit waveform resolution every 90.7 microseconds.

The data card's functions are coordinated by the VI, which initializes the DAS for data acquisition according to optional user inputs for number of samples, sample rate input channel number, and trigger level. The VI reads the external trigger input channel, and triggers when the voltage on the trigger channel falls below a user defined level. In each run, the last of four triggers starts the acquisition of actual data. The VI monitors operation of each module, detecting any operational errors and reporting them on the VI front panel for the user's information. Following acquisition, the VI scales the acquired data to volts, and plots the data versus time on the VI front panel (see Figure 1–13) for inspection by the user.

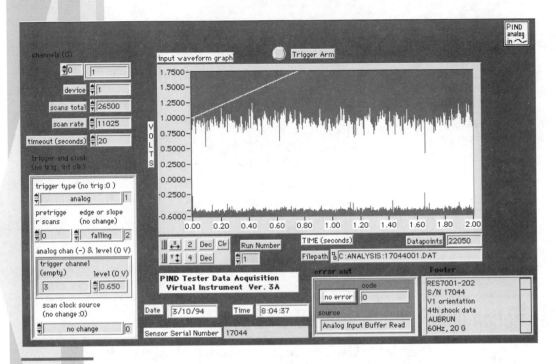

Figure 1–13
PIND3A Virtual Instrument (front panel)

The VI also provides the technician with an opportunity to accept or reject the data acquired after each run, and upon acceptance allows the user to specify the file name in which to store the acquired data. The voltage data and footer information (optional user definable data entered on the VI front panel) are stored together in each of the data files, along with file creation date and time determined from the computer system's internal clock. The data is stored in binary format, with time and voltage ordinates written in real single-precision pairs, and in chronological order. (See Figures 1–14, 1–15, 1–16, 1–17, and 1–18.)

The digitized waveform data files (LabVIEW output; **.dat** files), are processed later using PEAKS9B.EXE, a stand-alone FORTRAN program adapted from code previously developed at Rocketdyne for similar analytical purposes. This approach to post-processing was driven by the extremely heavy computational requirements and the advantage of batch operation over-

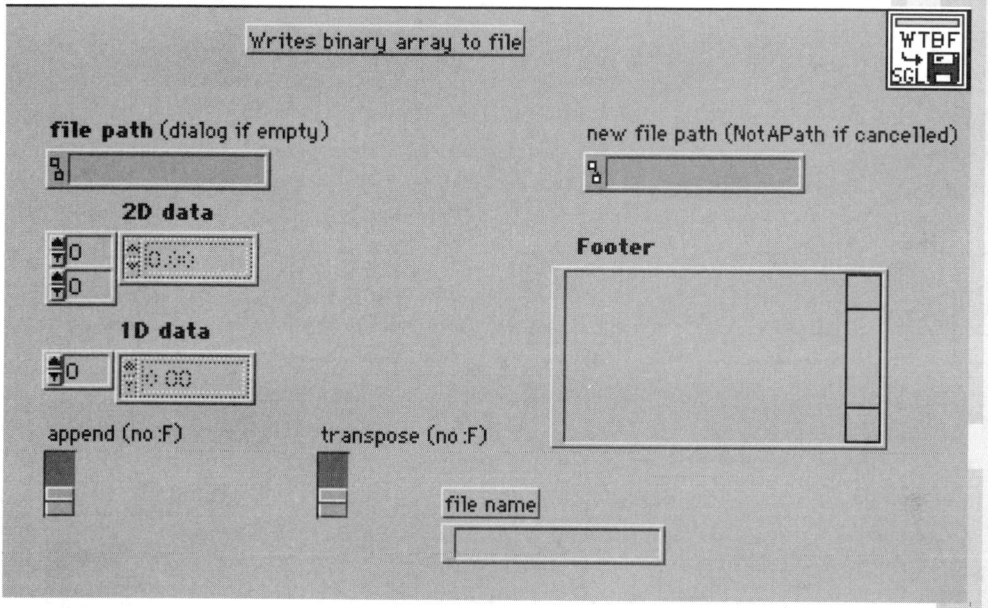

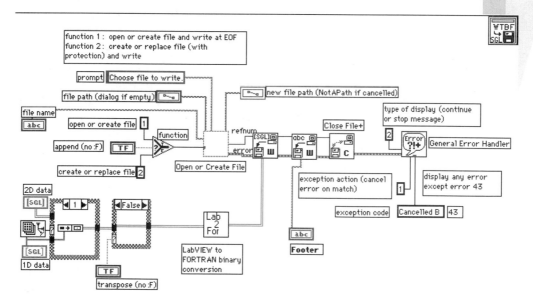

Figure 1–14
WTBF VI (front)

Figure 1–15
WTBF VI (diagram)

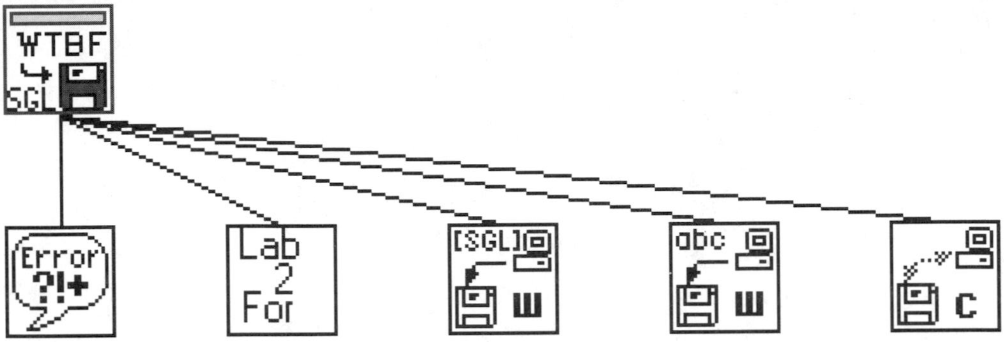

Figure 1–16
WTBF VI (hierarchy)

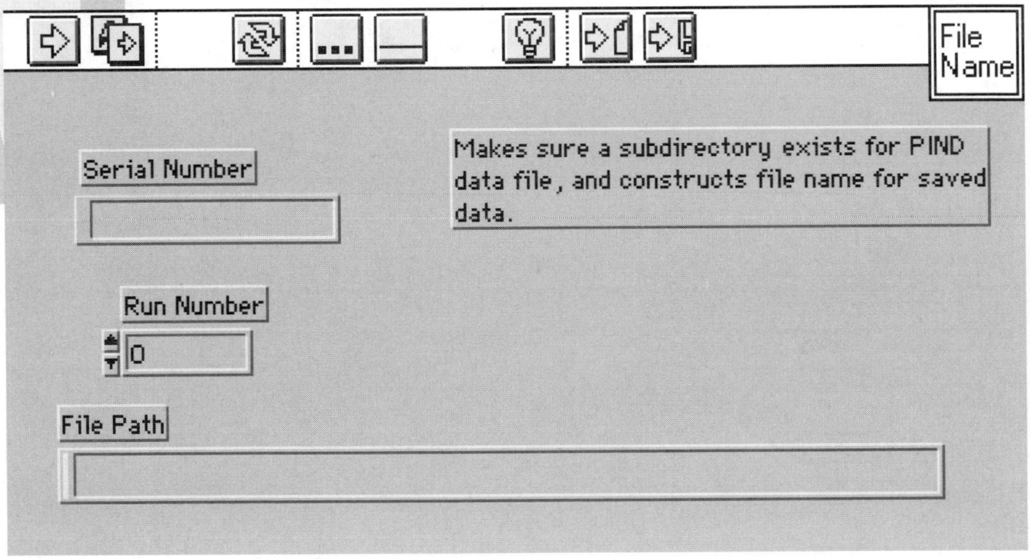

Figure 1–17
FNAME VI (front)

night on the numerous large data files acquired during the day. The post-processed data (**.mms** files) are then automatically imported into a spreadsheet for further interpretation and display using macros written specifically for that purpose.

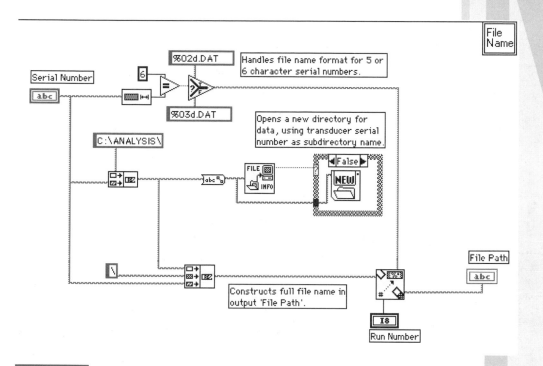

Figure 1–18
FNAME VI (diagram)

Development and Validation

A "top down" methodology and the techniques of structured programming were implemented in the development of the PIND VI and other developed modules, consisting of breaking the overall tasks into their major components, then breaking these into smaller components, and so on, until each component performs a single, easily manageable function. The LabVIEW environment allowed flexibility in the algorithm development, and the VI code was designed to be interactive so that the user could specify parameters such as number of data points to be analyzed and the degree of accuracy to be achieved. (See program hierarchy, Figure 1–20.)

Due to the particular workings of overhead routines outside of direct control, it was observed that sequencing of multiple

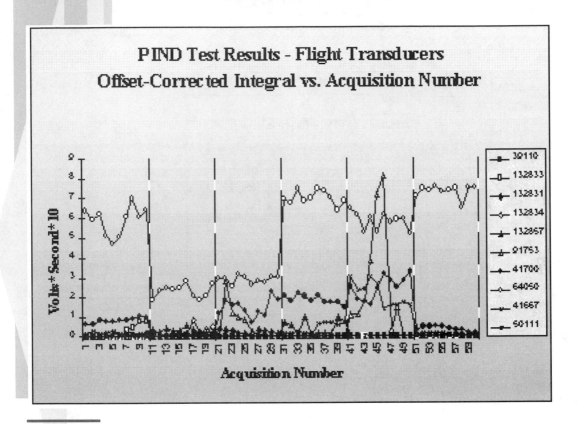

Figure 1–19
PIND test results

events did not guarantee the required acquisition delays in the desired sequence. Use of logical operators and appropriate trigger delays eliminated timeout errors and enabled data acquisition within the desired constraints.

After the VI was developed initially, formal validation and verification (V&V) was performed under the supervision of the Quality Assurance function through the testing of four pressure sensors chosen as benchmark units, and ongoing repeatability tests have been sustained using these units plus specialized (geometrically simple) reference standards. In addition to calibration of the shaker table (consistency of exitational vibrational energy being applied to the sensor), and of the microphone (consistent sensitivity of the transducer used to pick up the acoustical signal generated by particle impact noise within the sensor) of the PIND system, these benchmark tests have served as a good indicator of overall system "drift" on a time scale of months.

Structure and Operation

Upon execution of LabVIEW, the PIND3A front panel VI loads for execution, during which any needed subordinate VI's and library functions are called automatically. When started, PIND3A executes module AIWSA for four iterations. Each iteration monitors the PIND Tester trigger signal to count the PIND Tester vibration cycles. On the fourth iteration, module AIWSA acquires the PIND Tester data for further processing.

Following the fourth iteration of module AIWSA, PIND3A executes module FNAME ("Filename", Figures 1–17 and 1–18) which uses the serial number and run number entered by the user into the PIND3A front panel to construct the default path and file name under which to store the acquired PIND data. FNAME creates a subdirectory in the C:\ANALYSIS path in which to store this file (the default path and file name can be overridden by the user).

Following execution of module FNAME, PIND3A executes module WTBF ("Write To Binary File"; Figures 1–14, 1–15, and 1–16), which controls saving the PIND data in binary format (a format more efficient and compact than ASCII) onto the DAS computer hard disk. WTBF first uses a dialog box to obtain user confirmation of the path and file name. At this point, the user can change the file path and/or file name and save the data to the DAS computer hard disk, or the user can cancel the data save operation.

Upon choosing to save the acquired data, WTBF calls module LAB2FOR ("LabVIEW to FORTRAN"; Figures 1–22 and 1–23), which converts the LabVIEW single precision real data to FORTRAN binary format, meaning it converts the data from the

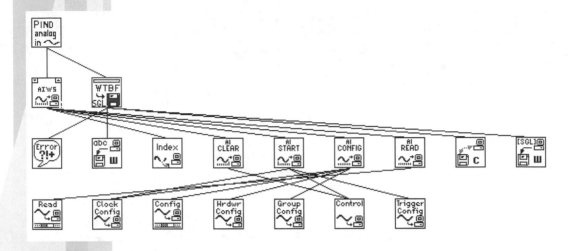

Figure 1–20
PIND3A Virtual Instrument (hierarchy)

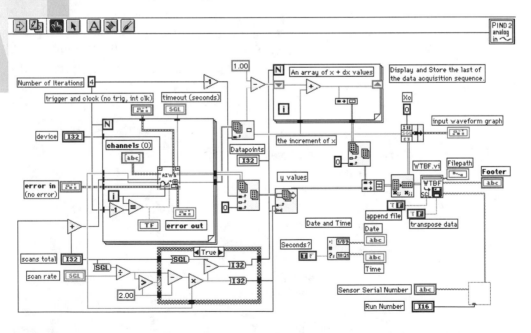

Figure 1–21
PIND3A Virtual Instrument (diagram)

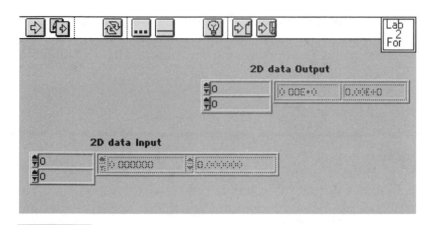

Figure 1–22
LAB2FOR (front)

general floating-point binary format to byte-ordered, platform-specific binary format readable by the external FORTRAN
post-processing application (in this case, "PEAKS"). LAB2FOR
then writes the data to the specified file, and terminates. Upon

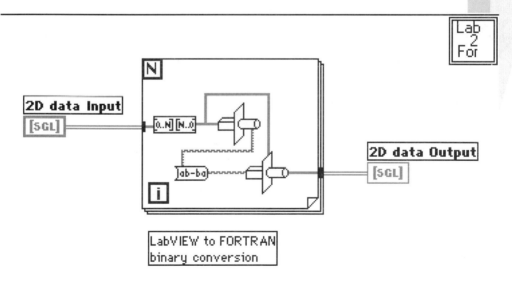

Figure 1–23
LAB2FOR (diagram)

choosing to cancel the data save operation, execution of PIND3A terminates and the acquired data is discarded. Otherwise, the data is saved for post-processing and the results accumulated for hardware allocation.

Acknowledgments

The PIND3A virtual instrument itself was developed by three individuals at Rocketdyne: John W. Reinert, Robert S. Nelson, and Robert B. Tarn. The collection and analysis of PIND data for assessment of flight sensor hardware involved dozens of others in a team effort.

■ **Contact Information**

Sonia Balcer
Instrumentation Development and Design
Rocketdyne Division, Boeing
Canoga Building 01, Dept-Grp 125-000, Mail stop AC83
Canoga Park, CA
818-586-0218 or 818-586-0200
Fax: 818-586-7197
Email: sonia.c.balcer@boeing.com

Overview

In this chapter, you will learn how to use the VIs in the Analysis Library to generate many different types of signals. Some of the applications for signal generation are

- *Simulating signals to test your algorithm when real-world signals are not available (for example, when you do not have a DAQ board for obtaining real-world signals).*

- *Generating signals to apply to a D/A converter (for example, in control applications such as opening or closing a valve).*

GOALS

- Learn about the concept of *normalized* frequency.
- Understand the difference between *Wave* and *Pattern* VIs (for example, the **Sine Wave** VI and the **Sine Pattern** VI).
- Build a simple function generator using the VIs in the **Signal Generation** subpalette.

KEY TERMS

- normalized frequency
- signal generator
- arbitrary wave
- sawtooth wave
- chirp

Signal Generation

Normalized Frequency

In the analog world, a signal frequency is measured in Hz or cycles per second. But the digital system often uses a digital frequency, which is the ratio between the analog frequency and the sampling frequency:

$$\text{digital frequency} = \frac{\text{analog frequency}}{\text{sampling frequency}}$$

This digital frequency is known as the *normalized* frequency. Its units are cycles/sample.

Some of the Signal Generation VIs use an input frequency control, f, that is assumed to use *normalized frequency* units of *cycles per sample*. This frequency ranges from 0.0 to 1.0, which corresponds to a real frequency range of 0 to the sampling frequency f_s. This frequency also wraps around 1.0, so that a normalized frequency of 1.1 is equivalent to 0.1. As an example, consider an analog signal of frequency $f_s/2$. If this signal is sampled at a frequency f_s, this will correspond to a normalized frequency of $1/2$ cycles/sample = 0.5 cycles/sample. The reciprocal of the normalized frequency, $1/f$, gives you the number of times that the signal is sampled in one cycle.

When you use a VI that requires the normalized frequency as an input, you must convert your frequency units to the normalized units of cycles/sample. You must use these normalized units with the following VIs:

- Sine Wave
- Square Wave
- Sawtooth Wave
- Triangle Wave
- Arbitrary Wave
- Chirp Pattern

If you are used to working in frequency units of cycles, you can convert cycles to cycles/sample by dividing cycles by the number of samples generated. Figure 2–1 shows the **Sine Wave** VI, which is being used to generate two cycles of a sine wave.

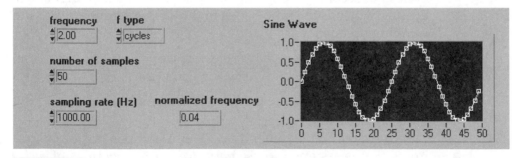

Figure 2–1
Frequency selection in units of cycles

Figure 2–2 shows the block diagram for converting cycles to cycles/sample.

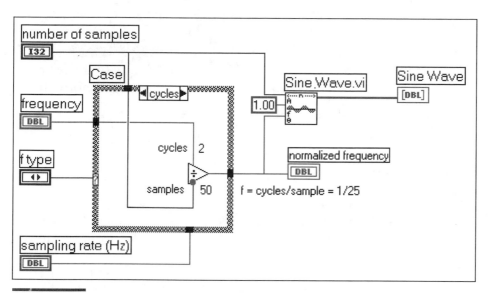

Figure 2–2
Conversion from cycles to cycles/samples

You simply divide the frequency (in cycles) by the number of samples. In the above example, the frequency of 2 cycles is divided by 50 samples, resulting in a normalized frequency of $f = 1/25$ cycles/sample. This means that it takes 25 (the reciprocal of f) samples to generate one cycle of the sine wave.

However, you may need to use frequency units of Hz (cycles/second). If you need to convert from Hertz (or cycles/second) to cycles/sample, divide your frequency in cycles/second by the sampling rate given in samples/second.

$$\frac{\text{cycles/second}}{\text{samples/second}} = \frac{\text{cycles}}{\text{sample}}$$

Figure 2–3 shows the **Sine Wave** VI used to generate a 60-Hz sine signal.

Figure 2–4 is a block diagram for generating a Hertz sine signal. You divide the frequency of 60 Hz by the sampling rate of

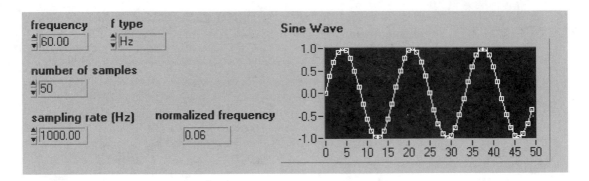

Figure 2–3
Frequency selection in units of Hertz

1000 Hz to get the *normalized* frequency of *f* = 0.06 cycles/sample. Therefore, it takes almost 17 (1/0.06) samples to generate one cycle of the sine wave.

The signal generation VIs create many common signals required for network analysis and simulation. You can also use

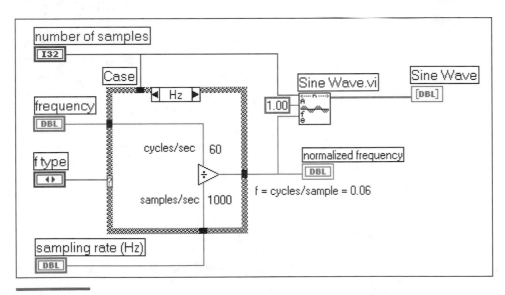

Figure 2–4
Conversion from Hertz to cycles/sample

the signal generation VIs in conjunction with National Instruments hardware to generate analog output signals.

Activity 2-1

Objective: To understand the concept of normalized frequency

1. Build the VI front panel and block diagram shown below.

■ Front Panel

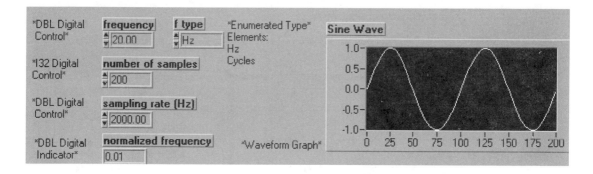

■ Block Diagram

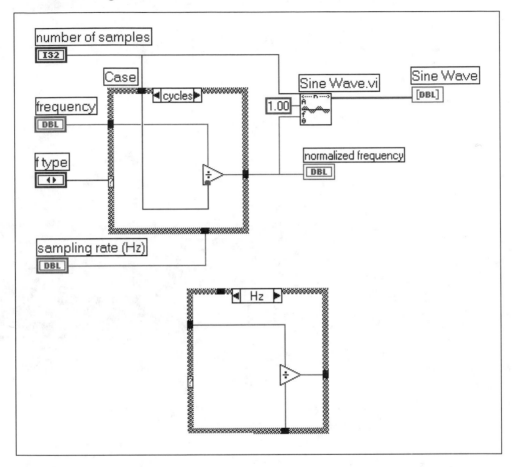

 Sine Wave VI (**Functions » Analysis » Signal Generation** subpalette).

2. Select a frequency of 2 cycles (**frequency** = 2 and **f type** = cycles) and **number of samples** = 100. Run the VI. Note that the plot will show 2 cycles. (The normalized frequency indicator tells you the normalized frequency.)

3. Increase the **number of samples** to 150, 200, and 250. How many cycles do you see?

4. Now keep the **number of samples** = 100. Increase the number of cycles to 3, 4, and 5. How many cycles do you see?

Thus, when you choose the frequency in terms of cycles, you will see that many cycles of the input waveform on the plot. Note that the sampling rate is irrelevant in this case.

5. Change **f type** to Hz and **sampling rate (Hz)** to 1000.
6. Keeping the **number of samples** fixed at 100, change the **frequency** to 10, 20, 30, and 40. How many cycles of the waveform do you see on the plot for each case? Explain your observations.
7. Repeat the above step by keeping the **frequency** fixed at 10 and change the **number of samples** to 100, 200, 300, and 400. How many cycles of the waveform do you see on the plot for each case? Explain your observations.
8. Keep the **frequency** fixed at 20 and the **number of samples** fixed at 200. Change the **sampling rate (Hz)** to 500, 1000, and 2000. Make sure you understand the results.
9. Save the VI as **Normalized Frequency.vi** in the library `Activities.llb`.

■ End of Activity 2-1

Wave and Pattern VIs

You will notice that the names of most of the signal generation VIs have the word *wave* or *pattern* in them. There is a basic difference in the operation of the two different types of VIs. It has to do with whether or not the VI can keep track of the phase of the signal that it generates each time it is called.

■ Phase Control

The *wave* VIs have a *phase in* control where you can specify the initial phase (in degrees) of the first sample of the generated waveform. They also have a *phase out* indicator that specifies what the phase of the next sample of the generated waveform is going to be. In addition, a *reset phase* control decides whether or not the phase of the first sample generated when the *wave* VI is called is the phase specified at the *phase in* control, or whether it is the phase available at the *phase out* control when the VI last executed. A TRUE value of *reset phase* sets the initial phase to *phase in*, whereas a FALSE value sets it to the value of *phase out* when the VI last executed.

The *wave* VIs are all reentrant (can keep track of phase internally) and accept frequency in normalized units (cycles/sample). The only *pattern* VI that presently uses normalized units is the **Chirp Pattern** VI. Setting the *reset phase* Boolean to FALSE allows for continuous sampling simulation.

Wave VIs are reentrant and accept the frequency input in terms of normalized units.

In Activity 2-3, you will generate a sine wave using both the **Sine Wave** VI and the **Sine Pattern** VI. You will see how in the **Sine Wave** VI you have more control over the initial phase than in the **Sine Pattern** VI.

Activity 2-2

Objective: To generate a sine wave of a particular frequency and see the effect of aliasing

■ Front Panel

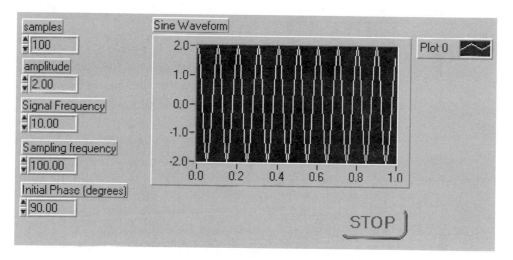

1. Open the **Generate Sine** VI from the library `Activities.llb`.

2. The front panel contains controls for the number of sample points to be generated, the amplitude, analog frequency, and initial phase (in degrees) of the sine wave to be generated, and the frequency at which this waveform is sampled.

3. Do not change the front panel default values. Switch to the block diagram.

■ Block Diagram

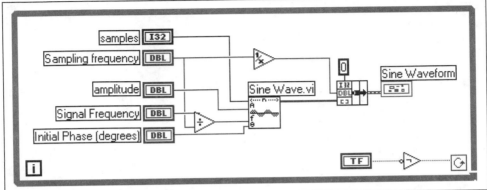

4. Examine the block diagram.

 Sine Wave VI (**Functions » Analysis » Signal Generation** subpalette). In this activity, this VI generates 100 points of a 10-Hz sine wave sampled at 100 Hz.

5. Notice in the block diagram that the signal frequency is divided by the sampling frequency *before* it is connected to the **Sine Wave** VI. This is because the **Sine Wave** VI requires the digital (normalized) frequency of the signal.

6. Run the VI. With the default front panel values, a 10-Hz sine wave should appear on the graph.

■ Sampling and Aliasing

7. Change the signal frequency on the front panel to 90 Hz and observe the waveform. The resulting signal looks just like the 10-Hz waveform.

As you saw in the previous chapter, this phenomenon is called *aliasing*, which occurs only in the digital domain. The famous Nyquist sampling theorem dictates that the highest representable useful frequency is at most half of the sampling frequency. In our case, the sampling frequency is 100 Hz, so the maximum representable frequency is 50 Hz. If the input frequency is over 50 Hz, as in our case of 90 Hz, it will be aliased back to $((n*50) - 90)$ Hz > 0, which is $(100 - 90)$ Hz, or 10 Hz. In other words, this digital system with a sampling frequency of 100 Hz cannot discriminate 10 Hz from 90 Hz, 20 Hz from 80 Hz, 51 Hz from 49 Hz, and so on.

The Importance of an Analog Antialiasing Filter

Therefore, in designing a digital system, you must make sure that any frequencies over half of the sampling frequency do not enter the system. *Once they are in, there is no way to remove them!* To prevent aliasing, you typically use an analog antialiasing lowpass filter. So, in this example, you can use an analog antialiasing filter to remove any frequencies over 50 Hz. After the signal is filtered, you are assured that whenever you see a 10-Hz signal with a 100-Hz sampling frequency, it is 10 Hz and not 90 Hz.

8. When you are done, stop the VI by clicking on the STOP button. Close the VI. Do not save any changes.

■ End of Activity 2-2

Activity 2-3

Objective:To generate a sinusoidal waveform using both the Sine Wave VI and the Sine Pattern VI and to understand the differences

1. Build the VI front panel and block diagram shown below.

■ Front Panel

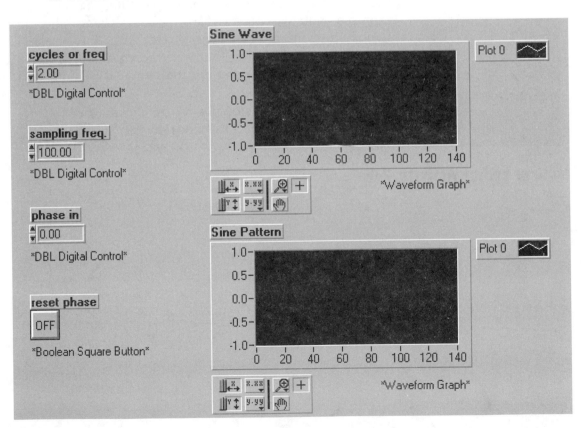

■ Block Diagram

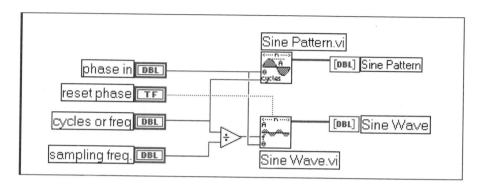

 Sine Pattern VI (**Functions » Analysis » Signal Generation** subpalette).

 Sine Wave VI (**Functions » Analysis » Signal Generation** subpalette).

2. Set the controls to the following values:

   cycles or freq: 2.00

   sampling freq: 100

   phase in: 0.00

   reset phase: OFF

   Run the VI several times.

   Observe that the **Sine Wave** plot changes each time you run the VI. Because **reset phase** is set to OFF, the phase of the sine wave changes with each call to the VI, being equal to the value of **phase out** during the previous call. However, the **Sine Pattern** plot always remains the same, showing two cycles of the sinusoidal waveform. The initial phase of the **Sine Pattern** plot is equal to the value set in the **phase in** control.

"Phase in" and "phase out" are specified in degrees.

3. Change **phase in** to 90 and run the VI several times. Just as before, the **Sine Wave** plot changes each time you run the VI. However, the **Sine Pattern** plot does not change; the initial phase of the sinusoidal pattern is 90 degrees—the same as that specified in the **phase in** control.

4. With **phase in** still at 90, set **reset phase** to ON and run the VI several times. The sinusoidal waveforms shown in both the **Sine Wave** and **Sine Pattern** plots start at 90 degrees but do not change with successive calls to the VI.

5. Keeping **reset phase** as ON, run the VI several times for each of the following values of **phase in**: 45, 180, 270, and 360. Note the initial phase of the generated waveform each time that the VI is run.

6. When you have finished, save the VI as **Wave and Pattern.vi** in the library `Activities.llb`.

■ End of Activity 2-3

Activity 2-4

Objective: To build a simple function generator

In this activity, you will build a very simple function generator that can generate the following waveforms:

- ■ Sine Wave
- ■ Square Wave
- ■ Triangle Wave
- ■ Sawtooth Wave

1. Build the VI front panel and block diagram shown below.

■ Front Panel

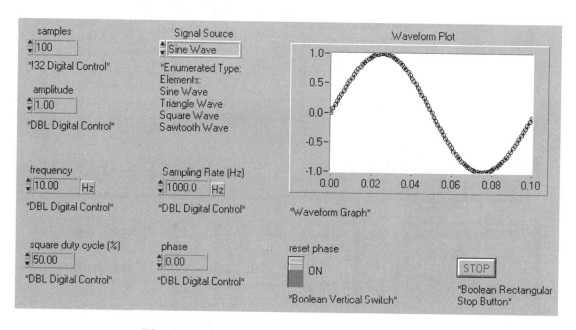

The **Signal Source** control selects the type of waveform that you want to generate.

The **square duty cycle (%)** control is used only for setting the duty cycle of the square wave.

The **samples** control determines the number of samples in the plot.

Note that these are all wave VIs, and therefore they require the frequency input to be the normalized frequency. So, you divide **frequency** by the **Sampling Rate (Hz)** and the result is the normalized frequency wired to the f input of the VIs.

■ Block Diagram

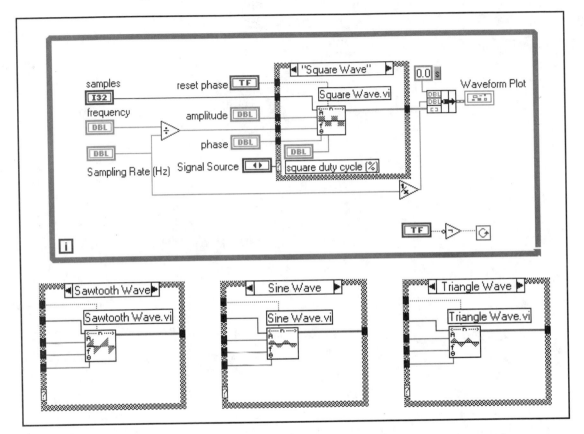

 Sine Wave VI (**Functions » Analysis » Signal Generation** subpalette) generates a sine wave of normalized frequency f.

 Triangle Wave VI (**Functions » Analysis » Signal Generation** subpalette) generates a triangular wave of normalized frequency f.

 Square Wave VI (**Functions » Analysis » Signal Generation** subpalette) generates a square wave of normalized frequency f with specified duty cycle.

Sawtooth Wave VI (**Functions » Analysis » Signal Generation** subpalette) generates a sawtooth wave of normalized frequency *f*.

2. Select a **Sampling Rate (Hz)** of 1000 Hz, **amplitude** = 1, **samples** = 100, **frequency** = 10, **reset phase** = ON, and **Signal Source** = Sine Wave. Note that because **Sampling Rate (Hz)** = 1000 and **frequency** = 10 Hz, every 100 samples corresponds to one cycle.

3. Run the VI and observe the resulting plot.

4. Change **samples** to 200, 300, and 400. How many cycles of the waveform do you see? Explain why.

5. With **samples** set to 100, change **reset phase** to OFF. Do you notice any difference in the plot?

6. Change **frequency** to 10.01 Hz. What happens? Why?

7. Change **reset phase** to ON. Now what happens? Explain why.

8. Repeat steps 4–7 for different waveforms selected in the **Signal Source** control. If you select the square wave, set the **square duty cycle (%)** to 50.00.

9. When you finish, stop the VI amd save it as **Function Generator.vi** in the library `Activities.llb`.

■ End of Activity 2-4

Wrap It Up!

In this chapter, you learned

- About the normalized frequency (*f*) that has units of cycles/sample.
- How to generate a sine wave of a particular frequency.
- That the wave VIs can keep track of the phase of the generated waveform.
- How to build a simple function generator that can generate a sine, square, triangular, and sawtooth wave.

■ Review Questions

1. Name two practical applications in which you would want to generate signals.

2. What is the normalized signal frequency for the following?

   a) sampling frequency = 100 Hz
      number of samples = 200
      signal frequency = 15 Hz

   b) sampling frequency = 100 Hz
      number of samples = 200
      signal frequency = 15 cycles

3. What are two main differences between the **Wave** and **Pattern** VIs?

4. Which of the following VIs require a normalized frequency input?

   a) Sine wave
   b) Sine pattern
   c) Chirp pattern
   d) Square wave

■ Additional Activities

1. Modify Activity 2-4 (**Function Generator.vi**) so that it can also generate an exponentially weighted sine wave as shown below:

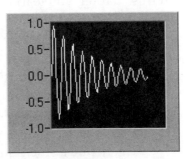

Such waveforms are found in the responses of damped oscillators, springs, RLC circuits, pendulums, and other oscillatory systems. They are extremely useful in providing stimulation to systems under test, and are commonly known as "ringdown."

2. Build a VI that can be used to generate arbitrary waveforms. For example, you can use a combination of the **Sine Pattern** and/or the **Ramp Pattern** VIs, along with some of the **Trigonometric (Numeric >> Trigonometric** subpalette) and **Logarithmic (Numeric >> Logarithmic** subpalette) VIs, in combination with the **Build Array (Array** subpalette) VI.

3. A chirp signal is one whose frequency increases (or decreases) linearly with time. The time waveform of a chirp with increasing frequency is as shown:

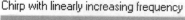

Chirp with linearly increasing frequency

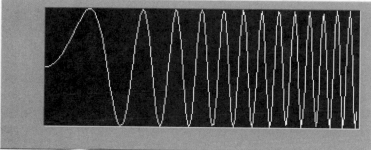

A chirp could be used to excite a physical structure and to measure the amplitude of its vibration throughout the corresponding frequency range. The purpose is to determine the resonant frequency of vibration. If necessary, the design of the structure may be modified so that the resonant frequency of vibration is outside that of the desired operating range. Build a VI that generates a chirp signal whose amplitude, and minimum and maximum frequencies, can be modified by controls on the front panel.

Use the Chirp Pattern (Analysis >> Signal Generation subpalette) VI.

4. Open and run the **Two Channel Oscilloscope** VI from the **Examples >> Apps >> Demos.llb** library in your LabVIEW folder. The front panel is as shown:

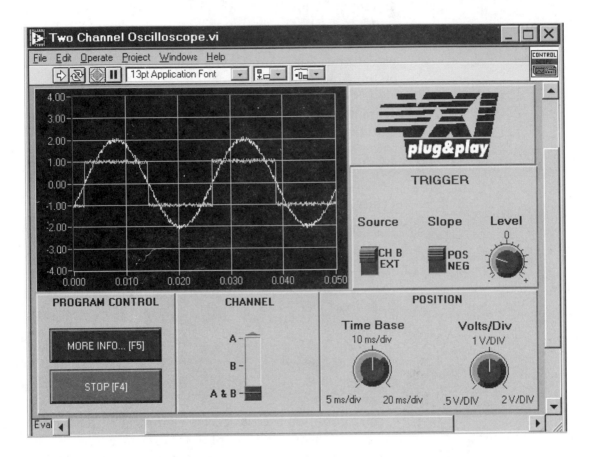

This VI uses some of the VIs from the Signal Generation library to simulate waveforms that can be seen on the front panel of the oscilloscope. It simulates a noisy sine wave on channel B and a noisy square wave on channel A.

a) The **Channel** control selects which waveform to display. Change the **Channel** control to its different settings (**A, B, A & B**) and observe the displayed waveform (s).

b) The **Position** controls (**Time Base** and **Volts/Div**) are used to adjust the scaling in the x and y axes respectively. Modify these controls and observe the corresponding changes in the displayed waveform (s).

c) The **Trigger** controls (**Source, Slope** and **Level**) select the triggering mode.

   i) Leave the **Source** control on **CH B** and change the **Slope** control between **POS** (positive) and **NEG** (negative). Observe the changes in the left edge of the display.

   ii) Leave the **Source** control on **CH B** and change the **Level** control between positive and negative values. Again, observe the left edge of the display.

d) From the menu bar, select **Project >> This VI's subVIs >> Channel A and/or B (demo).vi.** From the menu bar of the **Channel A and/or B (demo).vi** select **Windows >> Show Diagram.** The block diagram shows how the noisy signals are generated using the VIs in the **Signal Generation** subpalette.

e) When you finish, close the VI.

This activity can only be done with the LabVIEW full development system.

Signal Processing in Industrial Testing Using LabVIEW™

Graham Douglas
Systems Engineer—Controls
Delco Defense Systems Operations

The Challenge To design and implement a stabilization test for armament systems.

The Solution The LabView Analysis VIs and National Instruments' plug-in boards are used to build the system which would perform the stabilization test.

Figure 2–5
Rate stabilized 25-mm cannon industrial testing

Introduction

Delco Defense Systems Operations develops armament systems like the two-axis stabilized Bushmaster cannon shown on the vehicle above. Stabilization keeps the gun pointing in inertial space as the vehicle traverses rough terrain. All these systems undergo a rigorous series of tests at the end of assembly to ascertain their correct functioning. A test of major importance is the rate stabilization frequency response, also known as Bode testing. In exercising a stabilized axis, LabVIEW is first used to drive the system with sinusoidal excitation and then to analyze the results. Swept sine wave testing has several advantages over pseudorandom noise in nonlinear systems. Each discrete frequency can have its own input amplitude, low enough to stimulate the system without running afoul of nonlinearities, yet high enough to have sufficient quantization from output sensor A/D and D/A converters.

Signal Processing and Analysis

Figure 2–6 shows the use of the **Sine PatternVI** in conjunction with the **AO Continuous ReGenVI** to generate the stimulus. Notice also that the amplitudes are increased as the system response falls off at the higher frequencies.

National Instruments' computer plug-in boards are used to generate the stimulus and gather the response. The MIO-16L

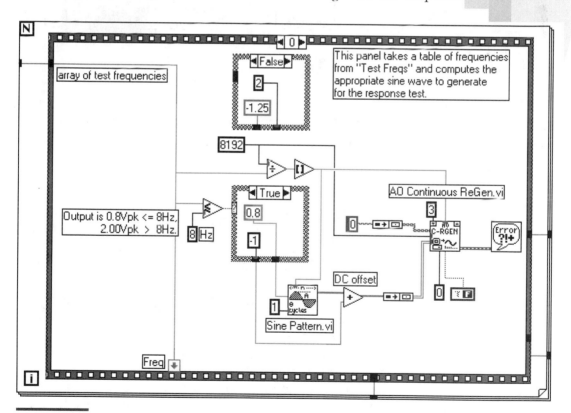

Figure 2–6
Stimulus generation

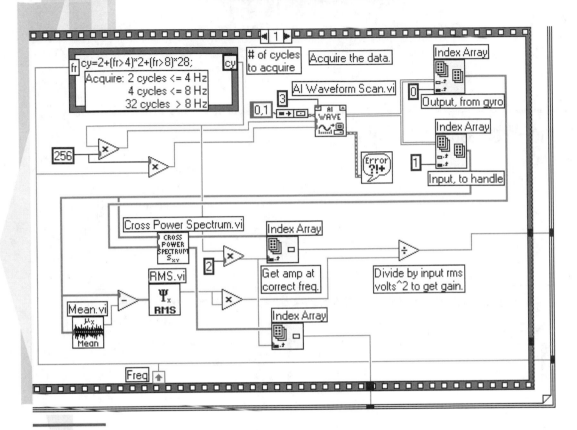

Figure 2–7
Extracting the response/stimulus

board can output a continuous signal while using Direct Memory Access (DMA) to gather the system's response.

Finally we arrive at the most interesting part from a signal processing view—the extraction of magnitude and phase for use in the Bode chart. The current frequency enters the sequence frame at the bottom and is used by a formula node to provide a table of test cycles versus frequency. At higher frequencies the system response is falling off and more cycles are used to extract the output in the presence of industrial noise. The AI Waveform Input Scan block has channels zero and one identified as signal sources. A good workhorse, the Index Array block is used to split the 2D array into a 1D array of singles. These arrays are operated on by the Cross Power Spectrum.vi which develops the cross-power spectrum. The upper output of this VI is in volts/rms squared while the lower output is the phase difference between the two inputs.

The gunner's method of controlling the rate stabilized gun is by moving a handle while observing the target through a telescopic sight mounted on the same structure as the weapon. Remember that in the generation of the stimulus a DC component was added to the sine wave. The real-world necessity for adding this component was to avoid velocity reversals and the attendant coulomb friction nonlinearities. So, if we follow the signal labeled "Input to handle" we see that the DC component is removed by using the Mean.vi. The remainder of the magnitude of signal extraction is straightforward. The indexed output of the Cross Power Spectrum VI is divided by the square of the RMS handle input to provide the ratio of gyro output to handle input. The array of magnitude and phase outputs leave the sequence on the For Loop for processing into a logarithmic Bode chart. The chart is archived and several figures of merit, gain and phase margin, are presented to the operator for pass/fail comparison.

■ Contact Information

Graham Douglas
Systems Engineer—Controls
Delco Defense Systems Operations
7410 Hollister Av, Goleta, CA 93117
805-961-7078
Fax: 805-961-7242
Email: lnusdo2.xzv1td@gmeds.com

Overview

In this chapter, you will learn the basics of transforming a signal from the time domain into the frequency domain.

Goals

- Learn about the Discrete Fourier Transform (DFT) and the Fast Fourier Transform (FFT).
- Determine the frequency spacing between the samples of the FFT (that is, the relationship between the sampling frequency fs, number of samples N, and the frequency spacing Δf).
- Familiarize yourself with the power spectrum and how it differs from both the DFT and the FFT.
- Understand how to interpret the information in the frequency domain for the DFT/FFT and the power spectrum, for both even and odd N.

Key Terms

- time domain
- frequency domain
- Fourier Transform
- sampling frequency
- frequency spacing
- power spectrum
- zero padding

Signal Processing

3

The Discrete Fourier Transform (DFT) and the Fast Fourier Transform (FFT)

The samples of a signal obtained from a DAQ board constitute the *time-domain* representation of the signal. This representation gives the amplitudes of the signal at the instants of *time* during which it had been sampled. However, in many cases you want to know the frequency content of a signal rather than the amplitudes of the individual samples. The representation of a signal in terms of its individual frequency components is known as the *frequency-domain* representation of the signal. The frequency-domain representation could give more insight about the signal and the system from which it was generated.

The algorithm used to transform samples of the data from the time domain into the frequency domain is known as the *Discrete Fourier Transform* or DFT. The DFT establishes the relationship between the samples of a signal in the time domain and their representation in the frequency domain. The DFT is widely used in the fields of spectral analysis, applied mechanics, acoustics, medical imaging, numerical analysis, instrumentation, and telecommunications.

Suppose you have obtained N samples of a signal from a DAQ board. If you apply the DFT to N samples of this time-domain representation of the signal, the result is also of length N samples, but the information it contains is of the frequency-domain representation. Figure 3–1 shows both the time domain & frequency domain representations of 32 samples of a sine wave. The relationship between the N samples in the time domain and the N samples in the frequency domain is explained below.

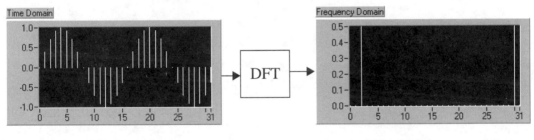

time-domain representation   frequency-domain representation

Figure 3–1
Time domain and frequency domain representations of a sampled signal

If the signal is sampled at a sampling rate of f_s Hz, then the time interval between the samples (that is, the sampling interval) is Δt, where

$$\Delta t = \frac{1}{f_s}$$

The sample signals are denoted by $x[i]$, $0 \leq i \leq N - 1$ (that is, you have a total of N samples). When the discrete Fourier transform, given by

$$X[k] = \sum_{i=0}^{N-1} x[i]e^{-j2\pi ik/N} \text{ for } k = 0, 1, 2, ..., N-1 \quad (3.1)$$

is applied to these N samples, the resulting output ($X[k]$, $0 \leq k \leq N - 1$) is the frequency-domain representation of $x[i]$. Note that both the time domain **x** and the frequency domain **X** have a total of N samples. Analogous to the *time spacing* of Δt between the samples of x in the time domain, you have a frequency spacing of

$$\Delta f = \frac{f_s}{N} = \frac{1}{N\Delta t}$$

between the components of **X** in the frequency domain. Δf is also known as the *frequency resolution*. To increase the frequency resolution (smaller Δf) you must either increase the number of samples N (with f_s constant) or decrease the sampling frequency f_s (with N constant).

In the following example, you will go through the mathematics of Equation 3.1 to calculate the DFT for a DC signal.

■ DFT Calculation Example

In the next section, you will see the exact frequencies to which the N samples of the DFT correspond. For the present discussion, assume that $X[0]$ corresponds to DC, or the average value, of the signal. To see the result of calculating the DFT of a waveform with the use of Equation 3.1, consider a DC signal having a constant amplitude of +1 V. Four samples of this signal are taken, as shown in Figure 3–2.

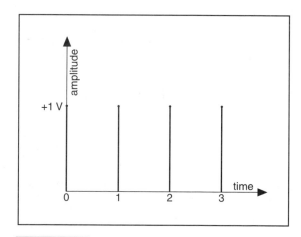

Figure 3–2
Samples of a 1V DC signal

Each of the samples has a value +1, giving the time sequence

$$x[0] = x[1] = x[2] = x[3] = 1$$

Using Equation 3.1 to calculate the DFT of this sequence and making use of Euler's identity,

$$\exp(-j\theta) = \cos(\theta) - j\sin(\theta) \quad \text{where } j = \sqrt{-1}$$

you get

$$X[0] = \sum_{i=0}^{N-1} x[i]e^{-j2\pi i 0/N} = x[0] + x[1] + x[2] + x[3] = 4$$

$$X[1] = x[0] + x[1]\left(\cos\left(\frac{\pi}{2}\right) - j\sin\left(\frac{\pi}{2}\right)\right) + x[2](\cos(\pi) - j\sin(\pi)) +$$

$$x[3]\left(\cos\left(\frac{3\pi}{2}\right) - j\sin\left(\frac{3\pi}{2}\right)\right) = (1 - j - 1 + j) = 0$$

$$X[2] = x[0] + x[1](\cos(\pi) - j\sin(\pi)) + x[2](\cos(2\pi) - j\sin(2\pi)) +$$
$$x[3](\cos(3\pi) - j\sin(3\pi)) = (1 - 1 + 1 - 1) = 0$$

$$X[3] = x[0] + x[1]\left(\cos\left(\frac{3\pi}{2}\right) - j\sin\left(\frac{3\pi}{2}\right)\right) + x[2](\cos(3\pi) -$$

$$j\sin(3\pi)) + x[3]\left(\cos\left(\frac{9\pi}{2}\right) - j\sin\left(\frac{9\pi}{2}\right)\right) = (1 + j - 1 - j) = 0$$

Therefore, except for the DC component, $X[0]$, all the other values are zero, which is as expected. However, the calculated value of $X[0]$ depends on the value of N (the number of samples). Because you had $N = 4$, this resulted in $X[0] = 4$. If $N = 10$, then you would have calculated $X[0] = 10$. This dependency of $X[.]$ on N also occurs for the other frequency components. Thus, you usually divide the DFT output by N, so as to obtain the correct magnitude of the frequency component.

■ Magnitude and Phase Information

You have seen that N samples of the input signal result in N samples of the DFT. That is, the number of samples in both the time and frequency representations is the same. From Equation 3.1, you see that regardless of whether the input signal $x[i]$ is real or complex, $X[k]$ is always complex (although the imaginary part may be zero). Thus, because the DFT is complex, it contains two pieces of information—the magnitude and the phase. It turns out that for real signals ($x[i]$ real) such as those obtained from the output of one channel of a DAQ board, the DFT is symmetric about the index $N/2$ (the Nyquist index) with the following properties:

$$| X[k] | = | X[N - k] | \text{ and phase}(X[k]) = - \text{phase}(X[N - k])$$

The terms used to describe this symmetry are that the magnitude of $X[k]$ is *even symmetric*, and phase($X[k]$) is *odd symmetric*. An even symmetric signal is one that is symmetric about the y-axis (like a cosine), whereas an odd symmetric signal is antisymmetric about the y-axis (like a sine). This is shown in Figure 3–3.

The net effect of this symmetry in the DFT of real signals is that there is repetition of information contained in the N complex samples of the DFT. Because of this repetition of information, only half of the samples of the DFT of real signals actually need to be computed or displayed, and the other half can be obtained from this assumption of symmetry.

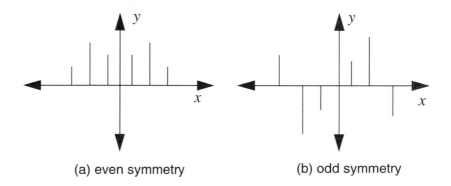

(a) even symmetry   (b) odd symmetry

Figure 3–3
Even and odd symmetry

If the input signal is complex, the DFT will be nonsymmetric and you cannot use this trick.

Frequency Spacing and Symmetry of the DFT/FFT

For a sampling interval of Δt seconds, with the first data sample corresponding to $t = 0$ seconds, the ith data sample is at $i\Delta t$ seconds. Similarly, a frequency resolution of Δf ($\Delta f = f_s/N$) means that the kth sample of the DFT occurs at a frequency of $k\Delta f$ Hz. (Actually, as you will soon see, this is valid for only up to half the number of samples. The other half represent negative frequency components.) Depending on whether the number of samples, N, is even or odd, you can have a different interpretation of the frequency corresponding to the kth sample of the DFT.

■ Even Number of Samples

For example, suppose N is even and let $p = N/2$. Table 3–1 shows the frequency to which each element of the complex output sequence **X** corresponds. For N even, $N/2$ is an integer, and so the DFT contains the exact Nyquist component at this index.

Table 3–1 *Correspondence between actual frequency and sample number for N even—note position of the Nyquist frequency*

Array Element	Corresponding Frequency
$X[0]$	DC component
$X[1]$	Δf
$X[2]$	$2\Delta f$
$X[3]$	$3\Delta f$
. . .	. . .

(Continued)

Table 3–1 *Correspondence between actual frequency and sample number for*
N *even—note position of the Nyquist frequency*

Array Element	Corresponding Frequency
$X[p-2]$	$(p-2)\Delta f$
$X[p-1]$	$(p-1)\Delta f$
$X[p]$	$p\Delta f$ (Nyquist frequency)
$X[p+1]$	$- (p-1)\Delta f$
$X[p+2]$	$- (p-2)\Delta f$
$\cdot$ $\cdot$ $\cdot$	$\cdot$ $\cdot$ $\cdot$
$X[N-3]$	$- 3\Delta f$
$X[N-2]$	$- 2\Delta f$
$X[N-1]$	$- \Delta f$

Again, note that the pth element, $X[p]$, corresponds to the
Nyquist frequency. The negative entries in the second column
beyond the Nyquist frequency represent *negative* frequencies.

For example, if $N = 8$, $p = N/2 = 4$, then each element corresponds to the frequencies shown in the following table.

Table 3–2 *Frequency correspondence for* N *even*

Element	Frequency
$X[0]$	DC
$X[1]$	Δf
$X[2]$	$2\Delta f$
$X[3]$	$3\Delta f$
$X[4]$	$4\Delta f$ (Nyquist freq.)
$X[5]$	$-3\Delta f$
$X[6]$	$-2\Delta f$
$X[7]$	$-\Delta f$

If this were a DFT of a real signal, $X[1]$ and $X[7]$ would have the
same magnitude, $X[2]$ and $X[6]$ would have the same magnitude,

and $X[3]$ and $X[5]$ would have the same magnitude. The difference is that whereas $X[1]$, $X[2]$, and $X[3]$ correspond to positive frequency components, $X[5]$, $X[6]$, and $X[7]$ correspond to negative frequency components. Note that $X[4]$ is at the Nyquist frequency.

Figure 3–4 represents this complex sequence for $N = 8$.

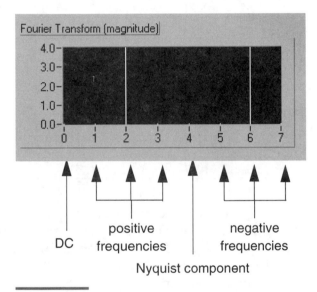

Figure 3–4
Two-sided transform for N even

Such a representation, where you see both the positive and negative frequencies, is known as the *two-sided* transform.

■ Odd Number of Samples

Now suppose that N is odd. Let $p = (N - 1)/2$. Table 3–3 shows the frequency to which each element of the complex output sequence **X** corresponds.

Note that when N is odd, $N/2$ is not an integer, and thus there is no component at the Nyquist frequency.

Table 3–3 *Correspondence between actual frequency and sample number for N odd—note absence of the Nyquist frequency.*

Array Element	Corresponding Frequency
$X[0]$	DC component
$X[1]$	Δf
$X[2]$	$2\Delta f$
$X[3]$	$3\Delta f$
$\vdots$	$\vdots$
$X[p-1]$	$(p-1)\Delta f$
$X[p]$	$p\Delta f$
$X[p+1]$	$-p\Delta f$
$X[p+2]$	$-(p-1)\Delta f$
$\vdots$	$\vdots$
$X[N-3]$	$-3\Delta f$
$X[N-2]$	$-2\Delta f$
$X[N-1]$	$-\Delta f$

For example, if $N = 7$, $p = (N-1)/2 = (7-1)/2 = 3$, and you have

Table 3–4 *Frequency correspondence for N odd*

Element	Frequency
$X[0]$	DC
$X[1]$	Δf
$X[2]$	$2\Delta f$
$X[3]$	$3\Delta f$
$X[4]$	$-3\Delta f$
$X[5]$	$-2\Delta f$
$X[6]$	$-\Delta f$

Now, if this were a DFT of a real signal, $X[1]$ and $X[6]$ have the same magnitude, $X[2]$ and $X[5]$ have the same magnitude, and

X[3] and X[4] have the same magnitude. However, whereas X[1], X[2], and X[3] correspond to positive frequencies, X[4], X[5], and X[6] correspond to negative frequencies. Because N is odd, there is no component at the Nyquist frequency.

Figure 3–5 represents the preceding table for N = 7.

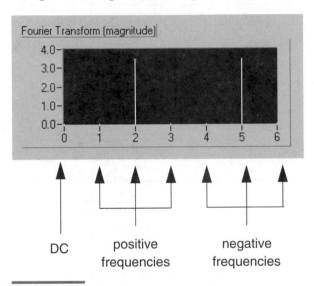

Figure 3–5
Two-sided transform for N odd

This is also a two-sided transform, because you have both the positive and negative frequencies.

■ Fast Fourier Transform

Direct implementation of the DFT (Equation 3.1) on N data samples requires approximately N^2 complex operations and is a time-consuming process. However, when the size of the sequence is a power of 2,

$$N = 2^m \text{ for } m = 1, 2, 3, \ldots$$

the computation of the DFT can be reduced to approximately $N \log_2(N)$ operations. This makes the calculation of the DFT much faster, and Digital Signal Processing (DSP) literature refers to these algorithms as Fast Fourier Transforms (FFTs). The FFT is nothing but a fast algorithm for calculating the DFT when the number of samples (N) is a power of 2. In some cases, Fast Fourier Transforms can be implemented for other special sequence lengths.

The advantages of the FFT include speed and memory efficiency. The size of the input sequence, however, must be a power of 2. The DFT can efficiently process any size sequence, but the DFT is slower than the FFT and uses more memory, because it must store intermediate results during processing.

■ Zero Padding

A technique employed to make the input sequence size equal to a power of 2 is to add zeros to the end of the sequence so that the total number of samples is equal to the next higher power of 2. For example, if you have 10 samples of a signal, you can add six zeros to make the total number of samples equal to 16 (= 2^4—a power of 2). This is shown in Figure 3–6.

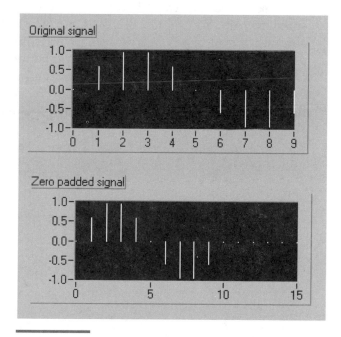

Figure 3–6
Effect of zero padding

In addition to making the total number of samples a power of two so that faster computation is made possible by using the FFT, zero padding also helps in increasing the frequency resolution (recall that $\Delta f = f_s/N$) by increasing the number of samples, N. This method, however, does not actually improve the resolution of the frequency estimate, but rather it provides frequency domain interpolation to increase resolution.

■ FFT VIs in the Analysis Library

The Analysis Library contains two VIs that compute the FFT of a signal. They are the **Real FFT** and **Complex FFT**.

The difference between the two VIs is that the **Real FFT** computes the FFT of a real-valued signal, whereas the **Complex FFT** computes the FFT of a complex-valued signal. However, keep in mind that the outputs of both VIs are complex.

Most real-world signals are real valued, and hence you can use the **Real FFT** for most applications. Of course, you could also use the **Complex FFT** by setting the imaginary part of the signal to zero. You should use the **Complex FFT** when the signal consists of both a real and an imaginary component. Such complex valued signals occur frequently in the field of telecommunications, where you modulate a waveform by a complex exponential. The process of modulation by a complex exponential results in a complex signal, as shown below:

$$e^{-j\omega t}$$

$$x(t) \longrightarrow \boxed{\times} \longrightarrow y(t) = x(t) \cos(\omega t) - jx(t) \sin(\omega t)$$

The block diagram in Figure 3–7 shows a simplified version of how you can generate a complex signal.

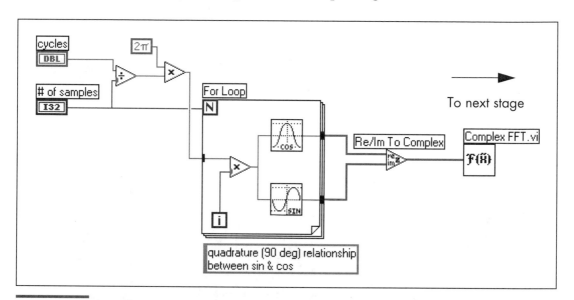

Figure 3–7
Generating a complex signal

Activity 3-1

Objective: To display the two-sided and the one-sided Fourier transform of a signal using the Real FFT VI, and to observe the effect of aliasing in the frequency spectrum

1. Build the VI front panel and block diagram as shown below.

■ Front Panel

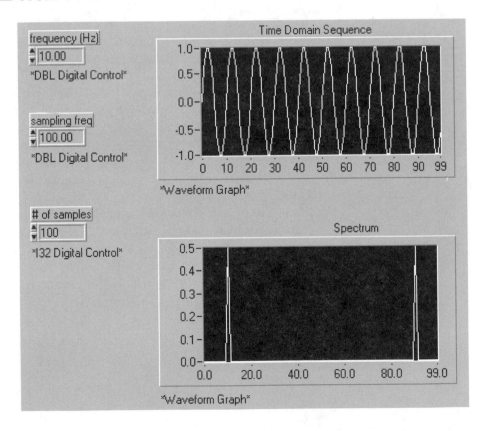

■ Block Diagram

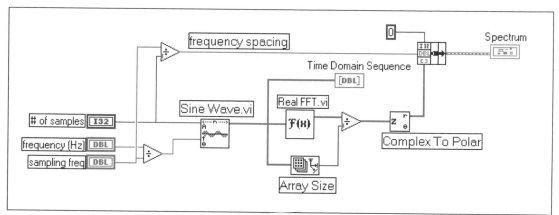

 The output of the **Array Size** function (**Functions » Array** subpalette) is used to scale the output of the FFT by the number of samples so as to obtain the correct amplitude of the frequency components.

Sine Wave VI (Functions » Analysis » Signal Generation subpalette) generates a time domain sinusoidal waveform.

Real FFT VI (Functions » Analysis » Digital Signal Processing subpalette) computes the FFT of the input data samples. The output of the **Real FFT VI** is divided by the FFT size (number of data points) to obtain the correct values.

Complex to Polar function (**Functions » Numeric » Complex** subpalette) separates the complex output of the FFT into its magnitude and phase parts. The phase information is in units of radians. Here you are displaying only the magnitude of the FFT.

The frequency spacing, Δf, is obtained by dividing the **sampling freq** by the **# of samples**.

2. Select **frequency (Hz)** = 10, **sampling freq** = 100, and **# of samples** = 100. Run the VI.

Notice the plots of the time waveform and the frequency spectrum. Because **sampling freq = # of samples** = 100, you are in effect sampling for 1 second. Thus, the number of cycles of the sine wave you see in the time waveform is equal to the **frequency** you select. In this case, you will see 10 cycles. (If you change the **frequency (Hz)** to 5, you will see 5 cycles.)

■ Two-Sided FFT

3. Examine the frequency spectrum (the Fourier transform). You will notice two peaks, one at 10 Hz and the other at 90 Hz. The peak at 90 Hz actually represents the frequency of –10 Hz. The plot you see is known as the *two-sided FFT* because it shows both the positive and the negative frequencies.

4. Run the VI with **frequency (Hz)** = 10 and then with **frequency (Hz)** = 20. For each case, note the shift in both peaks of the spectrum.

Also observe the time domain plot for frequency = 10 and 20. Which one gives a better representation of the sine wave? Why?

5. Because f_s = 100 Hz, you can accurately sample only signals having a frequency < 50 Hz (Nyquist frequency = $f_s/2$). Change **frequency (Hz)** to 48 Hz. You should see the peaks at ± 48 Hz on the spectrum plot.

6. Now change **frequency (Hz)** to 52 Hz. Is there any difference between the result of step 5 and what you see on the plots now? Because 52 > Nyquist, the frequency of 52 is aliased to $|100 - 52| = 48$ Hz.

7. Change **frequency (Hz)** to 30 Hz and 70 Hz and run the VI. Is there any difference between the two cases? Explain why.

8. Save this VI as **FFT_2sided.vi** in the library `Activities.llb`.

■ One-Sided FFT

9. Modify the block diagram of the VI as shown in the following diagram. You have seen that the FFT had repetition of information because it contained information about both the positive and the negative frequencies. This modification now shows only half the FFT points (only the positive frequency components). This representation is known as the *one-sided FFT*. The one-sided FFT shows only the positive frequency components. Note that you need to multiply the positive frequency components by two to obtain the correct amplitude. The DC component, however, is left untouched.

■ Block Diagram

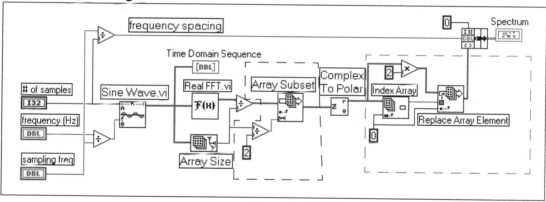

10. Run the VI with the following values: **frequency** = 30, **sampling freq** = 100, **# of samples** = 100.
11. Change the value of **frequency** to 70 and run the VI. Do you notice any difference between this and the result of step 10?
12. Save the VI as **FFT_1sided.vi** in the library `Activities.llb`.

■ End of Activity 3-1

The Power Spectrum

You have seen that the DFT (or FFT) of a real signal is a complex number, having a real and an imaginary part. The *power* in each frequency component represented by the DFT/FFT can be obtained by squaring the magnitude of that frequency component. Thus, the power in the kth frequency component (the kth element of the DFT/FFT) is given by $|X[k]|^2$. The plot showing the power in each of the frequency components is known as the *power spectrum*. Because the DFT/FFT of a real signal is symmetric, the power at a positive frequency of $k\Delta f$ is the same as the power at the corresponding negative frequency of $-k\Delta f$ (DC and Nyquist components not included). The total power in the DC and Nyquist (assuming N is even) components are $|X[0]|2$ and $|X[N/2]|2$, respectively.

■ Loss of Phase Information

Because the power is obtained by squaring the magnitude of the DFT/FFT, the power spectrum is always real and all the phase information is lost. If you want phase information, you must use the DFT/FFT, which gives you a complex output.

You can use the power spectrum in applications where phase information is not necessary (for example, to calculate the harmonic power in a signal). You can apply a sinusoidal input to a nonlinear system and see the power in the harmonics at the system output.

■ Frequency Spacing Between Samples

You can use the **Power Spectrum** VI in the **Analysis » Digital Signal Processing** subpalette to calculate the power spectrum of the time domain data samples. Just like the DFT/FFT, the number of samples from the **Power Spectrum** VI output is the same as the number of data samples applied at the input. Also, the frequency spacing between the output samples is $\Delta f = f_s/N$.

In Table 3–5, the power spectrum of a signal $x[i]$ is represented by Sxx. If N is even, let $p = N/2$. The table shows the format of the output sequence Sxx corresponding to the power spectrum.

Table 3–5 *Correspondence between actual frequency and sample number for* N *even—note position of the Nyquist frequency*

Array Element	Interpretation
$Sxx[0]$	Power in DC component
$Sxx[1] = Sxx[N-1]$	Power at frequency Δf
$Sxx[2] = Sxx[N-2]$	Power at frequency $2\Delta f$
$Sxx[3] = Sxx[N-3]$	Power at frequency $3\Delta f$
. . .	. . .
$Sxx[p-2] = Sxx[N-(p-2)]$	Power at frequency $(p-2)\Delta f$
$Sxx[p-1] = Sxx[N-(p-1)]$	Power at frequency $(p-1)\Delta f$
$Sxx[p]$	Power at Nyquist frequency

Figure 3–8 represents the information in the preceding table for a sine wave with amplitude = 2 V_{peak} (V_{pk}), and $N = 8$.

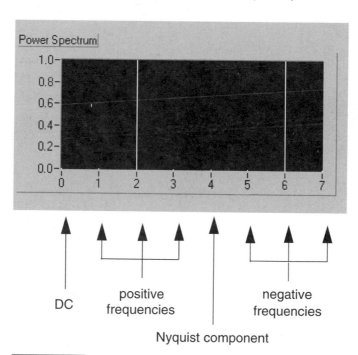

Figure 3–8
Power spectrum for *N* even

The output units of the **Power Spectrum** VI are in volts rms squared (V^2_{rms}). For example, if the peak amplitude (V_{pk}) of the input signal is 2 V_{pk}, its rms value is $V_{rms} = 2/\sqrt{2} = \sqrt{2}$, so V^2_{rms} = 2. This value is divided equally between the positive and negative frequency components, resulting in the plot shown above.

If N is odd, let $p = (N-1)/2$. Table 3–6 shows the format of the output sequence Sxx corresponding to the power spectrum.

Table 3–6 *Correspondence between actual frequency and sample number for N odd—note absence of the Nyquist frequency*

Array Element	Interpretation
$Sxx[0]$	Power in DC component
$Sxx[1] = Sxx[N-1]$	Power at frequency Δf
$Sxx[2] = Sxx[N-2]$	Power at frequency $2\Delta f$
$Sxx[3] = Sxx[N-3]$	Power at frequency $3\Delta f$
. . .	. . .
$Sxx[p-2] = Sxx[N-(p-2)]$	Power at frequency $(p-2)\Delta f$
$Sxx[p-1] = Sxx[N-(p-1)]$	Power at frequency $(p-1)\Delta f$
$Sxx[p] = Sxx[N-p]$	Power at frequency $p\Delta f$

Figure 3–9 represents the information in the preceding table for $N = 7$.

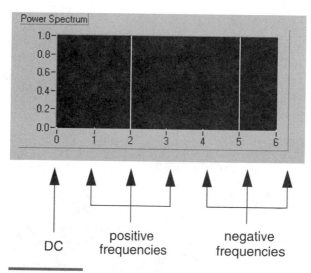

Figure 3–9
Power spectrum for *N* odd

Activity 3-2

Objective: To observe the difference between the FFT and the power spectrum representations

1. Open the **FFT_1sided** VI (from the library Activities.llb) that you built in the previous activity. Modify the block diagram and front panel as shown below.

■ Front Panel

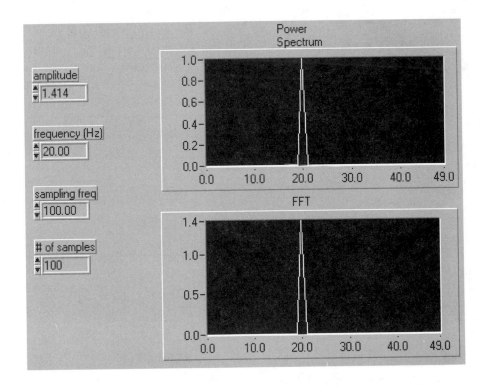

■ Block Diagram

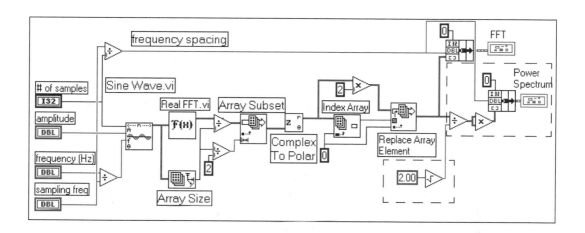

 Sine Wave VI (Functions » Analysis » Signal Generation subpalette) generates a time-domain sinusoidal waveform.

 Real FFT VI (Functions » Analysis » Digital Signal Processing subpalette) computes the FFT of the input data samples.

Array Subset function (Functions » Array subpalette) returns a portion of the array. Here you are selecting half the array.

 Complex to Polar function (**Functions » Numeric » Complex** subpalette) separates the complex output of the FFT into its magnitude and phase parts. The phase information is in radians. Here, you are displaying only the magnitude of the FFT.

The power spectrum is obtained by squaring the magnitude of the FFT. The division by $\sqrt{2}$ (1.414) makes the conversion from V_{pk} to V_{rms}.

You could also have wired the output of the Sine Wave VI directly to the input of the Power Spectrum VI (Analysis » Digital Signal Processing subpalette). The output of the Power Spectrum VI would indeed be the power spectrum of the signal. However, in that case, the phase information would be lost.

2. Enter the following values in the controls: **amplitude** = 1.414, **frequency (Hz)** = 20 Hz, **sampling freq** = 100, and **# of samples** = 100, and run the VI. Do you notice any difference in the FFT and power spectrum representations?
3. Change the amplitude to 1.00 and run the VI. What difference do you notice in the FFT and power spectrum representations?
4. Save the VI as **FFT and Power Spectrum.vi** in the library `Activities.llb`.

■ End of Activity 3-2

Wrap It Up!

- The time-domain representation (sample values) of a signal can be converted into the frequency-domain representation using the Discrete Fourier Transform (DFT).
- Fast calculation of the DFT is possible by using an algorithm known as the Fast Fourier Transform (FFT). You can use this algorithm when the number of signal samples is a power of two.
- The output of the conventional DFT/FFT is two-sided because it contains information about both the

positive and the negative frequencies. This output can be converted into a one-sided DFT/FFT by using only half the number of output points, as you saw in Activity 3-1.

■ The frequency spacing between the samples of the DFT/FFT is $\Delta f = f_s / N$.

■ The power spectrum can be calculated from the DFT/FFT by squaring the magnitude of the individual frequency components. The **Power Spectrum** VI in the advanced analysis library does this automatically for you. The units of the output of the **Power Spectrum** VI are V^2_{rms}. However, the power spectrum does not provide any phase information.

■ The DFT, FFT, and power spectrum are useful for measuring the frequency content of stationary or transient signals. The FFT provides the average frequency content of the signal over the entire time that the signal was acquired. For this reason, you use the FFT mostly for stationary signal analysis (when the signal is not significantly changing in frequency content over the time that the signal is acquired), or when you want only the average energy at each frequency line.

■ For measuring frequency information that changes during the acquisition, you should use the Joint Time-Frequency Analysis (JTFA) Toolkit or the Wavelet and Filter Banks Toolkit. You can obtain more information on these toolkits by contacting National Instruments.

■ Review Questions

1. Which of the following provides you with both the magnitude and phase information?

   a) FFT
   b) Power spectrum
   c) DFT
   d) Time-domain waveform

2. Which of the following are true?

   a) The magnitude spectrum is always even symmetric.
   b) The DFT is a fast algorithm for computing the FFT.
   c) The frequency spacing is given by

   $$\Delta f = \frac{f_s}{\text{number of samples}}$$

   where f_s is the sampling frequency.
   d) An even number of samples always results in a two-sided transform.

3. If you have 1024 samples, how many times faster is the FFT as compared to the DFT in calculating the Fourier transform?

■ Additional Activities

1. Build a VI that generates two sine waves, of amplitude A1 and A2, and frequencies f_1 and f_2, respectively, all of which can be controlled from the front panel. (Use the **Sine Wave.Vi** with additional controls for the sampling frequency and number of samples.) Combine (add) the sine waves and plot both the FFT and the power spectrum of the resulting signal. Set the sampling frequency = 1280 Hz, number of samples = 128, f_1 = 100 Hz, f_2 = 200 Hz, A1 = 1.0 = A2. Run the VI and observe the resulting amplitudes of the FFT and the power spectrum. Run the VI several times with A1 = 1, but A2 = 2, 5, and 10, respectively. Observe the differences in amplitudes of the two spectral lines. Note that the difference for the power spectrum plot is more, due to its squared nature, versus the FFT (where the relationship is linear).

2. A square wave consists of the sum of sine waves of odd harmonics, with the amplitudes of the higher harmonics falling as $1/p$, where $p = 1,3,5,...$, is the

harmonic number. Mathematically, with $p = 2n + 1$, we have

$$x[k] = \frac{2}{\pi} \sum_{n=0}^{N-1} \frac{1}{2n + 1} \sin(2\pi(2n + 1)fk)$$

where $x[k]$ is the square wave, and f is the fundamental frequency.

a) Build a VI that generates a square wave from a combination of sine waves specified on the front panel. Plot the one-sided DFT/FFT of the square wave, as well as the corresponding time-domain waveform.

b) Run the VI for the number of harmonics = 1, 2, 5, 10, 25, and 50. Observe the corresponding time- and frequency-domain plots. In particular, note that

i) The higher the number of harmonics, the closer the approximation is to an ideal square wave;

ii) The number of peaks in the approximation is equal to the number of harmonics;

iii) Even though the number of harmonics increases, there still exist overshoots and under-shoots at the points of transition of the square wave between positive and negative values. This is known as Gibb's phenomenon.

3. In this activity, you will learn a very important property of the Fourier transform. Follow the steps shown below:

Step 1: Generate an impulse waveform using the **Impulse Pattern** VI (**Functions >> Analysis >> Signal Generation** subpalette). Connect a Vertical Pointer Slide control to the input terminal **delay.** Using this control, you can vary the delay from the front panel.

Step 2: Compute the Fourier transform of the impulse signal using the **Complex FFT** VI (**Functions >> Analysis >> Digital Signal Processing** subpalette). One way to do this is to type cast the real input signal to a complex value (with zero imaginary part) before using the VI. Plot the magnitude of the frequency domain signal on a Waveform Graph.

Steps 1 and 2 perform transformation from the time-domain to the frequency domain. Now we will assume that the impulse signal generated in Step 1 above is in the frequency domain.

Step 3: Compute the inverse Fourier transform of the impulse signal in the frequency domain using the **Inverse Complex FFT** VI (**Functions >> Analysis >> Digital Signal Processing**). You will need to type cast similar to that done in Step 2 above. Plot the magnitude of the time domain signal on a Waveform Graph.

Compare the frequency domain and time domain signals plotted in Steps 2 and 3 above. You will observe that an impulse signal in the time domain corresponds to a flat waveform (square wave) in the frequency domain. On the other hand, an impulse signal in the frequency domain corresponds to a flat waveform in the time domain. This is well-known as the principle of duality. In other words, a sharp feature in the time domain corresponds to a smooth signal in the frequency domain and vice-versa.

Save the VI as **Duality.vi** in the library `Activities.11b` and close it.

The Statistical Properties of Acoustical Fields in a Reverberation Room

Dr. Matias H. W. Budhiantho
Satya Wacana Christian University, Salatiga, Indonesia

Dr. Elmer L. Hixson
University of Texas at Austin

Monica Obermier
University of Texas at Austin

The Challenge To determine the statistical properties of acoustical fields in a reverberation room using a four-microphone sensor.

The Solution Using the VIs from LabVIEW's Analysis Library, the statistical properties were calculated from data obtained using the National Instruments' AT-A2150C data acquisition board.

Introduction

The acoustic power emitted from a noisy machine can be determined by the spatial and temporal average of the energy density produced in a calibrated reverberation room. A reverberation room is one in which very hard walls absorb very little acoustic energy. The noise from a machine produces an energy density in the room that increases until the power "in" equals the power "lost" to the room. Then the total energy is energy density times room volume and power is total energy divided by the reverberation time of the room (T is the time for sound to decay 60 dB). The statistical properties of these random fields indicate the confidence level of the measurement. Total energy is the sum of potential and kinetic energy but they have different statistical properties. A special four-microphone sensor was designed from which potential and kinetic energy density can be determined.

The electrical signals from the four microphones must be averaged over time and space then squared to get potential energy density. The pressure difference in the X, Y, and Z direction must be time integrated, averaged, then the squares in each direction are added to get the kinetic energy density. Finally, the sensor is traversed along a 3-meter track with measurements taken at regular intervals. Then the mean value and statistical distributions are displayed.

LabVIEW IMPLEMENTATION

To produce the mean and standard deviation of the total energy density three virtual instruments were constructed from the standard LabVIEW package. These were CALIBRATE to calibrate the sensor, ACQUIRE to acquire and store the four microphone signals and ANALYSIS to calculate pressure, particle velocity components, potential and kinetic energy densities, their sum and the statistical properties of each. But first, the sensor had to be powered, signals amplified and converted to digital form.

■ Sensor

The sensor shown in Figure 3–10 uses four miniature microphones that include an FET preamp. There are microphones labeled "m" at positions r at the origin and at x, y, and z on their respective coordinate axes. The distance r to x, r to y and r to z was set at 40mm to optimize the response from 300 to 3000 kHz which was the expected frequency range. Each channel was band limited to 300–3000 Hz and 48 dB gain was provided to give maximum signals near the 2 VRMS limit of the A/D converters.

The Data acquisition board chosen was the NI AT-A2150C.

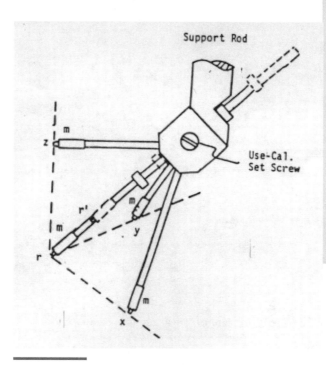

Figure 3–10
Sensor with four miniature microphones (m) at positions
x, y, z, and r

The 4-channel simulations sampling was required to get phase information. Available sampling rates were adequate for the 3 KHz upper frequency. The 93 dB signal to noise ratio and -94 dB distortion insured accurate data.

■ Calibrate

To calibrate the sensor it was placed in a plane wave tube with the r microphone retracted to, r^1, as shown in Figure 3–10. This placed r in the plane of the other microphones so that all received the same sound level as measured by a standard sound level meter.

Figure 3–11 shows a block diagram of the analysis system with calibration constants "G". After the gain in the Pr channel

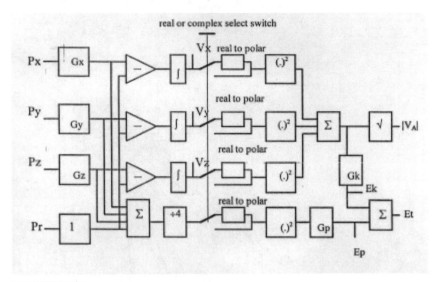

Figure 3–11
Block Diagram of the Analysis System

was set to give the correct Pr for a sound pressure level of 95 dB relative to 20 microPascals (the threshold of human hearing), the LabVIEW VI from the Analysis package took ratios of Px, Py and Pz to Pr to get Gx, Gy and Gz which were saved. The calculated values of potential and kinetic energy density (which are equal for a plane wave) were divided by the potential and kinetic channels to get Gp and Gk which were stored for use in analysis.

With microphone, r, restored to its original position, the probe was ready for measurement.

■ Acquire

This VI was required to start the transverse along the 3 meter track, calculate the spatial intervals, take data at these intervals and store the data (P_r, P_x, P_y, P_z) in 4 arrays. The VI was designed so that the experimenter could choose the intervals in fractions of the acoustic wavelength which is the distance between pressure maxima (sound speed divided by frequency) and the number of measurements at each position.

The position of the sensor was indicated by the potential from a linear potentiometer with 10.0 Volts dc across it. A NI Lab-PC+ DAQ board was used to measure this potential to identify the position of the sensor. The A/D converter sample rate was chosen and set by this VI.

■ Analyze

Several tasks were required from this VI. So several sub VI's from the LabVIEW analysis package were used. Since the raw data were identified and stored, the analysis could be performed at any later time.

As shown in Figure 3–11, the first calculation required average sound pressure and each velocity component. The average pressure was obtained by simultaneously summing the four stored arrays and dividing by 4. Thus the instantaneous pressure was obtained.

A velocity component required the time integral of the pressure gradient in that direction. Thus the difference of the Pr data array and each of the P_x, P_y and P_z arrays were taken and integrated. Each of the velocity components and pressure were squared and average to produce time averaged mean square values. The mean square pressure multiplied by Gp gives potential energy density (Ep). The mean square velocity components are summed and multiplied by G_k to give kinetic energy density (Ek). These two are summed to give total energy density (Et).

Because of the great versatility of the LabVIEW System and the use of all of the quantities determined above, many functions could be calculated and displayed. The spatial distribution functions were determined and histograms were plotted. The histograms showing the statistical properties of potential, kinetic and total energy density are shown in Figure 3–12. This was the basic purpose of this measurement program. However, many more important properties of the acoustic field were determined.

The frequency spectrum of each function was readily determined by the FFT routine. The phase relations between velocity components was determined by the Fast Hilbert Transform rou-

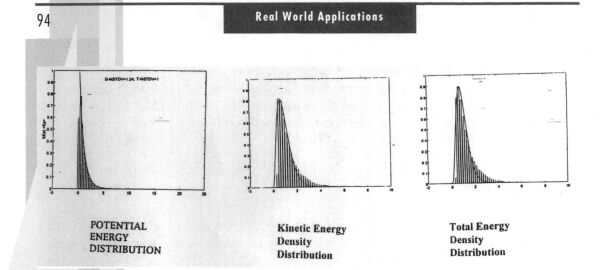

POTENTIAL
ENERGY
DISTRIBUTION

Kinetic Energy
Density
Distribution

Total Energy
Density
Distribution

Figure 3–12
Statistical distributions

tine to get the direction cosines of the intensity vector. The product of the pressure and velocity components gives the acoustic intensity vector and its 3 components. And of course the instantaneous time waveforms of each quantity can be displayed. Because data were taken along a transverse, a plot of each function versus distance can be obtained. And finally all analysis output can be printed in spread sheet form.

Conclusions

Traditionally acoustic fields have been investigated with a scanned single microphone or an array of microphones. With the development of the 4-microphone sensor, a great deal more information can be obtained. The implementation is more complicated than single or array microphone systems, however, LabVIEW allowed an easy, quick and powerful implementation of the new system.

With LabVIEW we have an automated calibration procedure for the sensor. Then the measurement parameters are calculated, the measurements are carried out and the data stored automatically. The analysis and display capabilities of LabVIEW provides much more information about acoustic fields than have previously been obtainable. Also our experiences has been that acoustic power measurements can be performed faster and more accurately than by traditional methods.

■ References

1. Moryl, J. A., and Hixson, E. L. (1989). "A total acoustical energy density sensor with application to energy density measurement in a reverberation room," *Proc. Inter-Noise* 87, vol. II, pp. 1195–1198.

2. Budhiantho, M. H. W., and Hixson, E. L. (1996). "Acoustic velocity related distributions," *J. Acoust. Soc. Am.* Vol. 100, No. 4 Pt. 2, p. 2838.

■ Contact Information

Dr. Elmer L. Hixson, Professor
Department of Electrical & Computer Engineering
University of Texas at Austin,
Austin, TX 78712.
512-471-1294
Fax: 512-471-5553
Email: ehixson@mail.utexas.edu

Overview

In this chapter, you will learn about windows and how they affect the spectral characteristics of a signal.

GOALS

- Understand spectral leakage and the need for smoothing windows.
- Identify the difference (both time and frequency domains) between a windowed and a nonwindowed signal.
- Learn about the differences between the various types of windows in the Analysis Library and their applications.
- Separate two sine waves that are closely spaced in the frequency domain but have a large amplitude difference.

KEY TERMS

- smoothing windows
- spectral leakage
- window applications

Windowing

4

About Spectral Leakage and Smoothing Windows

In practical applications, you can obtain only a finite number of samples of the signal. When you use the DFT/FFT to find the frequency content of a signal, it is inherently assumed that the data that you have is a single period of a periodically repeating waveform. This is shown in Figure 4–1. The first period shown is the one sampled. The waveform corresponding to this period is then repeated in time to produce the periodic waveform.

As seen in the figure, because of the assumption of periodicity of the waveform, discontinuities between successive periods occur that were not present in the original continuous signal. This happens when you sample a noninteger number of cycles of the periodic continuous signal. These "artificial" discontinuities turn up as frequencies that were not present in the original signal. The spectrum you get by using the DFT/FFT, therefore, will not be the actual spectrum of the original signal but will be a smeared version. It appears as if the energy at one frequency

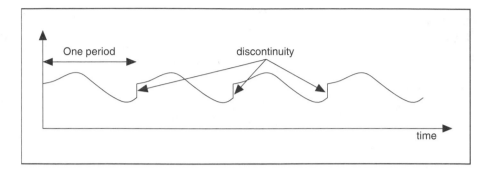

Figure 4–1
Periodic waveform created from sampled period

has "leaked out" into all the other frequencies. This phenomenon is known as *spectral leakage*.

Figure 4–2 shows a sine wave and its corresponding Fourier
transform. The sampled time-domain waveform is shown in
Graph 1. Because the Fourier transform assumes periodicity, you
repeat this waveform in time, and the periodic time waveform

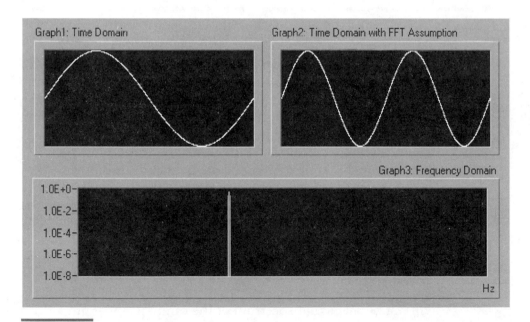

Figure 4–2
Sine wave and corresponding Fourier transform

of the sine wave of Graph 1 is shown in Graph 2. The corresponding spectral representation is shown in Graph 3. Because the time record in Graph 2 is periodic, with no discontinuities, its spectrum is a single line showing the frequency of the sine wave. The reason that the waveform in Graph 2 does not have any discontinuities is because you have sampled an integer number of cycles (in this case, 1) of the time waveform.

In Figure 4–3, you see the spectral representation when you sample a noninteger number of cycles of the time waveform (namely 1.25). Graph 1 now consists of 1.25 cycles of the sine wave. When you repeat this periodically, the resulting waveform, as shown in Graph 2, consists of discontinuities. The corresponding spectrum is shown in Graph 3. Notice how the energy is now spread over a wide range of frequencies. This smearing of the energy is *spectral leakage.* The energy has leaked out of one of the FFT lines and smeared itself into all the other lines.

Leakage exists because of the finite time record of the input signal. To overcome leakage, one solution is to take an infinite time record, from $-\infty$ to $+\infty$. Then the FFT would calculate a single spectral line at the correct frequency. Waiting for infinite time

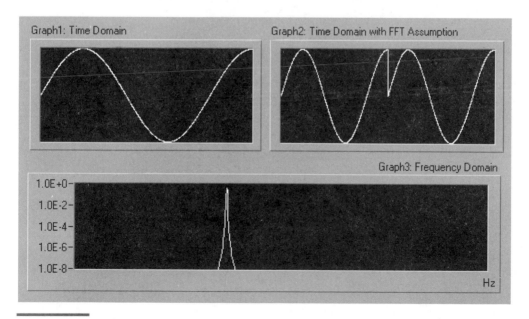

Figure 4–3
Spectral representation when sampling a nonintegral number of samples

is, however, not possible in practice. Because you are limited to having a finite time record, another technique, known as *windowing*, is used to reduce the spectral leakage.

The amount of spectral leakage depends on the amplitude of the discontinuity. The larger the discontinuity, the more the leakage. You can use windowing to reduce the amplitude of the discontinuities at the boundaries of each period and thus reduce the spectral leakage. It consists of multiplying the time record by a finite-length window whose amplitude tapers smoothly and gradually toward zero at the edges. This is shown in Figure 4–4,

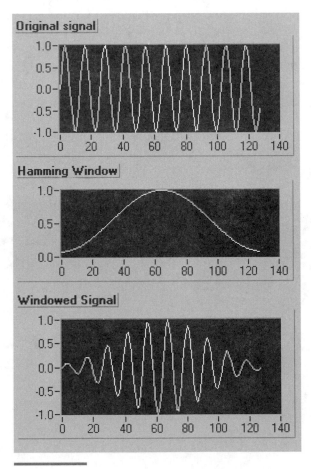

Figure 4–4
Time signal windowed using a Hamming window

where the original time signal is windowed using a *Hamming* window. Notice that the time waveform of the windowed signal gradually tapers to zero at the ends. Therefore, when performing Fourier or spectral analysis on finite-length data, you can use windows to minimize the transition edges of your sampled waveform. A smoothing window function applied to the data sequence before it is transformed into the frequency domain minimizes spectral leakage.

Note that if the time record contains an integral number of cycles, as shown in Figure 4–2, the assumption of periodicity does not result in any discontinuities, and thus there is no spectral leakage. The problem arises only when you have a nonintegral number of cycles.

Windowing Applications

There are several reasons to use windowing. Some of these are

- To define the duration of the observation.
- To reduce spectral leakage.
- To separate a small amplitude signal from a larger amplitude signal with frequencies very close to each other.

Characteristics of Different Types of Window Functions

Applying a window to a signal in the time domain is equivalent to multiplying the signal by the window function. Because multiplication in the time domain is equivalent to convolution in the frequency domain, the spectrum of the windowed signal is a convolution of the spectrum of the original signal with the spectrum of the window. Thus, windowing changes the shape of the signal in the time domain, as well as affecting the spectrum that you see.

Many different types of windows are available in the LabVIEW/ BridgeVIEW Analysis Library. Depending on your application, one may be more useful than others. Some of these windows are

1. *Rectangular (no window):* The rectangular window, also known as the uniform window, has a value of 1.0 over its entire time interval. Mathematically, it can be written as

$$w[n] = 1.0 \quad \text{for } n = 0, 1, 2, \ldots, N\text{--}1$$

where N is the length of the window. Applying a rectangular window is equivalent to not using any window. This is because the rectangular function just truncates the signal to within a finite time interval. The rectangular window has the highest amount of spectral leakage. The rectangular window for $N = 32$ is in Figure 4–5.

Figure 4–5
Rectangular window

The rectangular window is useful for analyzing transients that have a duration shorter than that of the window. It is also useful in the special case where the width of the window is equal to an integral number of cycles of the waveform.

2. *Exponential:* The shape of this window is that of a decaying exponential. It can be mathematically expressed as

$$w[n] = \exp\left(\frac{n}{N-1} \times \ln(f)\right) \text{ for } n = 0, 1, 2, \ldots, N-1$$

where f is the final value. The initial value of the window is 1, and it gradually decays toward the final value f. The final value of the exponential can be adjusted to between 0 and 1. The exponential window for $N = 32$, with the final value specified as 0.1, is shown in Figure 4–6.

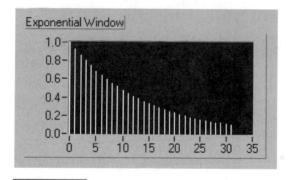

Figure 4–6
Exponential window

This window is useful in analyzing transients (signals that exist only for a short time duration) whose duration is longer than the length of the window. This window can be applied to signals that decay exponentially, such as the response of structures with light damping that are excited by an impact (for example, a hammer). Other forms of exponential weighting are used in making standard acoustic measurements.

3. *Hanning:* This window has the shape of a sinusoid. Its defining equation is

$$w[n] = 0.5 - 0.5 \cos(2\pi n/N) \quad \text{for } n = 0, 1, 2, \ldots, N-1$$

A Hanning window with $N = 32$ is shown in Figure 4–7.

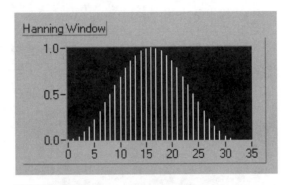

Figure 4–7
Hanning window

The Hanning window is useful for analyzing transients longer than the time duration of the window, and also for general-purpose applications.

4. *Hamming:* This window is a modified version of the Hanning window. Its shape is also that of a sinusoid. It can be defined as

$$w[n] = 0.54 - 0.46 \cos(2\pi n/N) \quad \text{for } n = 0, 1, 2, \ldots, N-1$$

A Hamming window with $N = 32$ is shown in Figure 4–8.

You see that the Hanning and Hamming windows are somewhat similar. However, note that in the time domain, the Hamming window does not get as close to zero near the edges as does the Hanning window.

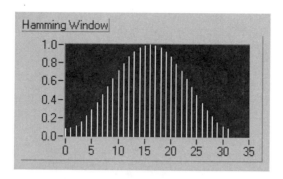

Figure 4–8
Hamming window

5. *Flat Top*: The shape of this window is given by

$$w[n] = 0.2810639 - 0.5208972 \cos(2\pi n/N) +$$
$$0.1980399 \cos(4\pi n/N) \quad \text{for } n = 0, 1, 2, ..., N-1$$

This window has the best amplitude accuracy of all the window functions. The increased amplitude accuracy (± 0.02 dB for signals exactly between integral cycles) is at the expense of frequency selectivity. The flat top window is most useful in accurately measuring the amplitude of single frequency components when there are no significant frequency components nearby.

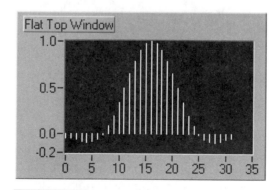

Figure 4–9
Flat Top window

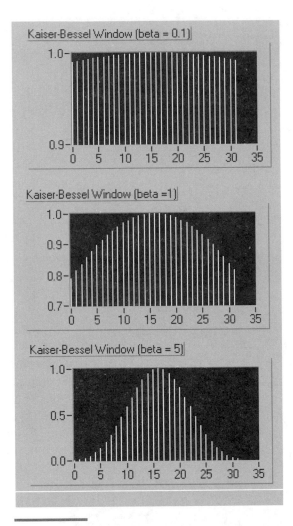

Figure 4–10
Kaiser-Bessel window for = β, 0.1, 1.0, and 5.0

6. *Kaiser-Bessel:* This window is a "flexible" window whose shape the user can modify by adjusting the parameter *beta*. Thus, depending on your application, you can change the shape of the window to control the amount of spectral leakage. The Kaiser-Bessel window for different values of *beta* is shown in Figure 4–10.

Note that for small values of beta, the shape is close to that of a rectangular window. Actually, for beta = 0.0, you do get a rectangular window. As you increase beta, the window tapers off more to the sides.

This window is good for detecting two signals of almost the same frequency, but significantly different amplitudes.

7. *Triangle:* The shape of this window is that of a triangle. It is given by

$$w[n] = 1 - |\ (2n - N)\ /\ N\ |\ \text{for } n = 0, 1, 2, ..., n - 1$$

A triangle window for $N = 32$ is shown in Figure 4–11.

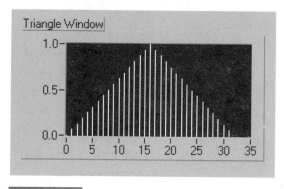

Figure 4–11
Triangle Window

■ What Type of Window Do I Use?

Now that you have seen several of the many different types of windows that are available, you may ask, "What type of window should I use?" The answer depends on the type of signal you have and what you are looking for. Choosing the correct window requires some prior knowledge of the signal that you are analyzing. In summary, Table 4–1 shows the different types of signals and the appropriate windows that you can use with them.

Table 4-1   *Applications of different windows*

Type of signal	Window
Transients whose duration is shorter than the length of the window	Rectangular
Transients whose duration is longer than the length of the window	Exponential, Hanning
General-purpose applications	Hanning
System analysis (frequency response measurements)	Hanning (for random excitation), rectangular (for pseudorandom excitation)
Separation of two tones with frequencies very close to each other, but with widely differing amplitudes	Kaiser-Bessel
Separation of two tones with frequencies very close to each other, but with almost equal amplitudes	Rectangular
Widely spaced components where amplitude estimation must be accurate	Flat Top

In many cases, you may not have sufficient prior knowledge of the signal, so you need to experiment with different windows to find the best one.

Table 4-2 summarizes the different windows.

Table 4-2 *Brief summary of different windows*

Windows	Equation	Shape	Applications
Rectangular (None)	$w[n] = 1.0$		Detecting transients whose duration is shorter than the length of the window; separating two tones with frequencies and amplitudes very close to each other; system response
Exponential	$w[n] = e^{\frac{n}{N-1}\ln(f)}$ where f = final value		Transients whose duration is longer than the length of the window
Hanning	$w[n] = 0.5 - 0.5\cos\left(\frac{2\pi n}{N}\right)$		General-purpose applications; system analysis; transients whose duration is longer than the length of the window
Hamming	$w[n] = 0.54 - 0.46\cos\left(\frac{2\pi n}{N}\right)$		Speech signal processing

(Continued)

Table 4–2 *Brief summary of different windows*

Windows	Equation	Shape	Applications
Flat Top	$w[n] = 0.2810639$ $- 0.5208972\cos\left(\dfrac{2\pi n}{N}\right)$ $+ 0.1980399\cos\left(\dfrac{4\pi n}{N}\right)$		Accurate measurement of single-frequency components when there are no significant frequency components nearby.
Kaiser-Bessel	$w[n] = \dfrac{I_0\left(\beta\sqrt{1 - a^2}\right)}{I_0(\beta)}$ where $a = 1 - 2n/N$ $I_0() = $ Bessel function of the first kind		Separation of two tones with frequencies very close to each other, but with significanlty different amplitudes.
Triangle	$w[n] = 1 - \left|\dfrac{2n - N}{N}\right|$		No special application

Activity 4-1

Objective: To see the effect of windowing on spectral leakage

1. Open the **Spectral Leakage** VI from the library `Activities.llb`. The VI is running when it opens. Using this VI, you can see the effect of windowing on spectral leakage.

■ Front Panel

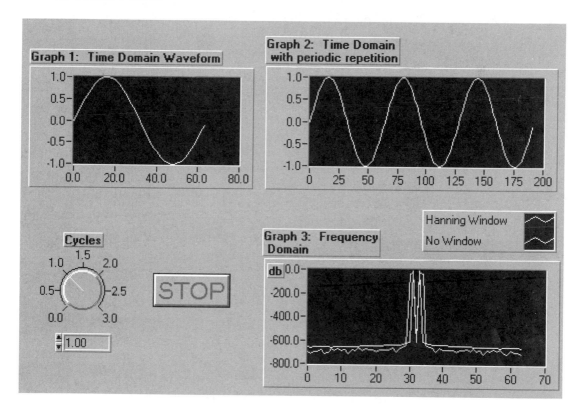

The Spectral Leakage VI searches for the Nyquist Shift VI. The Nyquist Shift VI is in LabVIEW » Examples » Analysis » dspxmpl.llb.

You can see three plots on the front panel:

Graph 1 shows the time record of the signal that has been sampled.

Graph 2 shows the repeated time record (assuming periodicity).

Graph 3 shows the frequency spectrum (in dB). The white line shows the spectrum without windowing and the yellow line shows the spectrum by windowing using a Hanning window.

You can use the **Cycles** dial to control the number of time-domain waveform cycles that have been sampled. The display below the **Cycles** dial tells you the exact number of cycles (to two decimal places). You can also type a specific value in this display.

2. First, you will see the effect of windowing when you sample an integral number of cycles.

Set the **Cycles** dial to 1.0. (or type 1.0 in the display beneath it.)

As you can see in **Graph 1**, you have exactly one cycle of the time waveform. **Graph 2** shows the repeated time record. Notice the absence of any discontinuities, and the peak corresponding to the sine wave in the frequency domain (**Graph 3**). You see two peaks because you have the two-sided spectrum.

Notice the spreading of the frequency components. This spreading is the effect of using a window function. In this case, you used the Hanning window. Different windows have different amounts of spreading.

3. Now see what happens when you sample a noninte-gral number of cycles. Set the **Cycles** dial to 1.3 and observe the difference in the plots in **Graph 3**. Exper-iment by changing the **Cycles** dial and observing the waveforms in graphs 2 and 3.

In the white plot corresponding to **No Window**, the energy in the frequency of interest spreads out across the spectrum. Hence the frequency of interest is sometimes not clearly distinguishable. In the yellow plot corresponding to **Hanning Window**, the spectral leakage across the spectrum is reduced and the energy is more concentrated around the frequency of interest.

4. Stop the VI by pressing the **STOP** button.
5. Close the VI. Do not save any changes.

■ End of Activity 4-1

Activity 4-2

Objective:To see the difference (both time and frequency domains) between a windowed and nonwindowed signal

1. Build the VI front panel and block diagram as shown below.

■ Front Panel

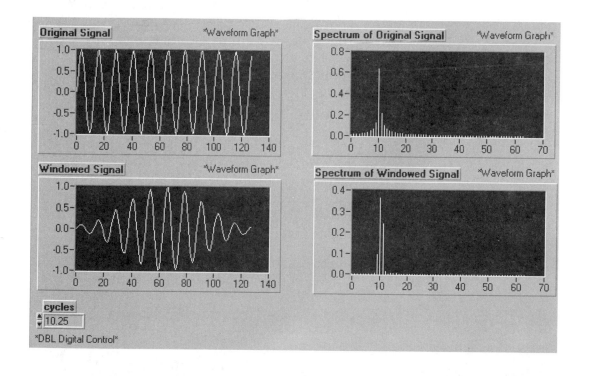

■ Block Diagram

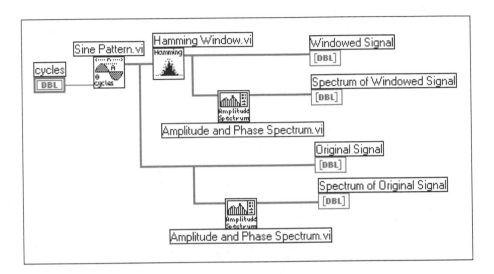

 The **Sine Pattern** VI (**Functions » Analysis » Signal Generation** subpalette) generates a sine wave with the number of cycles specified in the **cycles** control.

The time waveform of the sine wave is windowed using the **Hamming Window** VI (**Functions » Analysis » Windows** subpalette), and both the windowed and nonwindowed time waveforms are displayed on the left two plots on the front panel.

The **Amplitude and Phase Spectrum** VI (**Functions » Analysis » Measurement** subpalette) obtains the amplitude spectrum of the windowed and nonwindowed time waveforms. These waveforms are displayed on the two plots on the right side of the front panel.

2. Pop up on the Legend of the Waveform Graph on the Front Panel and select Common Plots to show only vertical lines.

3. Set **cycles** to 10 (an integral number) and run the VI. Note that the spectrum of the windowed signal is broader (wider) than the spectrum of the nonwindowed signal. But both the spectra are concentrated near 10 on the x-axis.

4. Change **cycles** to 10.25 (a nonintegral number) and run the VI. Note that the spectrum of the nonwindowed signal is now more spread out than it was before. This is because now you have a noninteger number of cycles, and when you repeat the waveform to make it periodic, you get discontinuities. The spectrum of the windowed signal is still concentrated, but that of the nonwindowed signal has now smeared all over the frequency domain. (This is spectral leakage.)

5. Change **cycles** to 10.5 and observe the frequency-domain plots. Spectral leakage of the original signal is clearly apparent.

6. When you finish, save the VI as **Windowed and Unwindowed Signal.vi** in the library `Activities.llb`.

7. Close the VI.

■ End of Activity 4-2

Activity 4-3

Objective:To learn about the different windows in the Analysis Library

1. Open the **Window Plots** VI from the library `Activities.llb`. It is running when it opens.

■ Front Panel

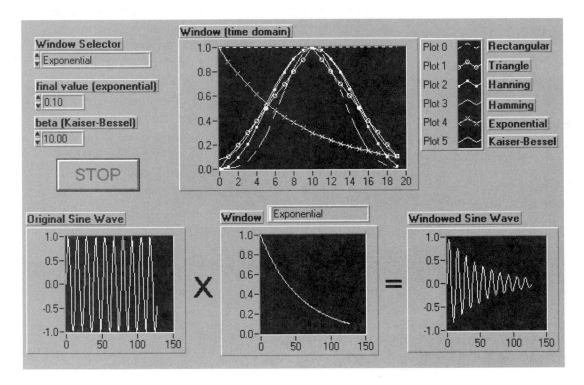

The topmost plot shows you the shapes (in the time domain) of six different types of windows in the Analysis Library. They are all shown on the same plot for comparison purposes.

The bottom three plots show the effect of multiplying a time-domain signal (a sine wave) by the window. The left plot shows the original time signal, the middle plot shows the shape of the window being applied, and the right plot shows the resulting signal.

In the **Window Selector** control, you can select one of the six different types of windows.

The **final value (exponential)** control specifies the value to which the exponential window should decay. This value is normally between 0 and 1.

The value in the **beta (Kaiser-Bessel)** control can be adjusted to change the shape of the Kaiser-Bessel window. The higher the value of *beta*, the lower the spectral leakage, and vice versa.

2. Select different windows in the **Window Selector** control and observe their shapes.

In particular, select the **Rectangular (None)** window. What difference do you see in the first and the third plots shown on the bottom? Explain.

Select the **Exponential** window. Observe the shape when you change the value in the **final value (exponential)** control. What happens as the final value increases? Decreases? Is equal to 1.0?

Select the **Kaiser-Bessel** window. Observe the shapes when you change the value of beta between –10 and +10. What happens when *beta* = 0.0?

3. When you finish, stop the VI by pressing the **STOP** button.

4. Close the VI. Do not save any changes.

■ End of Activity 4-3

Activity 4-4

Objective:To use windows to separate two sine waves of almost the same frequency, but widely differing amplitudes

In this activity, two sine waves of different amplitudes are summed together and then transformed into the frequency domain. Sine wave 1 has a much smaller amplitude than sine wave 2. Without windowing, it is not possible to distinguish between the two sine waves in the frequency domain. With an appropriate choice of a window, you can clearly separate the peaks in the frequency domain corresponding to the two sine waves. The frequency domain plot shows the results so that you can compare the effect of different window functions.

1. Open and run the **Window Comparison** VI from the library `Activities.llb`.

■ Front Panel

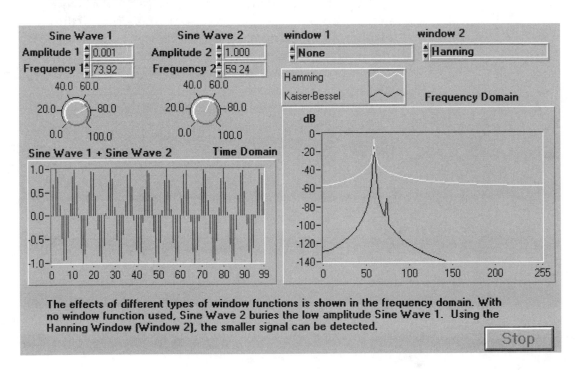

The effects of different types of window functions is shown in the frequency domain. With no window function used, Sine Wave 2 buries the low amplitude Sine Wave 1. Using the Hanning Window (Window 2), the smaller signal can be detected.

Stop

- The frequency of each sine wave is adjustable with either the knob or digital controls.

- The amplitude of each sine wave is adjustable with the digital controls.

- You can select a different window function from the **window 1** and **window 2** controls.

2. Using the digital controls, set the amplitude of Sine Wave 1 as 0.001, and that of Sine Wave 2 as 1.000. With the knob controls, set the frequency of Sine Wave 1 to near 70, and that of Sine Wave 2 to near 60. In effect, you are adjusting the frequency of Sine Wave 2 using the knob control, so that the smaller amplitude is nearer the larger amplitude in the frequency-domain plot.

3. Notice in the **Frequency Domain** graph that when the frequency of the smaller amplitude signal (Sine Wave 1) is closer to that of the larger amplitude signal (Sine Wave 2), the peak corresponding to the smaller signal is not detected. The discontinuity is what causes the spectrum to spread out. Signals at smaller amplitudes are lost in the sidelobes of the larger amplitude signal. Applying a window function is the only way to detect the smaller signal.

4. Compare different window functions by choosing another window from the **window 1** and **window 2** controls. Which one(s) can distinguish between the two frequency components?

5. When you are done, stop the VI by clicking on the **STOP** button. Close the VI. Do not save any changes.

■ End of Activity 4-4

Wrap It Up!

- The DFT/FFT assumes that the finite time waveform that you obtain is one period of a periodic signal that exists for all time. This assumption of periodicity could result in discontinuities in the periodic signal and gives rise to a phenomenon known as spectral leakage, whereby the energy at a particular frequency leaks throughout the spectrum.

- To reduce the spectral leakage, the finite time waveform is multiplied by a "window" function.

- Windows can be used to distinguish between closely spaced frequency components or improve amplitude accuracy of spectral measurements.

■ Review Questions

1. Why does spectral leakage occur?
2. Name four applications of using windows.
3. Which window(s) would you use for the following?

   a) Separation of two tones with frequencies very close to each other.
   b) System analysis.
   c) Transients.

■ Additional Activities

1. In this activity, you will experiment with windows whose parameters can be modified by controls on the front panel. These are the force, exponential, and Kaiser-Bessel windows. As you have learnt, the shape of the force window can be adjusted by varying its *duty cycle*, the shape of the exponential window depends on *the final value* chosen, and the shape

of the Kaiser-Bessel window is controlled by the value of the parameter *beta*.

a) Build a VI whose front panel and block diagram are as shown:

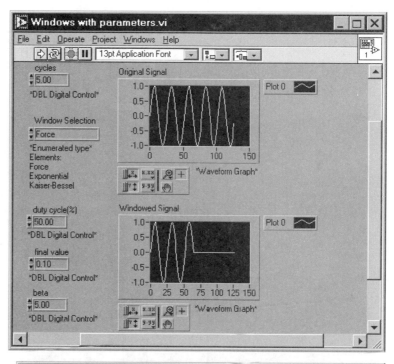

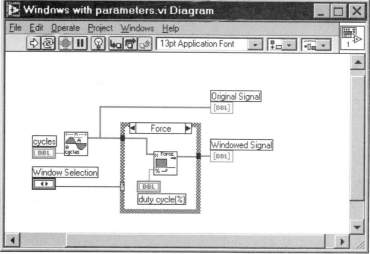

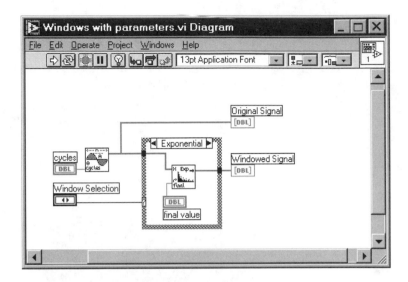

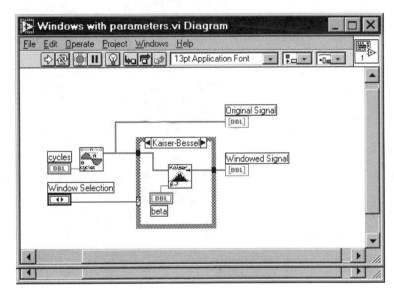

b) **Set cycles** = 5, **Window Selection** = Force, **duty cycle(%)** = 50, and run the VI. Observe the shape of the **Original Signal** and the **Windowed Signal.** Change the **duty cycle(%)** from 10 to 90 in increments of 20, and observe the shape of the **Windowed Signal.** For what value of **duty cycle(%)** is the **Windowed Signal** the same as the **Original**

Signal? What happens if you set **duty cycle(%)** to more than 100%?

c) Set **Window Selection** = Exponential, **final value** = 0.10, and run the VI. Increase **final value** in increments of 0.2 from 0.1 to 0.9, and run the VI each time. Observe the shape of the **Windowed Signal.** For what setting of **final value** is the **Windowed Signal** the same as the **Original Signal**? What happens if you set of **final value** to a value higher than 1.00?

d) Set **Window Selection** = Kaiser-Bessel, **beta** = 1.00, and run the VI. Increase **beta** in increments of 1.0 from 1.0 to 10.0, and run the VI each time. Observe the shape of the **Windowed Signal.** For what setting of **beta** is the **Windowed Signal** the same as the **Original Signal**? What happens if you set **beta** to a value less than 0.00?

e) Save the VI as **Windows with parameters.vi** in the library `Activities.llb` and close it.

Extracting Features from CTFM Sonar

Danny Ratner
Phillip McKerrow
Intelligent Robotics Laboratory
School of Information Technology and Computer Science
University of Wollongong
NSW 2522
Australia
email: dr19@uow.edu.au
email: Phillip_McKerrow@uow.edu.au

Figure 4–12
4wd mobile robot and sonar head

Figure 4–13
Landmarks along the robot's path

The Challenge We are developing an autonomous mobile robot equipped with CTFM sonar. The outdoor 4wd robot (Figure 4–12) is navigating along outdoor walkways (Figure 4–13) using landmark recognition. The problem is to build a fast prototype of a complicated navigation system that interacts with the sonar, the robot, and other sensors mounted on it.

The Solution We are using LabVIEW for programming the signal processing of the sonar and the control behaviour of the mobile robot. We found that the combination of simple graphical programming with simple interface to data acquisition cards enables us to develop modules of fast prototypes for our application.

In this article we focus on the aspect of using LabVIEW for the CTFM sonar signal processing.

CTFM Ultrasonic Signals

The ultrasonic sensor (Figure 4–14) Continuously Transmits a Frequency Modulated (CTFM) signal. The signal is swept down from 100Khz to 50Khz. The echoes detected by the receivers are demodulated with the transmitted signal to produce tones proportional to range (Figure 4–15). The time domain audio signal is converted to a frequency spectrum using a fast Fourier transform (FFT).

For this research project, we developed a 20 element phased array transmitter. The transmitter insonifies the environment with a vertical sheet of ultrasonic energy that can be steered electronically. The echoes from the 3D field of audition are detected with a quadaural receiver. This sensor is mounted on a pan and tilt unit at the front of the robot.

There is a delay (t_a) from the start of the sweep until an echo returns from an object (a plant or a post in Figure 4–13) due to the speed of sound. After the time to an echo from maximum range (t_d) there is a period of continuous demodulated frequencies (t_0) until the start of the next sweep. During this period any changes in the demodulated frequencies represent motion of the object. The sampling of the signal for the FFT should be done during this period for best results.

We developed a system to capture the demodulated echo, display it and calculate the power spectrum of the echo using

Figure 4–14
Scanning array

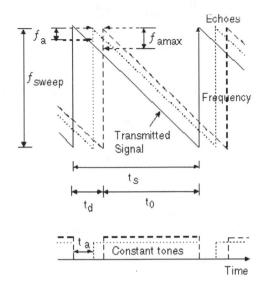

Figure 4–15
CTFM signals, where
f_{sweep} is the frequency sweep range,
f_a is the audio frequency of the demodulated echo,
f_{amax} is the audio frequency for maximum range,

t_s is the sweep time,
t_a is the time to an echo,
t_d is the time to an echo at the maximum range, and
t_0 is the time during which audio tones are constant.

LabVIEW. The echo spectra in the Figure 4–17 were obtained using this system.

Problem

The CTFM sonar produces a spectrum where each spectral line represents the amount of energy reflected at that range. A geometrically complex object like a plant has a complex spectrum. In order to recognise this object, or even to classify it into a class of objects, requires signal processing to extract features from the spectrum. These signal features represent geometric information about the object. Some recognition can be done by matching the

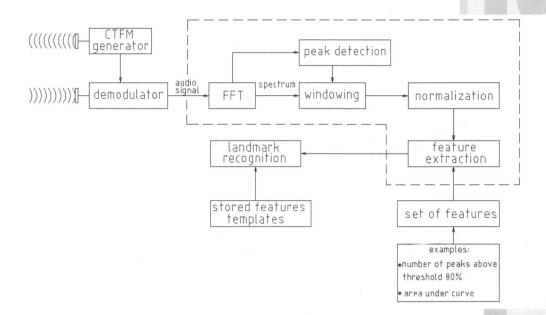

Figure 4–16
System block diagram

sensed spectrum with a previously recorded spectrum. But, the specular nature of ultrasonics in air coupled with many small reflectors (leaves) results in significant changes in the spectrum with small changes in the location of the plant relative to the sensor.

The FFT converts the audio signal output from the sensor into a spectrum where each spectral line has a frequency and an amplitude. The frequency is proportional to range and the amplitude is proportional to the area of the reflecting surfaces at that range. The width of the spectrum is proportional to the acoustic depth of the object. So a flat surface gives a very narrow spectrum with high amplitude. A plant is a complex object and gives a complex set of tones as shown in Figure 4–17.

We have found that matching features results in classification that is much more robust to changes in plant orientation and location. Thus, the task of the signal processing software is to extract the features, which are then used by the recognition software to determine which landmark the robot is currently sensing.

The key to using features to recognise objects is understanding the mapping from geometric feature into signal feature. Obtaining

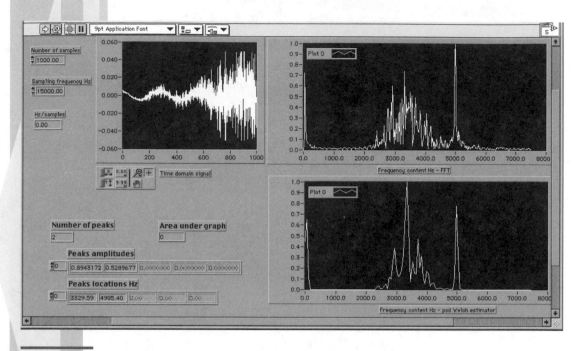

Figure 4–17
Time domain audio signal (left) is represented in the frequency domain (top right). Applying Welch method power spectrum estimator is shown in the bottom right plot. Details of the two major peaks from the estimated spectrum is presented on the front panel.

this understanding is part of an research project where we have identified 19 significant features. In this article, we will develop the signal processing to extract two of the features.

Acquiring the Audio Signal

The complete sensing system can be decomposed into 4 stages (Figure 4–16): Sensor, analog to digital converter, signal processing and landmark recognition. The output of the sensor is an audio signal with bandwidth of 5 kHz, where 0 .. 5 kHz is proportional to a range of 0 .. 1 m, 0 .. 2 m or 0 .. 4 m depending on the length of the FM sweep. This output is connected to an analog to digital converter in a PCMCIA data acquisition card plugged into a Macintosh Powerbook.

At the start of each FM sweep a trigger pulse starts the conversion process. The data capture is done with a vi derived from the **Acquire N Scans—Strig.VI** located in the **Analog Input VI** library. The modifications to this VI include a plot of the audio signal, plot of the spectrum and a button to save the audio signal as an ascii file when desired.

Signal Processing of the Audio Signal

The signal processing is composed of two stages: audio processing and feature extraction. In the audio processing stage the signal is transformed into a known state to make feature extraction robust. The signal is transformed from the time domain to the frequency domain using the **Real FFT** VI from the digital signal processing library. The time domain signal has a low frequency modulation due to the direct coupling from the transmitter to the receiver which results in a spike at the left of the spectrum. The spectrum in Figure 4–17 represent a reflection off two objects. A plant in the foreground (broad section of the spectrum) and a wall in the background (narrow tall spike on the right).

As we are interested in recognising the plant, we will select the spectral lines for the plant and discard the rest. The location of the plant is determined by finding the maximum amplitude spectral line in the plant. Then n spectral lines from either side of this peak are selected to form a plant window. The spectrum in the window is considered to be the plant spectrum. This step also makes the recognition range independent, because we are left with relative ranges in the window.

The final step before feature extraction is to normalise the signal by scaling all spectral lines so that the amplitude of the peak line is 1.0. This normalisation step minimises the effect of spreading loss with range. Also, it gives all spectra the same parameters, which makes feature extraction consistent, and hence recognition more robust.

The **Peak Detector** VI from the **Measurement** library is used to locate the objects in the estimated spectrum (bottom right, Figure 4–17) to identify the location of the legitimate landmark and the additional object in the background. The next stage is normalization of the landmark spectrum and applying a window around the peak for feature extraction.

Feature extraction

The features represent geometric information about the plant and show less variation with rotation then the spectra. To illustrate the process we will discuss two features: number of lines above a threshold, and area under the curve. The number of lines above a threshold gives us a measure of the acoustic density of the plant. We have written a subVI that thresholds the spectrum in the window and counts the number of lines above the threshold. The area under the curve is a measure of the total reflecting area in the plant. It is calculated using the **Numeric Integration** VI included in the **Additional Numeric Methods** library.

Recognition

Signal processing only performs part of the processing required by an intelligent system. It transforms the signal into data. Following stages convert the data into information, interpret the information and make decisions. In this system, the next stage is landmark recognition. Statistical classification is used to match the extracted features to previously recorded feature sets in order to determine which landmark is being sensed.

Conclusion

A CTFM system has been designed for landmark navigation. Initial experiments in landmark detection show that there is sufficient information in the echoes to distinguish between different path surfaces and between different landmarks. We have tested several components of the system and are currently integrating them. Also, we have found visual programming in LabVIEW to be considerably faster than programming in C++.

Acknowledgment

The CTFM ultrasonic system was built by Prof. Leslie Key of BAT Systems in New-Zealand.

Suggested exercises:

- Use standard VIs to generate an audio signal with several frequencies of different amplitudes.
- Use another VI to find the spectrum of this signal.
- Build a VI to find the frequency with the maximum amplitude.

■ **Contact Information**

Danny Ratner
Intelligent Robotics Laboratory
School of It & CS
University of Woolongong
NSW 2522
Australia
++612-422-3772
Fax: ++612-4221-4170
Email: dr19@uow.edu.au

Overview

You will now learn about some of the VIs that are already available in the Analysis Library to perform various real-world instrument measurement tasks. These VIs are collectively referred to as measurement VIs.

GOALS

- Learn about the measurement VIs and how they can perform various signal processing operations common in many benchtop spectral analyzers.
- Calculate the frequency (amplitude and phase) spectrum of a time-domain signal, with the appropriate units.
- Become familiar with the Coherent Gain (CG) and Equivalent Noise Bandwidth (ENBW) window constants.
- Determine the total harmonic distortion present in a signal.

KEY TERMS

- frequency spectrum
- amplitude
- phase
- coherent gain
- equivalent noise bandwidth

- total harmonic distortion
- impulse response
- AC components
- DC components
- pulse parameters

Measurement

5

The Measurement VIs

The measurement VIs perform specific measurement tasks such as

- Calculating the total harmonic distortion present in a signal.
- Determining the impulse response, or transfer function, of a system.
- Estimating pulse parameters such as the rise time, overshoot, and so on.
- Computing the amplitude and phase spectrum of a signal.
- Estimating the AC and DC components of a signal.

In the past, these computations have traditionally been performed by benchtop instruments. The measurement VIs make these measurements possible in the G programming language on your desktop computer. They are built on top of the digital signal processing VIs and have the following characteristics:

- The input time-domain signal is assumed to be real valued.
- Outputs are in magnitude and phase, scaled, and in the appropriate units, ready for immediate graphing.
- The spectrums calculated are single-sided and range from DC to Nyquist (sampling frequency/2).
- Wherever appropriate, corrections are automatically applied for the windows being used. The windows are scaled so that each window gives the same peak spectrum amplitude result within its amplitude accuracy constraints.

In general, you can directly connect the ouput of the data acquisition VIs to the inputs of the measurement VIs. The outputs of the measurement VIs can be connected to graphs for an appropriate visual display:

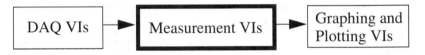

Several measurement VIs perform commonly used time domain-to-frequency domain transformations such as calculation of the amplitude and phase spectrum, the power spectrum, the network transfer function, and so on. Other measurement VIs perform functions such as scaled time-domain windowing and power and frequency estimation.

Calculating the Frequency Spectrum of a Signal

In many applications, knowing the frequency content of a signal provides insight into the system that generated the signal. The information thus obtained can be used in the design of bridges, for calibration purposes, for estimating the amount of noise and vibration generated by parts of machines, and so on. The next activity demonstrates how to use the **Amplitude and Phase Spectrum** VI to identify two frequency components.

Activity 5-1

Objective: To compute the frequency spectrum of a signal

■ Front Panel

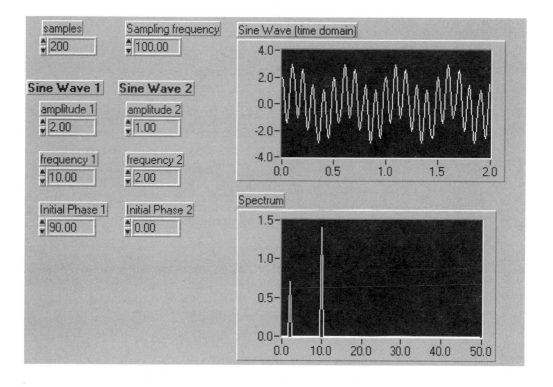

1. Open the **Compute Frequency Spectrum** VI found in the library `Activities.llb`. Two sine waves of frequencies 2 Hz and 10 Hz are superimposed. The 10-Hz sine wave has an amplitude of 2 V, and the 2-Hz sine wave has an amplitude of 1 V. The sampling frequency is 100 Hz and 200 points of data are generated.

2. Switch to the block diagram.

■ Block Diagram

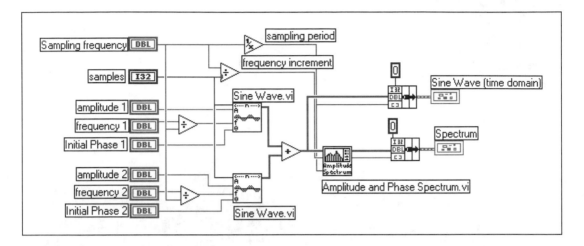

3. Examine the block diagram.

    The **Amplitude and Phase Spectrum** VI
   (**Functions** » **Analysis** » **Measurement**
   subpalette) calculates the amplitude spectrum
   and the phase spectrum of a time domain signal. The
   connections to this VI are shown below.

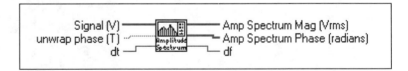

   The time domain signal is applied at the **Signal (V)**
   input. The magnitude and phase of the input signal
   spectrum are available at the **Amp Spectrum Mag
   (Vrms)** and **Amp Spectrum Phase (radians)** outputs,
   respectively.

   The initial phase input to the Sine Wave VI is specified in degrees.

If the units of the input time domain signal are in volts peak (V_p), the units of the magnitude of the amplitude spectrum is in volts rms (V_{rms}). The relationship between the units is $V_{rms} = V_p/\sqrt{2} = 0.707V_p$.

4. Run the VI.

    The graph should display two peaks, one at 2 Hz and the other at 10 Hz. The amplitude of the 2-Hz sine wave is 0.707 V, and that of the 10-Hz waveform is 1.414 V, which are the rms values for sine waveforms of peak amplitudes 1 and 2 V, respectively.

5. Change the phase of the sine waves by adjusting the **Initial Phase 1** and **Initial Phase 2** controls and run the VI.

    Do you notice any change in the time waveform? The spectrum?

6. Make the parameters of both the sine waves equal. That is, set **amplitude 1** = **amplitude 2** = 2, **frequency 1** = **frequency 2** = 10, **Initial Phase 1** = **Initial Phase 2** = 0, and **sampling frequency** = 200, and run the VI.

    Is the amplitude of the peak in the power spectrum the value that you would expect?

7. When you are done, close the VI. Do not save any changes.

■ **End of Activity 5-1**

Coherent Gain (CG) and Equivalent Noise Bandwidth (ENBW)

You saw in Chapter 4 that many different types of windows can be applied to a signal. These windows have different shapes, and they affect the signal in different ways. Two parameters that are useful in comparing various types of windows are the *Coherent Gain* (CG) and the *Equivalent Noise Bandwidth* (ENBW).

■ Coherent Gain

The coherent gain of a window is the *zero frequency* gain (or the *DC gain*) of the window. It is calculated by normalizing the maximum amplitude of the window to one and then summing the values of the window amplitudes over the duration of the window. The result is then divided by the length of the window (that is, the number of samples).

For example, consider the rectangular window shown below with amplitude equal to *A* and nine samples:

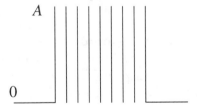

You first normalize (divide) all the heights by *A* to get the maximum height equal to 1:

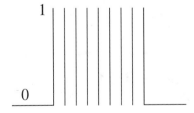

Then you add all the heights to get nine (nine lines each with a height equal to 1). This sum is then divided by the number of samples (nine) to get a value of 1. Thus, the CG of the rectangular window is equal to 1. Mathematically, the CG is given by

$$CG = \frac{1}{N} \sum_{n=0}^{N-1} w[n]$$

where N is the total number of samples over the duration of the window and $w[n]$ are the normalized amplitudes of the samples.

For comparison purposes, Table 5–1 below shows the CG of several commonly used windows. Note that the rectangular (uniform) window has the highest CG, whereas the CG of other windows is lower than that of the rectangular window. Can you explain why this is so?

Table 5–1   *Coherent gains of several windows*

Window	CG
Uniform (None)	1.00
Hamming	0.54
Hanning	0.5
Triangle	0.5
Exact Blackman	0.43
Blackman	0.42
Blackman-Harris	0.42
4-Term Blackman-Harris	0.36
Flat Top	0.28
7-Term Blackman-Harris	0.27

■ Equivalent Noise Bandwidth (ENBW)

You can use the ENBW to compare frequency responses having different shapes. An ideal frequency response is supposed to be rectangular in shape (see the chapter on digital filters). However, in practice, the frequency response differs from the ideal. Because the frequency response of each window has a different shape, each response will pass different amounts of noise power. The ENBW for a particular window is equal to the width of a frequency response having an ideal rectangular shape that will pass the same amount of noise power as the frequency response of that window.

Suppose that the solid line in Figure 5–1 below shows the frequency response of a window, and the dashed line shows the ideal rectangular response. The responses are first adjusted to have a gain of unity at zero frequency (DC).

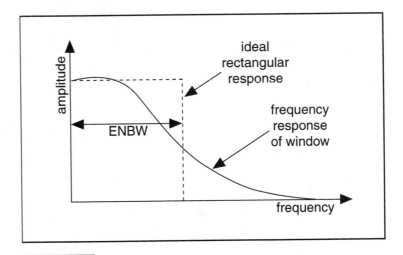

Figure 5–1
Calculation of ENBW

To calculate the ENBW, the width of the rectangular response is adjusted so that it has the same area as that of the nonideal response of the window. This width is then equal to the ENBW for that window. It is found that the ENBW is the smallest for the uniform window, and larger for the other windows. Table 5–2 shows the ENBW for different windows, relative to the ENBW for the uniform window.

Table 5–2   *Equivalent noise bandwidth of several windows*

Window	ENBW
Uniform (None)	1.00
Hamming	1.36
Hanning	1.5
Triangle	1.33
Exact Blackman	1.69
Blackman	1.73
Blackman-Harris	1.71
4-Term Blackman-Harris	2.00
Flat Top	2.97
7-Term Blackman-Harris	2.63

Activity 5-2

Objective: To learn about the coherent gain and equivalent noise bandwidth properties of windows

This activity will familiarize you with different window shapes and their CG and ENBW. The values of the CG and ENBW are used in other measurement VIs such as the **Power and Frequency Estimate** VI and **Spectrum Unit Conversion** VI.

1. Open the **Time Domain Windows** VI found in the library `Activities.llb`.
2. Open and examine the block diagram.

■ Block Diagram

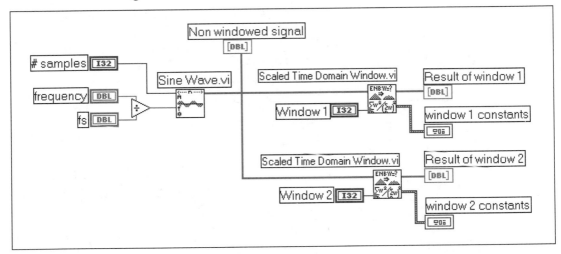

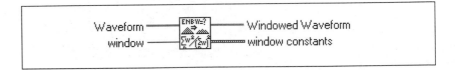 The VI that windows the input signal (a sine wave generated by the **Sine Wave** VI) and gives as the output the resulting windowed waveform is the **Scaled Time Domain Window** VI. The connections to this VI are shown below:

The time waveform is applied at the **Waveform** input, and the window selection is done by the **window** control. The VI also outputs the CG and the ENBW of the selected window at the **window constants** terminal.

The resulting output waveform is automatically scaled so that when you compute the amplitude or power spectrum of the windowed waveform, all the windows will give the same value.

You will pass the sine wave through two of these VIs and compare the resulting windowed waveforms.

3. Switch to the front panel.

■ Front Panel

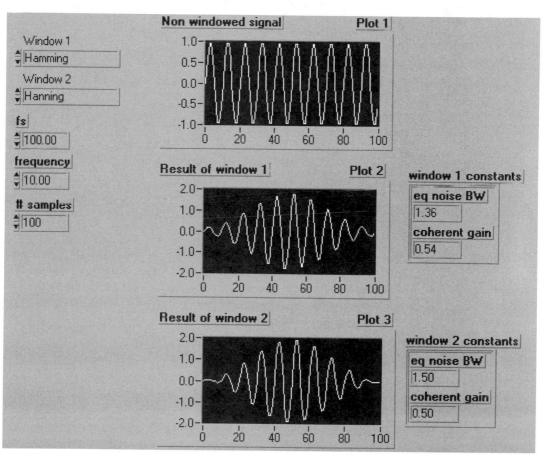

The **Window 1** and **Window 2** controls select the two types of windows that you want to apply to your signal (the sine wave).

The number of samples and the frequency of the sine wave are controlled by the **# samples** and **frequency** controls, respectively. The sampling frequency is adjusted by the **fs** control.

The topmost plot shows you the original time-domain waveform (without windowing). The lower two plots show you the signal after application of the two windows specified in the **Window 1** and **Window 2** controls.

4. Select **Window 1** as *None (Uniform)* and **Window 2** as *Hanning* and run the VI. Leave all the controls at their default values.

   Compare the waveforms in **Plot 1** and **Plot 2** and observe that using the uniform window is equivalent to not using any window.

   Observe from **Plot 2** and **Plot 3** the difference in the windowed time domain waveform due to the application of the *uniform* and *Hanning* windows, respectively.

   Note the differences in the CG and ENBW of the two windows.

5. The shapes of the *Hamming* and *Hanning* windows are very close to each other. Choose these windows in the **Window 1** and **Window 2** controls. Run the VI and compare the waveforms in **Plot 2** and **Plot 3**.

   Can you notice any difference? Which one is wider? In particular, compare the CG values of these two windows.

6. As mentioned before, the CG was the same as the DC gain. Choose different windows and run the VI. Observe that multiplying the maximum amplitude of the windowed signal by the CG of the window gives unity.

7. When you finish, close the VI. Do not save any changes.

■ End of Activity 5-2

Harmonic Distortion

When a signal, $x(t)$, of a particular frequency (for example, f_1) is passed through a nonlinear system, the output of the system consists of not only the input frequency (f_1), but also its harmonics ($f_2 = 2f_1, f_3 = 3f_1, f_4 = 4f_1$, and so on). The number of harmonics, and their corresponding amplitudes that are generated depends on the degree of nonlinearity of the system. In general, the more the nonlinearity, the higher the harmonic frequencies, and vice versa:

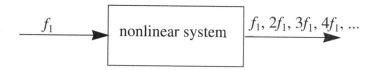

An example of a nonlinear system is a system where the output $y(t)$ is the cube of the input signal $x(t)$:

$$\xrightarrow{\cos(\omega t)} \boxed{y(t) = f(x) = x^3(t)} \xrightarrow{\cos^3(\omega t)}$$

So, if the input is

$$x(t) = \cos(\omega t),$$

the output is

$$x^3(t) = 0.5\cos(\omega t) + 0.25[\cos(\omega t) + \cos(3\omega t)]$$

Therefore, the output contains not only the input fundamental frequency of ω, but also the third harmonic of 3ω.

■ Total Harmonic Distortion

To determine the amount of nonlinear distortion that a system introduces, you need to measure the amplitudes of the harmonics that were introduced by the system relative to the amplitude of the fundamental. Harmonic distortion is a relative measure of the amplitudes of the harmonics as compared to the amplitude of the fundamental. If the amplitude of the fundamental is A_1, and the amplitudes of the harmonics are A_2 (second harmonic), A_3 (third harmonic), A_4 (fourth harmonic), ..., A_N (Nth harmonic), the Total Harmonic Distortion (THD) is given by

$$THD = \frac{\sqrt{A_2^{\,2} + A_3^{\,2} + ...A_N^{\,2}}}{A_1}$$

and the percentage total harmonic distortion (% THD) is

$$\%THD = \frac{100 \times \sqrt{A_2^{\,2} + A_3^{\,2} + ...A_N^{\,2}}}{A_1}$$

In the next activity, you will generate a sine wave and pass it through a nonlinear system. The block diagram of the nonlinear system is shown in Figure 5–2.

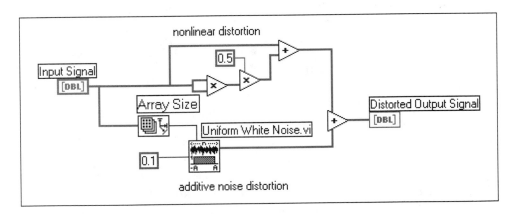

Figure 5–2
A nonlinear system with added uniform white noise

Verify from the block diagram that if the input is $x(t) = \cos(\omega t)$, and the noise is $n(t)$, then the output is

$$y(t) = \cos(\omega t) + 0.5 \cos^2(\omega t) + 0.1n(t)$$
$$= \cos(\omega t) + [1 + \cos(2\omega t)]/4 + 0.1n(t)$$
$$= 0.25 + \cos(\omega t) + 0.25 \cos(2\omega t) + 0.1n(t)$$

Therefore, this nonlinear system generates an additional DC component as well as the second harmonic of the fundamental.

■ Using the Harmonic Analyzer VI

You can use the **Harmonic Analyzer** VI to calculate the %THD present in the signal at the output of the nonlinear system. It finds the fundamental and harmonic components (their amplitudes and corresponding frequencies) present in the power spectrum applied at its input and calculates the percentage of total harmonic distortion (%THD) and the percentage of total harmonic distortion plus noise (%THD + Noise). The connections to the **Harmonic Analyzer** VI are shown in Figure 5–3.

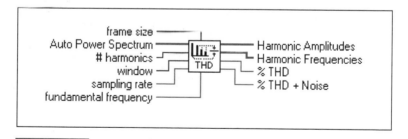

Figure 5–3
The Harmonic Analyzer VI

To use this VI, you need to give it the power spectrum of the signal whose THD you want it to calculate. Thus, in this example, you need to make the following connections:

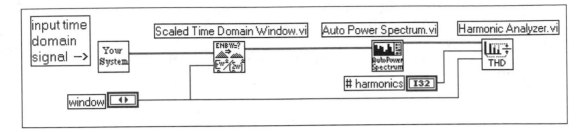

Figure 5–4
An example of using the harmonic analyzer VI

The **Scaled Time Domain Window** VI applies a window to the output $y(t)$ of the nonlinear system (**Your System**). This is then passed on to the **Auto Power Spectrum** VI, which sends the power spectrum of $y(t)$ to the **Harmonic Analyzer** VI, which then calculates the amplitudes and frequencies of the harmonics, the THD, and the %THD.

You can specify the number of harmonics you want the VI to find in the **# harmonics** control. Their amplitudes and corresponding frequencies are returned in the **Harmonic Amplitudes** and **Harmonic Frequencies** array indicators.

The number specified in the # harmonics control includes the fundamental. So, if you enter a value of 2 in the # harmonics control, the Harmonic Analyzer will find the fundamental (say, of freq f_1) and the second harmonic (of frequency $f_2 = 2f_1$). This would correspond to measuring the 2nd harmonic distortion. If you enter a value of N, the VI will find the fundamental and (N – 1) harmonics.

The following are explanations of some of the other controls.

Fundamental frequency is an estimate of the frequency of the fundamental component. If left as zero (the default), the VI uses the frequency of the non-DC component with the highest amplitude as the fundamental frequency.

Window is the type of window you applied to your original time signal. It is the window that you select in the **Scaled Time Domain Window** VI. For an accurate estimation of the THD, it is recommended that you select a window function. The default is the uniform window.

Sampling rate is the input sampling frequency in Hz.

The **% THD + Noise** output requires some further explanation. The calculations for **% THD + Noise** are almost similar to that for **% THD**, except that the noise power is also added to that of the harmonics. It is given by

$$\%THD + Noise = 100 \times \frac{\sqrt{sum(APS)}}{A_1}$$

where $sum(APS)$ is the sum of the Auto Power Spectrum elements minus the elements near DC and near the index of the fundamental frequency.

Activity 5-3

Objective: To use the Harmonic Analyzer VI for harmonic distortion calculations

1. Open the **THD Example** VI from the library `Activities.llb`.
2. Switch to the block diagram.

■ Block Diagram

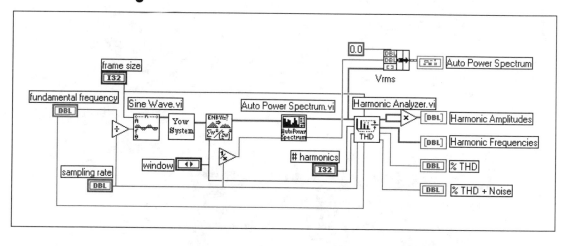

Some of this will already be familiar to you. **Your System** is the nonlinear system that you saw previously. Its output is windowed, and the power spectrum calculated and given to the **Harmonic Analyzer** VI.

The **Sine Wave** VI generates a fundamental of frequency specified in the **fundamental frequency** control.

The output of the **Harmonic Analyzer** VI is in V_{rms} (if the input from the Auto Power Spectrum is in V^2_{rms}). This output is then squared to convert it to V^2_{rms}.

3. Switch to the front panel.

At the bottom, you see a plot of the auto power spectrum of the output of the nonlinear system. On the top-right side are the array indicators for the frequencies and amplitudes of the fundamental and its harmonics. The size of these arrays depends on the value entered in the **# harmonics** control.

■ Front Panel

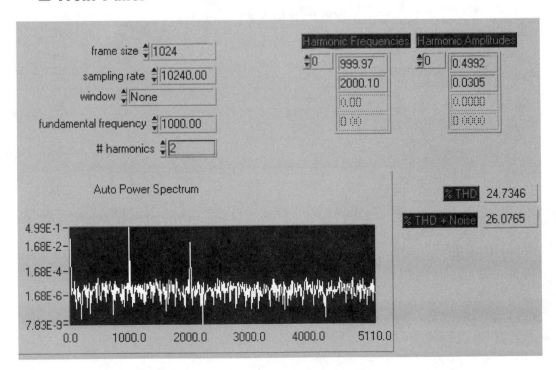

4. Change the **fundamental frequency** to 1000, **# harmonics** to 2, and run the VI several times. Each time, note the values in the output indicators (**Harmonic Frequencies**, **Harmonic Amplitudes**, **% THD**, and **% THD + Noise**).

Why do you get different values each time you run the VI?

Which of the values, % THD or % THD + Noise, is larger? Can you explain why?

5. Run the VI with different selections of the window control and observe the peaks in the power spectrum.

Which window gives the narrowest peaks? The widest? Can you explain why?

See the values of the ENBW for each window in Table 5–2.

6. Change the fundamental frequency to 3000 and run the VI.

Why do you get an error?

Consider the relationship between the Nyquist frequency and the frequency of the harmonic(s).

7. When you finish, close the VI. Do not save any changes.

■ End of Activity 5-3

Wrap It Up!

- The ready-made VIs to perform common measurements are available in the **Analysis » Measurements** subpalette.
- Some of these measurement tasks include calculating the amplitude and phase spectrum of a signal and the amount of harmonic distortion. Other VIs calculate properties of a system such as its transfer function, its impulse response, the cross-power spectrum between the input and output signals, and so on.
- Because a real-world signal is time limited by a window function, the study of certain properties of these window functions is an important consideration in interpreting the results of your measurements.
- The CG is a measure of the DC gain of the window, whereas the ENBW is a measure of the amount of noise power that a window introduces into a measurement.

■ Review Questions

1. Why is it important to know the coherent gain of a window?
2. Which **Measurement** VIs calculate both the amplitude and phase spectrum of the input waveform?
3. Name some applications where you would use the **Measurement** VIs.
4. When a **Measurement** VI calculates the spectrum of a signal, is it a one-sided or a two-sided spectrum?
5. What is the peak value corresponding to $2V_{rms}$?
6. Which window has the highest coherent gain? Why?
7. What is the difference between %THD and %THD + Noise? Which is larger?

■ Additional Activities

1. Open and run the **Vibration Analysis** VI from **the Examples >> Apps >> Demos.llb** library in your LabVIEW folder. This Virtual Instrument simulates a vibration analysis example. You can set the velocity using the **Set Velocity (km/hr)** control on the front panel. As the velocity of the engine is increased, the displacement of the component also increases and the peak in the **Power Spectrum (dB)** plot also moves towards a higher frequency. If the acquisition rate is not high enough, the component frequencies of the displacement waveform cannot be calculated accurately.

   a) Keep the **Acquisition Rate** at 250 kHz. While observing the peak in the power spectrum, gradually increase the velocity to beyond 150 km/hr using the **Set Velocity (km/hr)** control. Observe the effect of aliasing as the peak first moves towards the right and then moves back towards the left.

   b) Increase the **Acquisition Rate** to 500 kHz, and then vary the **Set Velocity (km/hr)** control from 0 km/hr to beyond 150 km/hr. Observe that in this case there is no aliasing.

This activity can only be done with the LabVIEW Full Development System.

2. A common measurement made in the field of digital communications, or in control systems, is to estimate the parameters of a (possibly noisy) pulse. The information obtained can be used to determine the level (1 or 0, High or Low) of a signal, or the characteristics of a system. The **Pulse Parameters** VI from the **Measurement** subpalette analyzes an input sequence for a pulse pattern and determines the best set of pulse parameters that describes the pulse. Some common

parameters of a pulse are overshoot, undershoot, rise time, top, base, and amplitude.

a) Build a VI having the block diagram shown below:

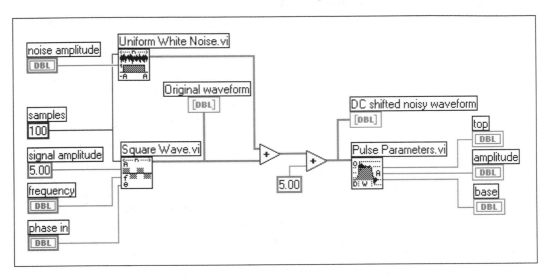

The **Square Wave** and **Uniform White Noise** VIs can be found in the **Analysis >> Signal Generation** subpalette, whereas the **Pulse Parameters** VI can be found in the **Analysis >> Measurement** subpalette. The **frequency** and **phase in** digital controls are used to adjust the frequency and initial phase of the square wave. The **noise amplitude** digital control adjusts the amplitude of the noise added to the square wave. A DC value of +5V is then added to the noisy square wave. **Original waveform** and **DC shifted noisy waveform** are **Waveform Graphs**. **Top, amplitude,** and **base** are digital indicators. **Top** is the line that best represents the values when the pulse is active, high, or on; **base** is the line that best represents the values when the pulse is inactive, low, or off; **amplitude** is the difference between the **top** and the **base**.

b) Set the controls to the following values... **frequency** = 0.02, **phase in** = 180.0, **noise amplitude** =

0.5, and run the VI several times. Each time, observe the estimated values of **top, base,** and **amplitude.**

c) Change the value of **noise amplitude** in increments of 0.5 starting from 0.5 upto 5.0. Each time, run the VI and observe the estimated values of **top, base, and amplitude.** Note that the error in the estimation increases as the amplitude of the noise increases.

d) Set **noise amplitude** back to 0.5. Change the values of **phase in** to 0, 45, 90, 135, and 180. Observe the initial phase of the generated square wave in the **Original waveform** plot.

e) Save the VI as **My Pulse Parameters** VI in the library `Activities.11b` and close it.

3. In many experimental systems, a small DC offset gets added to the output waveform. This DC offset is so small that it can go undetected by the user. The figure below shows a sinusoidal waveform of amplitude 2.0 V offset by a small DC voltage of 0.01 V.

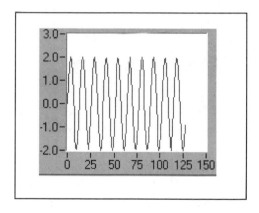

Sinusoidal signal with a small DC offset

In this activity, you will see how the small offset voltage can cause significant errors when the waveform is supplied as an input to other VIs and how the error can be corrected.

Select the **Test System** VI from the library `Activities.llb` and place it on the block diagram of your VI. The output of this VI is a sinusoidal signal with a small DC offset. Your aim is to integrate this output. Connect the **System Output** terminal of this VI to the **X** input of the **Integral x(t)** VI (**Analysis** >> **Digital Signal Processing** subpalette). Use default values for all the other terminals of this VI. Plot the integrated output on a waveform graph. Run the VI.

Normally, you would expect the integral of a sinusoidal signal as shown in the figure above to be a cosine waveform. However, you will observe that the integrated output is tilted and it appears like a cosine waveform on an increasing ramp

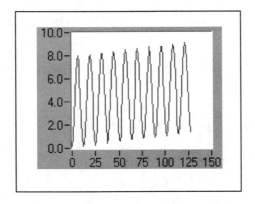

Integrated signal with small DC component

This shows that a small DC offset can introduce significant errors in your measurement. We will now see how the error can be corrected.

Select the **AC & DC Estimator** VI (**Analysis** >> **Measurement** subpalette). Connect the **System Output** terminal of the **Test System** VI to the input terminal of the **AC & DC Estimator** VI.

This VI will output the estimated AC and DC value of the input waveform. Subtract the DC value from each element of the output of the **Test System** VI. Then pass this waveform to the **Integral x(t)** VI and plot the integrated output on a waveform graph.

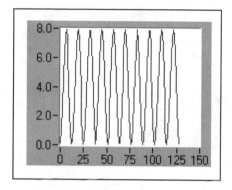

Integrated signal after removing the DC component

As shown in the figure above, the integrated output is now a proper cosine waveform. Thus detecting the DC component in the output of your system and removing it, yields correct results in the later processing stages.

Save the VI as **Offset Voltage Effect**.vi in the library Activities.llb and close it.

You can find the solution to this activity in the library Solutions.llb.

Periodic Signal Analyzer

Daniel Ch. von Grünigen
Professor of Signal Processing
Burgdorf School of Engineering and Architecture
Burgdorf, Switzerland

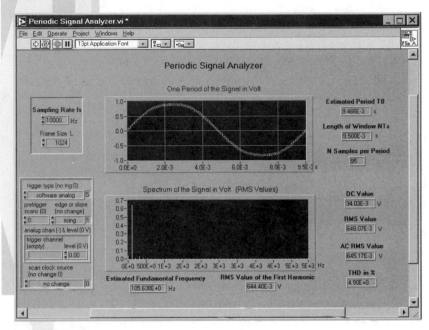

Figure 5–5
Front panel of the Periodic Signal Analyzer VI

The Challenge To determine several parameters (such as the fundamental frequency, RMS values, and the total harmonic distortion) of a periodic signal.

The Solution Use LabVIEW to build a VI that will acquire the signal and calculate these parameters.

Introduction

The Periodic Signal Analyzer VI (found on the enclosed CD) samples a periodic signal and represents one period and its harmonics to determine several parameters of the periodic signal. The front panel of the Periodic Signal Analyzer VI is shown above.

Periodic signals are signals whose waveform repeats periodically. Mathematically a periodic signal $x(t)$ can be defined by the equation $x(t) = x(t + T_0)$, where T_0 is the period and $f_0 = 1/T_0$ is the fundamental frequency. Examples are a sine wave and a triangular wave produced by a signal generator, the AC line voltage, the voltage at the output of a microphone driven by a tone, and so forth. According to Fourier theory, a periodic signal can be decomposed into a weighted sum of sine waves, called the harmonics, whose frequencies are an integral multiple of the fundamental frequency:

$$x(t) = A_0 + \sum_{k=1}^{\infty} A_k \cos(2\pi k f_0 t + \varphi_k)$$

The DC value A_0 and the amplitudes A_k of the harmonics can easily be determined by the DFT which is defined as

$$X(k) = \sum_{n=0}^{N-1} x(n) e^{-j2\pi kn/N}$$

where $x(0), x(1), x(2), ..., x(N-1)$ are N values of $x(t)$, sampled with the sample interval T_s. If the length of the measuring time NT_s, also called rectangular window length, is equal to the period T_0, then the DC value X_{DC} and all the other parameters of the periodic signal [1], [2] are given as:

$$X_{DC} = A_0 = \frac{1}{N} X(0)$$

The RMS values, denoted by X_k, are

$$X_k = \frac{A_k}{\sqrt{2}} = \frac{\sqrt{2}}{N} |X(k)| \qquad\qquad k = 1, 2, ..., N/2$$

The RMS value, the AC RMS value, and the total harmonic distortion THD

$$X_{\text{RMS}} = \sqrt{\sum_{k=0}^{N/2} X_k^2}, \quad \text{with} \qquad X_0 = X_{DC}$$

$$X_{\text{AC RMS}} = \sqrt{\sum_{k=1}^{N/2} X_k^2}, \text{ and}$$

$$\text{THD}_{\text{in}\%} = 100 \frac{\sqrt{\sum_{k=2}^{N/2} X_k^2}}{X_1}$$

When N is not even, then $N/2$ is rounded toward the nearest lower integer.

■ Application

The Periodic Signal Analyzer VI can be used to measure the harmonic content of periodic signals, such as

- the line voltage,
- the output of an amplifier driven by a sine wave,
- a thyristor-controlled AC current,
- the sound of a flute,
- and so forth

■ Input

One channel of a DAQ device (default device number 1 and default channel number 0) has to be driven by a periodic voltage. The user can choose the sampling frequency $f_s = 1/T_s$ and and the frame size L, which should be a power of 2 for fast computation.

■ Output

The output on the front panel depicts one period of the periodic waveform and its harmonics. It also displays an estimate of the fundamental frequency f_0 and the period T_0 of the signal, the number N of samples per period, and the length NT_s of the measuring rectangular window. The VI determines and displays the RMS value of the first harmonic, the DC value, the RMS value, the AC RMS value, and the total harmonic distortion.

■ Method

Based on the L input samples, the Periodic Signal Analyzer VI estimates the fundamental frequency f_0, which is the frequency of the first harmonic. The estimation assumes that the power at the first harmonic is bigger than the power of the remaining harmonics (if this assumption is not fullfilled, the estimation of the fundamental frequency fails and the VI produces meaningless results). The period T_0 of the signal can then be computed by $1/f_0$. The number N of samples per period can subsequently be determined by $N = \text{round}(T_0/T_s)$. The first N samples of the input signal are finally DFT transformed and used as a basis to calculate the remaining parameters.

■ Example

Choose the sampling rate $f_s = 10000$ Hz and the frame size $L = 1024$. Input a 1-kHz sine wave of 1-V amplitude and then a 1-kHz triangle of 1-V amplitude. Watch how a bigger frame size improves the estimation of the period T_0 and a higher sampling rate increases the number N of samples per period. Both enhancements improve the results. Examine if the length of the window NT_s is equal to the period T_0. For precise results this condition should also be fullfilled as well as possible. Remember that a triangular wave has only odd harmonics.

■ References

1. E. Oran Brigham, *The Fast Fourier Transform and its Applications*, Prentice Hall, 1988.
2. Daniel Ch. von Grünigen, *Digitale Signalverarbeitung*, AT-Verlag, Aarau, 1993.

■ Contact Information

Daniel Ch. von Grünigen
Dr. Sc. Techn. ETH, Professor for Signal Processing
Burgdorf School of Engineering and Architecture
Department of Electrical Engineering
Jlcoweg 1
CH-3400 Burgdorf, Switzerland
Phone: ++41 (0) 34 426 68 11
Fax: ++41 (0) 34 426 68 13
Email: daniel.vongruenigen@isburg.ch

Voltage Fluctuation on High Voltage Network

Andres Thorarinsson
Chief Engineer, BS
Gudmundur Eiriksson
Application Engineer, BS
Vista Engineers, Reykjavik, Iceland

The Challenge Measure voltage and frequency fluctuation, and overtones, on a 220 kV network due to switch closure of 100 MW load of new Aluminum factory.

The Solution Use DAQCard and LabVIEW to sample voltage, frequency and current transducers at 5.5 kHz.

Introduction

To most people, a steady clean voltage supply for work and home is taken for granted. However, there is a steady ongoing battle between users and suppliers, where the users pollute the clean 60Hz (50Hz for part of the word) sinewave originating from generators.

Another problem facing suppliers is the variable load on the network. This is because users are switching the power on or off, like lights, heating or cooling, or maybe the whole factory is switching in or out. The supplier has to keep a steady 60Hz frequency (50Hz) and steady voltage (120V at users level) all the time. To keep a steady frequency is like driving a car at constant speed, no matter whether it is traveling uphill or downhill, or if the car is empty or heavily loaded. The operator has to be alert at the throttle all the time.

To make the problem more visible to the suppliers staff, it is requested that voltage fluctuations and overtones originating from a users network are measured and displayed for analyzers. The system described in this article was put together in only a few days and was found to work very nicely, replacing specialized equipment costing thousands of dollars.

System Requirements

It was requested that a measurement system would be installed at a High Voltage feeder station for a new aluminum factory. The purpose was to measure fluctuations on voltage, frequency and current of the feeder when part of the load was switched in and out.

As overtones of 60Hz (50Hz) were found interesting, it was requested that a sampling frequency of 5.5 kHz be implemented, which would allow easy separation of 11 overtones.

In order to use disk space sparsely, the system was only to store on disk time marked fluctuations but otherwise store only one sample per second.

In order to measure fluctuations on 220kV 3 phase feeder line, the measurement system was to be connected to isolation amplifiers for 220kV to 10V, ±1000A to ±20mA, 50Hz to 10V, and ±50 Mvar to ±10V.

The front page of the system was to show in real time the measurements for voltage, current, Mvar and frequency, and the size of the first 11 overtones.

System Design

The system design was relatively straight forward. It was decided to use the DAQ-card PCI-1200 for measurements, for its low price, ease-of-use and stability in all aspects. To connect to the isolation amplifiers, a 5B isolation amplifiers were connected to the PCI-1200. In order to measure up to the 11th overtone, it was decided to sample 10 times as high as the overtone. For a

50Hz system, the 11th overtone is 550Hz. So, in order to have a good waveform, a sampling frequency of 5500 samples/sec was used. Then, thought was given to the man-machine-interface (MMI). The operators wanted to be able to decide which signal was connected to each channel of the DAQ-card. Then, they wanted to select one or more channels to trigger measurements flushed to disk, and finally, they wanted to select the threshold for the channels, which when stepped over would trigger measurements. This was all done with a single MMI, that would also allow the operator to start and to stop the application.

System Performance

The system is capable of doing what was expected. It is left unattended for long periods of time, and as it is networked, it is not necessary to visit the site for disk maintenance.

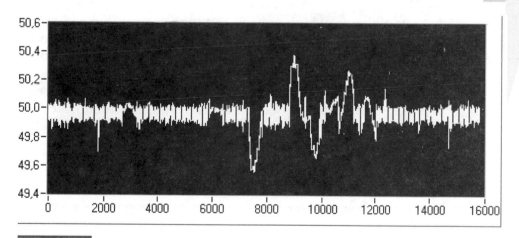

Figure 5–6
Fluctuations on 50Hz-network frequency due to load disturbance

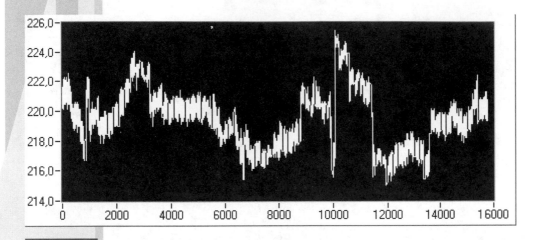

Figure 5–7
The voltage on 220kV feeder jumps up and down as load varies, and load capacitors are switched in and out.

Results

Now, all kinds of high voltage and high current fluctuations are visible to operators and analyzers of the supplier, e.g. the Power Company. Because of this system, highly specialized equipment worth thousands of dollars is not used, but has been replaced by this relatively inexpensive and simple solution. Furthermore, a project that started only to measure voltage fluctuations on the aluminum factory, is now being expanded to hydro generators and switching stations in different parts of the country.

Conclusion

With off-the-shelf DAQ-boards and LabVIEW software, it was possible to produce state-of-the-art measurement and diagnosis tools in a short time, in a field that before was only for highly specialized vendors.

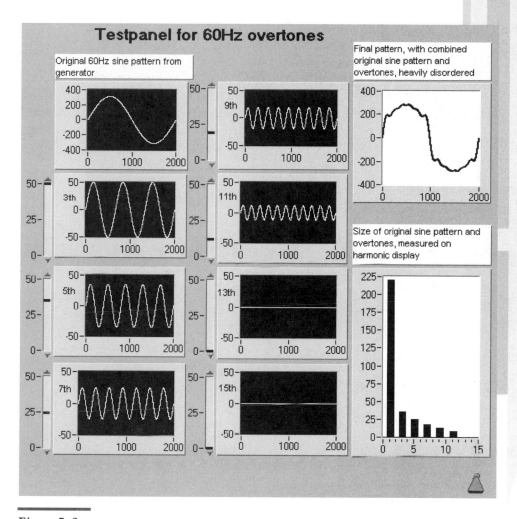

Figure 5–8

A testpanel done with LabVIEW. The purpose is to better the understanding of overtones. In general, all loads, except for purely resistive loads (like heaters), produce some overtones. For example, electronic speed drives for variable speed of motors are heavy polluters.

Summary

For many complicated DAQ problems are simple solutions based on off-the-shelf hardware and software. With basic training and a determined mind, an engineer is able to compete with bigger vendors with flexible and splendid solutions at relatively lower costs.

■ Contact Information

Andres Thorarinsson
Email: vista@centrum.is

Overview

In this chapter, you will learn about the characteristics of different types of digital filters and how to use them in practical filtering applications.

GOALS

- Master the basics of filtering and why it is needed (applications).
- Learn about the frequency response characteristics of different types of ideal filters—lowpass, highpass, bandpass, bandstop.
- Become familiar with the differences between ideal filters and practical filters.
- Know the advantages of digital filters over analog filters.
- Understand the differences between Infinite Impulse Response (IIR) and Finite Impulse Response (FIR) filters.
- Learn about the characteristics of different types of IIR filters.
- Learn about the transient response of IIR filters.
- Learn about the characteristics of FIR filters.

KEY TERMS

- digital filters
- analog filters
- passband ripple
- stopband attentuation
- lowpass
- highpass
- bandpass
- bandstop
- FIR filters
- IIR filters
- transient response
- impulse response

Digital Filtering

What Is Filtering?

Filtering is the process by which the frequency content of a signal is altered. It is one of the most commonly used signal processing techniques. Common everyday examples of filtering are the bass and treble controls on your stereo system. The bass control alters the low-frequency content of a signal, and the treble control alters the high-frequency content. By varying these controls, you are actually filtering the audio signal. Some other applications are for removing noise and performing decimation (lowpass filtering the signal and reducing the sampling rate).

Ideal Filters

Filters alter or remove unwanted frequencies. Depending on the frequency range that they either pass or attenuate, they can be classified into the following types:

- *A lowpass filter* passes low frequencies but attenuates high frequencies.
- A *highpass filter* passes high frequencies but attenuates low frequencies.
- A *bandpass filter* passes a certain band of frequencies.
- A *bandstop filter* attenuates a certain band of frequencies.

The ideal frequency response of these filters is shown in Figure 6–1.

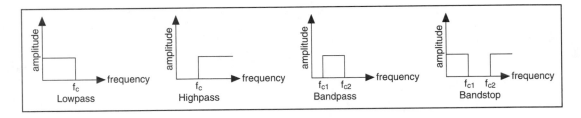

Figure 6–1
Frequency response of ideal filters

You see that the lowpass filter passes all frequencies below f_c, whereas the highpass filter passes all frequencies above f_c. The bandpass filter passes all frequencies between f_{c1} and f_{c2}, whereas the bandstop filter attenuates all frequencies between f_{c1} and f_{c2}. The frequency points f_c, f_{c1} and f_{c2} are known as the *cutoff frequencies* of the filter. When designing filters, you need to specify these cutoff frequencies.

The frequency range that is passed through the filter is known as the *passband* (PB) of the filter. An ideal filter has a gain of one (0 dB) in the passband so that the amplitude of the signal neither increases nor decreases. The *stopband* (SB) corresponds to that range of frequencies that do not pass through the filter at all and are rejected (attenuated). The passband and the stopband for the different types of filters are shown in Figure 6–2.

Note that whereas the lowpass and highpass filters have one passband and one stopband, the bandpass filter has one pass-

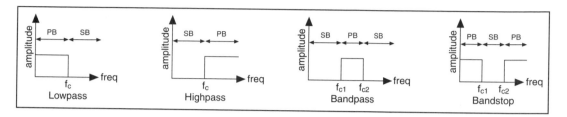

Figure 6–2
Passband (PB) and stopband (SB) of ideal filters

band, but two stopbands, and the bandstop filter has two pass-bands, but one stopband.

■ How Filters Affect Signal Frequency Content

Suppose you have a signal containing frequencies of 10 Hz, 30 Hz, and 50 Hz. This signal is passed through a lowpass, high-pass, bandpass, and bandstop filter. The lowpass and highpass filters have a cutoff frequency of 20 Hz, and the bandpass and bandstop filters have cutoff frequencies of 20 Hz and 40 Hz. The output of the filter in each case is shown in Figure 6–3.

Practical (Nonideal) Filters
■ The Transition Band

Ideally, a filter should have a unit gain (0 dB) in the passband, and a gain of zero (– ∞dB) in the stopband. However, in a real implementation, not all of these criteria can be fulfilled. In prac-tice, there is always a finite transition region between the pass-band and the stopband. In this region, the gain of the filter changes gradually from 1 (0 dB) in the passband to 0 (– ∞dB) in

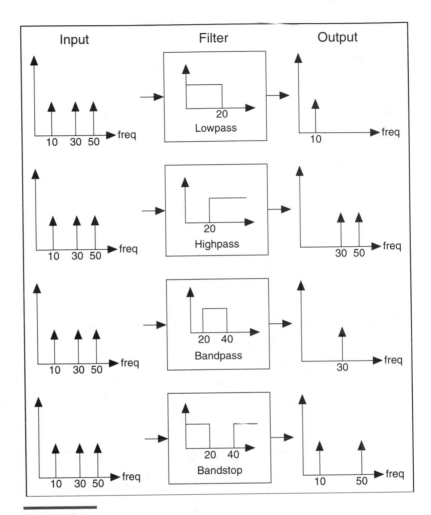

Figure 6–3
Effect of ideal filters on frequency content

the stopband. The diagrams in Figure 6–4 show the passband, the stopband, and the Transition Region (TR) for the different types of nonideal filters. Note that the passband is now the frequency range within which the gain of the filter varies from 0 dB to –3 dB. Although the –3 dB range is most commonly used, depending on the application, other values (–0.5 dB, –1 dB, etc.) may also be considered.

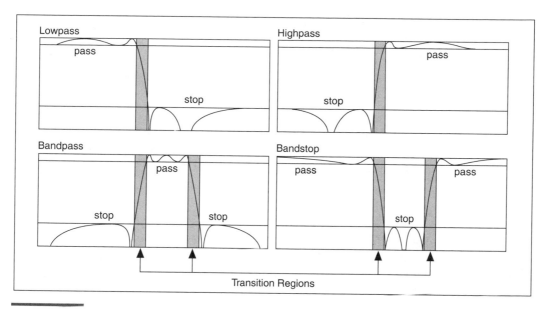

Figure 6–4
Frequency response of practical (nonideal) filters

■ Passband Ripple and Stopband Attenuation

In many applications, it is acceptable to allow the gain in the passband to vary slightly from unity. This variation in the passband is called the *passband ripple* and is the difference between the actual gain and the desired gain of unity. The attenuation in the stopband (also known as the *stopband attenuation*), in practice, cannot be infinite, and you must specify a value with which you are satisfied. Both the passband ripple and the stopband attenuation are measured in decibels or dB, defined by

$$dB = 20\log_{10}(A_o(f)/A_i(f))$$

where $\log_{10}$ denotes the logarithm to the base 10, and $A_i(f)$ and $A_o(f)$ are the amplitudes of a particular frequency f before and after the filtering, respectively.

For example, for -0.02 dB passband ripple, the formula gives

$$-0.02 = 20\log_{10}(A_o(f)/A_i(f))$$

$$A_o(f)/A_i(f) = 10^{-0.001} = 0.9977$$

which shows that the ratio of output and input amplitudes is close to unity.

If you have -60 dB attenuation in the stopband, you have

$$-60 = 20\log_{10}(A_o(f)/A_i(f))$$

$$A_o(f)/A_i(f) = 10^{-3} = 0.001$$

which means the output amplitude is 1/1000 of the input amplitude. Figure 6–5, though not drawn to scale, illustrates this concept.

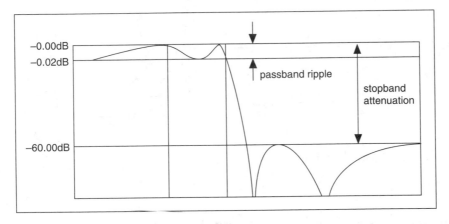

Figure 6–5
Passband ripple and stopband attenuation of a lowpass filter

Attenuation is usually expressed in decibels without the word "minus," but a negative dB value is normally assumed.

Advantages of Digital Filters over Analog Filters

An analog filter has an analog signal at both its input and its output. Both the input, *x(t)*, and output, *y(t)*, are functions of a continuous variable *t* and can take on an infinite number of values. Analog filter design is about 50 years older than digital filter design. Thus, analog filter design books featuring simple, well-tested filter designs exist and can be found extensively in the literature. However, this type of filter design is often reserved for specialists because it requires advanced mathematical knowledge and understanding of the processes involved in the system affecting the filter.

Modern sampling and digital signal processing tools have made it possible to replace analog filters with digital filters in applications that require flexibility and programmability. These applications include audio, telecommunications, geophysics, and medical monitoring. The advantages of digital filters over analog filters are

- They are software programmable and so are easy to "build" and test.

- They require only the arithmetic operations of multiplication and addition/subtraction and so are easier to implement.

- They are stable (do not change with time or temperature) and predictable.

- They do not drift with temperature or humidity or require precision components.

- They have a superior performance-to-cost ratio.

- They do not suffer from manufacturing variations or aging.

IIR and FIR Filters

Another method of classification of filters is based on their impulse response. But what is an impulse response? The response of a filter to an input that is an impulse ($x[0] = 1$ and $x[i] = 0$ for all $i \neq 0$) is called the *impulse response* of the filter (see Figure 6–6). The Fourier transform of the impulse response is known as the *frequency response* of the filter. The

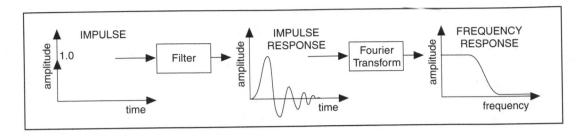

Figure 6–6
Impulse response and frequency response of a filter

frequency response of a filter tells you what the output of the filter is going to be at different frequencies. In other words, it tells you the gain of the filter at different frequencies. For an ideal filter, the gain should be 1 in the passband and 0 in the stopband. So, all frequencies in the passband are passed "as is" to the output, but there is no output for frequencies in the stopband.

If the impulse response of the filter falls to zero after a finite amount of time, it is known as a *Finite Impulse Response (FIR)* filter. However, if the impulse response exists indefinitely, it is known as an *Infinite Impulse Response (IIR)* filter. Whether the impulse response is finite or not (that is, whether the filter is FIR or IIR) depends on how the output is calculated.

The basic difference between FIR and IIR filters is that for FIR filters, the output depends only on the current and past input values, whereas for IIR filters, the output depends not only on the current and past input values, but also on the past output values.

As an example, consider a cash register at a supermarket. Let $x[k]$ be the cost of the kth item that a customer buys, where $1 \leq k \leq N$, and N is the total number of items. The cash register adds the cost of each item to produce a "running" total. This "running" total $y[k]$, up to the kth item, is given by

$$y[k] = x[k] + x[k-1] + x[k-2] + x[k-3] + \ldots + x[1] \qquad (6.1a)$$

Thus, the total for N items is $y[N]$. Because $y[k]$ is the total up to the kth item, and $y[k-1]$ is the total up to the $(k-1)$st item, you can rewrite Equation 6.1a as

$$y[k] = y[k-1] + x[k] \qquad (6.1b)$$

If you add a sales tax of 8.25%, Equations 6.1a and 6.1b can be rewritten as

$$y[k] = 1.0825x[k] + 1.0825x[k-1] +$$
$$1.0825\ x[k-2] + 1.0825x[k-3] +$$
$$\dots + 1.0825x[1] \tag{6.2a}$$

$$y[k] = y[k-1] + 1.0825x[k] \tag{6.2b}$$

Note that both Equations 6.2a and 6.2b are identical in describing the behavior of the cash register. The difference is that whereas 6.2a is implemented only in terms of the inputs, 6.2b is implemented in terms of both the input and the output. Equation 6.2a is known as the *nonrecursive*, or FIR, implementation. Equation 6.2b is known as the *recursive*, or IIR, implementation.

■ Filter Coefficients

In Equation 6.2a, the multiplying constant for each term is 1.0825. In Equation 6.2b, the multiplying constants are 1 (for $y[k-1]$) and 1.0825 (for $x[k]$). These multiplying constants are known as the *coefficients* of the filter. For an IIR filter, the coefficients multiplying the inputs are known as the *forward coefficients*, and those multiplying the outputs are known as the *reverse coefficients*.

Equations of the form 6.1a, 6.1b, 6.2a, or 6.2b that describe the operation of the filter are known as *difference* equations.

■ Advantages and Disadvantages of FIR and IIR Filters

Comparing IIR and FIR filters, the advantage of digital IIR filters over FIR filters is that IIR filters usually require fewer coefficients to perform similar filtering operations. Thus, IIR filters execute much faster and do not require extra memory, because they execute in place.

The disadvantage of IIR filters is that the phase response is nonlinear. If the application does not require phase information, such as simple signal monitoring, IIR filters may be appropriate. You should use FIR filters for those applications requiring linear phase responses. The recursive nature of IIR filters makes them more difficult to design and implement.

Infinite Impulse Response Filters

Infinite Impulse Response filters are digital filters whose output is calculated by adding a weighted sum of past output values with a weighted sum of past and current input values. Denoting the input values by $x[.]$ and the output values by $y[.]$, the general difference equation characterizing IIR filters is

$$a_0 y[i] + a_1 y[i-1] + a_2 y[i-2] + \ldots + a_{N_y-1} y[i-(N_y-1)] =$$
$$b_0 x[i] + b_1 x[i-1] + b_2 x[i-2] + \ldots + b_{N_x-1} x[i-(N_x-1)]$$

$$a_0 y[i] = -a_1 y[i-1] - a_2 y[i-2] + \ldots - a_{N_y-1} y[i-(N_y-1)] +$$
$$b_0 x[i] + b_1 x[i-1] + b_2 x[i-2] + \ldots + b_{N_x-1} x[i-(N_x-1)]$$

$$y[i] = \frac{1}{a_0} \left(-\sum_{j=1}^{N_y-1} a_j y[i-j] + \sum_{k=0}^{N_x-1} b_k x[i-k] \right) \qquad (6.3)$$

where N_x is the number of *forward* coefficients (b_k) and N_y is the number of *reverse* coefficients (a_j). The output sample at the present sample index i is the sum of scaled present and past inputs ($x[i]$ and $x[i-k]$ when $k \neq 0$) and scaled past outputs ($y[i-j]$ when $j \neq 0$). Usually, N_x is equal to N_y and the *order* of the filter is equal to $N_x - 1$.

In all of the IIR filters implemented in LabVIEW/BridgeVIEW, the coefficient a_0 is 1.

■ Practical IIR Filters

A lower order reduces arithmetic operations and therefore reduces computation error. A problem with higher-order filtering is that you quickly run into precision errors with orders much greater than 20–30. This is the main reason for the "cascade" implementations over the "direct" form. (We will discuss these implementations in more detail in Chapter 11 on the Digital Filter Design Toolkit.) It is recommended that the orders of 1–20 are reasonable, with 30 being an upper limit. A higher order also means more filter coefficients and hence longer processing time.

The impulse response of the filter described by Equation 6.3 is of infinite length for nonzero coefficients. In practical filter applications, however, the impulse response of stable IIR filters decays to near zero in a finite number of samples.

In practice, the frequency response of filters differs from that of ideal filters. Depending on the shape of the frequency response, the IIR filters can be further classified into

- Butterworth filters
- Chebyshev filters
- Chebyshev II or inverse Chebyshev filters
- Elliptic or Cauer filters

The characteristics of each filter type are described below.

■ Butterworth Filters

A Butterworth filter has no ripple in either the passband or the stopband. Due to the lack of ripple, it is also known as the *maximally flat filter*. Its frequency response is characterized by a smooth response at all frequencies. Figure 6–7 shows the response of a lowpass Butterworth filter of different orders—the x-axis scaling is in terms of f/f_s (where f_s is the sampling frequency), whereas the y-axis is scaled so that the gain in the passband is unity.

The region where the output of the filter is equal to 1 (or very close to 1) is the passband of the filter. The region where the output is 0 (or very close to 0) is the stopband. The region in between the passband and the stopband where the output gradually changes from 1 to 0 is the transition region.

The advantage of Butterworth filters is a smooth, monotonically decreasing frequency response in the transition region. As seen from the figure, the higher the filter order, the steeper the transition region.

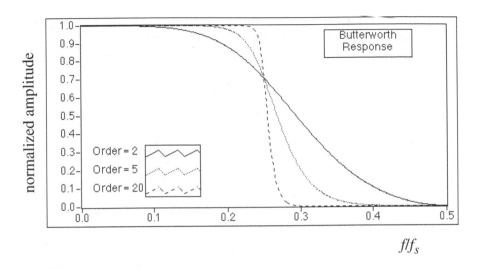

Figure 6–7
Frequency response of Butterworth filters of order 2, 5, and 20

■ Chebyshev Filters

The frequency response of Butterworth filters is not always a good approximation of the ideal filter response because of the slow rolloff between the passband (the portion of interest in the spectrum) and the stopband (the unwanted portion of the spectrum). On the other hand, Chebyshev filters have a smaller transition region than a Butterworth filter of the same order. However, this is achieved at the expense of ripple in the passband. Using LabVIEW or BridgeVIEW, you can specify the maximum amount of ripple (in dB) in the passband for a Chebyshev filter. The frequency response characteristics of Chebyshev filters have an equiripple (ripples all have the same magnitude) magnitude response in the passband, monotonically decreasing magnitude response in the stopband, and a sharper rolloff in the transition region as compared to Butterworth filters of the same order.

The graph in Figure 6–8 shows the response of a lowpass Chebyshev filter of different orders. In this case, the y-axis scaling is in decibels. Once again, note that the steepness of the transition region increases with increasing order. Also, the number of ripples in the passband increases with increasing order.

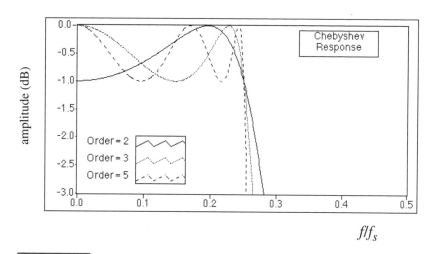

Figure 6–8

Frequency response of Chebyshev filters of order 2, 3, and 5

The advantage of Chebyshev filters over Butterworth filters is the sharper transition between the passband and the stopband with a lower-order filter. As mentioned before, this produces smaller absolute errors and higher execution speeds.

■ Chebyshev II or Inverse Chebyshev Filters

Chebyshev II, also known as inverse Chebyshev or Type II Chebyshev filters, are similar to Chebyshev filters, except that Chebyshev II filters have ripple in the stopband and are maximally flat in the passband. For Chebyshev II filters, you can specify the amount of attenuation (in dB) in the stopband. The frequency response characteristics of Chebyshev II filters are equiripple magnitude response in the stopband, monotonically decreasing magnitude response in the passband, and a rolloff sharper than Butterworth filters of the same order. The graph in Figure 6–9 plots the response of a lowpass Chebyshev II filter of different orders.

The advantage of Chebyshev II filters over Butterworth filters is that Chebyshev II filters give a sharper transition between the passband and the stopband with a lower-order filter. This difference corresponds to a smaller absolute error and higher execution speed. One advantage of Chebyshev II filters over regular

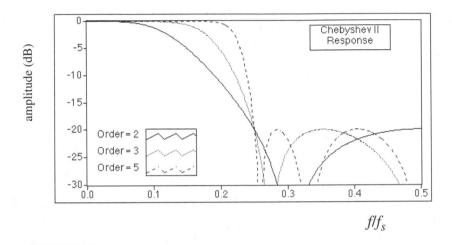

Figure 6–9
Frequency response of inverse Chebyshev filters of order 2, 3, and 5

Chebyshev filters is that Chebyshev II filters have the ripples in the stopband instead of the passband.

■ Elliptic Filters

You saw that Chebyshev filters have a sharper transition region than a Butterworth filter of the same order. This is because they allowed ripple in either the passband or the stopband. Elliptic filters distribute the ripple over both the passband as well as the stopband. Equiripples in the passband and the stopband characterize the magnitude response of elliptic filters. Therefore, compared with the same order Butterworth or Chebyshev filters, the elliptic filter provides the sharpest transition between the passband and the stopband. For this reason, elliptic filters are quite popular in applications where short transition bands are required and where ripple can be tolerated. The graph in Figure 6–10 plots the response of a lowpass elliptic filter of different orders. The x-axis scaling is in terms of f/f_s, whereas the y-axis is scaled so that the gain in the passband is unity.

Notice the sharp transition edge for even low-order elliptic filters. For elliptic filters, you can specify the amount of ripple (in dB) in the passband as well as the attenuation (in dB) in the stopband.

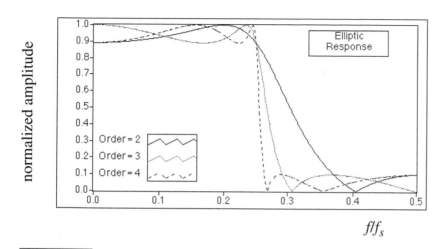

Figure 6–10
Frequency response of elliptic filters of order 2, 3, and 4

■ IIR Filter Comparison

A comparison of the lowpass frequency responses for the four different IIR filter designs, all having the same order (five), is shown in Figure 6–11. The elliptic filter has the narrowest transition region, whereas the Butterworth filter has the widest.

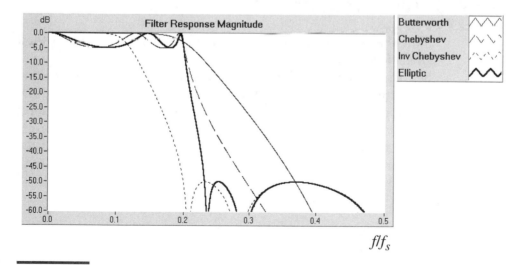

Figure 6–11
Comparison of frequency responses of IIR filters

Table 6–1   *Comparison of IIR Filters*

IIR Filter Design	Response Characteristics	Width of Transition Region for a Fixed Order	Order Required for Given Filter Specifications
Butterworth	No ripple	Widest	Highest
Chebyshev	Ripple in passband		
Inverse Chebyshev	Ripple in stopband		
Elliptic	Ripple in passband and stopband	Narrowest	Lowest

The filter designs of the various IIR filters are compared in Table 6–1.

The LabVIEW and BridgeVIEW digital filter VIs handle all the design issues, computations, memory management, and actual data filtering internally and are transparent to the user. You do not need to be an expert in digital filters or digital filter theory to process the data. All you need to do is to specify the control parameters such as the filter order, cutoff frequencies, amount of ripple, and stopband attenuation.

■ How Do I Decide Which Filter to Use?

Now that you have seen the different types of filters and their characteristics, the question arises as to which filter design is best suited for your application. In general, some of the factors affecting the choice of a suitable filter are whether you require linear phase, whether you can tolerate ripples, and whether a narrow transition band is required. The flowchart in Figure 6–12 is expected to serve as a guideline for selecting the correct filter. Keep in mind that in practice you may need to experiment with several different options before finally finding the best one.

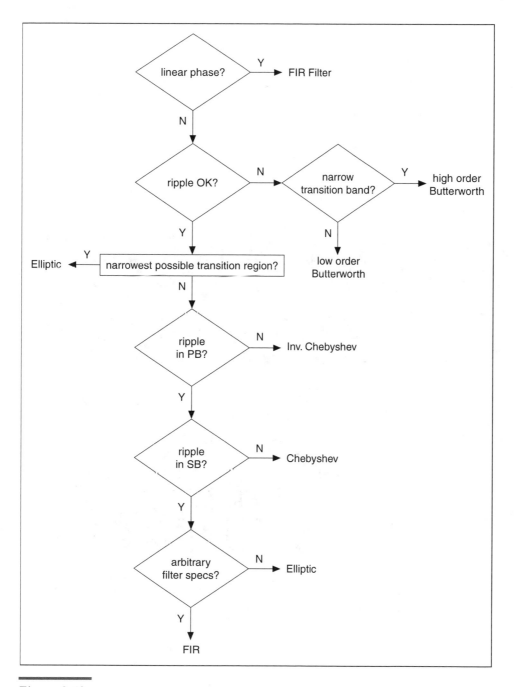

Figure 6–12
Choosing the right filter for your application

Activity 6-1

Objective: To filter data samples that consist of both high-frequency noise and a sinusoidal signal

In this activity, you combine a sine wave generated by the **Sine Pattern** VI with high-frequency noise. (The high-frequency noise is obtained by highpass filtering uniform white noise with a Butterworth filter.) The combined signal is then lowpass filtered by another Butterworth filter to extract the sine wave.

■ Front Panel

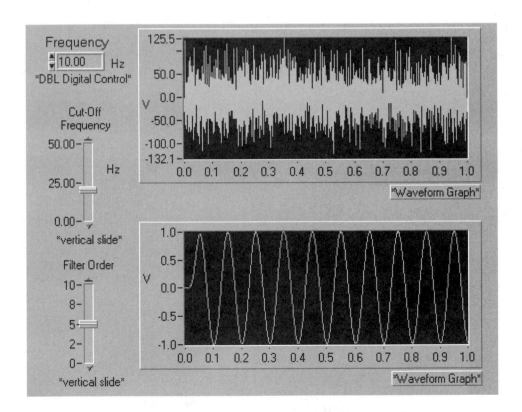

1. Open a new VI and build the front panel as shown above.

a) Select a **Digital Control** from the **Numeric** [...]
and label it **Frequency.**

b) Select **Vertical Slide** from the **Numeric** palette and
label it **Cut-Off Frequency.**

c) Select another **Vertical Slide** from the **Numeric**
palette and label it **Filter Order.** Change the Representation to I32.

d) Select a **Waveform Graph** from the **Graph** palette
for displaying the noisy signal, and another **Waveform Graph** for displaying the original signal.

2. Build the block diagram as shown below.

■ Block Diagram

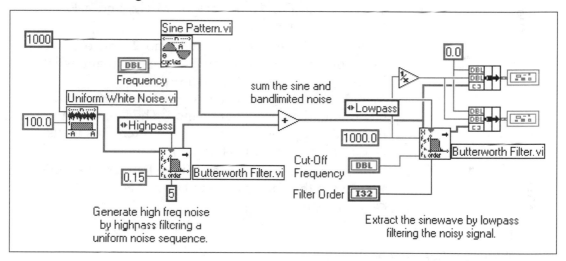

Sine Pattern VI (**Functions » Analysis » Signal Generation** subpalette) generates a sine wave of the desired frequency.

Uniform White Noise VI (**Functions » Analysis » Signal Generation** subpalette) generates uniform white noise that is added to the sinusoidal signal.

Butterworth Filter VI (**Functions » Analysis » Filters** subpalette) is used to highpass filter the noise.

Note that you are generating 10 cycles of the sine wave, and there are 1000 samples. Also, the sampling frequency to the **Butterworth Filter** VI on the right side is specified as 1000 Hz. Thus, effectively you are generating a 10-Hz signal.

You can easily create the Filter Type constants (Enumerated - Lowpass, Highpass, Bandpass and Bandstop) by clicking the right mouse button on the Filter Type terminal and selecting Create Constant.

3. Switch back to the front panel. Select a **Frequency** of 10 Hz, a **Cut-Off Frequency** of 25 Hz, and a **Filter Order** of 5. Run the VI.

4. Reduce the **Filter Order** to 4, 3, and 2, and observe the difference in the filtered signal. Explain what happens as you lower the filter order.

     In particular, observe the filtered waveform. At the beginning, there is a "flat" region. The length of this region depends on the order of the filter. This is discussed further in the next section.

5. When you finish, save the VI as **Extract the Sine Wave.vi** in the library `Activities.llb`.

6. Close the VI.

■ End of Activity 6-1

Activity 6-2

Objective: To compare the frequency response characteristics of various IIR filters

1. Open the **IIR Filter Design** VI from the library `Activities.llb`.

■ Front Panel

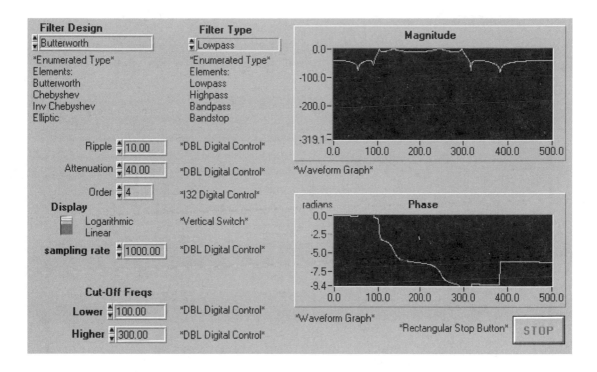

The front panel offers the choice of different types of filters.

The **Filter Design** control selects one of the four different designs of filters: Butterworth, Chebyshev, Inverse Chebyshev, or Elliptic.

The **Filter Type** control selects one of four different types of filters: *highpass*, *lowpass*, *bandpass*, or *band-stop*.

The **Display** control selects the display of the magnitude response (magnitude of the frequency response) to be either *linear* or *logarithmic*.

An easy way to create the desired control on the front panel is to pop up on the corresponding terminal of the subVI on the block diagram and select Create Control.

2. Switch to the block diagram. It appears as shown below.

■ Block Diagram

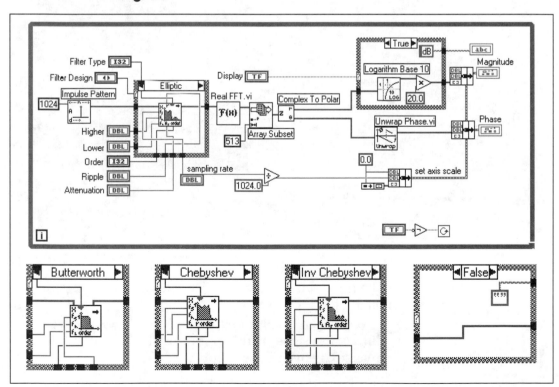

As seen in the block diagram, note that the frequency response is obtained by applying an impulse at the input of a filter and calculating the Fourier transform of the output.

The **Impulse Pattern** VI (**Functions » Analysis » Signal Generation** subpalette) generates an impulse that is given to the selected filter. The number of sample points is equal to 1024.

The **Real FFT** VI (**Functions » Analysis » Digital Signal Processing** subpalette) computes the Fourier transform of the output of the filter.

The **Array Subset** VI (**Functions » Array** subpalette) selects 513 FFT points out of 1024, so as to generate a one-sided spectrum.

The **Complex to Polar** VI (**Functions » Numeric » Complex** subpalette) converts the complex output of the **Real FFT VI** to its polar (magnitude and phase) representation. The magnitude can then be plotted in either the linear or dB scales.

Normally, the phase shown is limited to between $-\pi$ and $+\pi$. So, even if the phase lies outside the range $-\pi$ and $+\pi$, it is "wrapped" around to lie between these values. The **Unwrap Phase VI** (**Functions » Analysis » Digital Signal Processing** subpalette) is used to "unwrap" the phase to its true value, even if its absolute value exceeds π.

3. Switch to the front panel and select a **Filter Design** of Butterworth, **Filter Type** = Lowpass, **Ripple** = 10, **Attenuation** = 40, **Order** = 4, **Display** = Logarithmic, **sampling rate** = 1000, **Lower Cut-Off Frequency** = 100, and **Higher Cut-Off Frequency** = 300. Run the VI.

4. Increase the filter **Order** to 5, 10, 15, and 20, and note the difference in the magnitude and phase responses.

In particular, what changes do you notice in the transition region?

5. Keep the filter **Order** fixed at 5, and change the **Filter Design** to select different IIR filters. Note the changes in the magnitude and phase plots. For a given filter order, which of the four different filter designs has the smallest transition region?

6. When you finish, stop the VI by clicking on the **STOP** button in the lower-right corner.

7. Close the VI. Do not save any changes.

■ End of Activity 6-2

Activity 6-3

Objective: To use a digital filter to remove unwanted frequencies

In this activity, you will add two sine waves of different frequencies and then filter the resulting waveform using a Butterworth lowpass filter to obtain only one of the sine waves.

1. Open the **Low Pass Filter** VI from the library `Activities.llb`. This VI shows how to design a lowpass Butterworth filter to remove a 10-Hz signal from a 2-Hz signal.

■ Front Panel

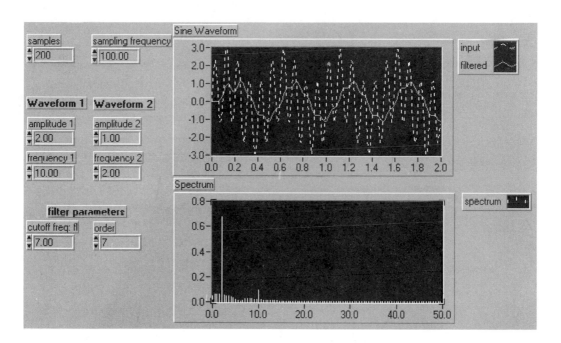

The number of samples to be generated and the sampling frequency are controlled by the **samples** and **sampling frequency** controls, respectively.

The amplitude and frequency of the two sine waves can be controlled by the **amplitude 1**, **ampli-**

tude 2, **frequency 1**, and **frequency 2** controls on the front panel.

The **cutoff freq:f1** control controls the cutoff frequency of the lowpass filter, whose order is adjusted by the **order** control.

2. Switch to the block diagram and examine it.

■ Block Diagram

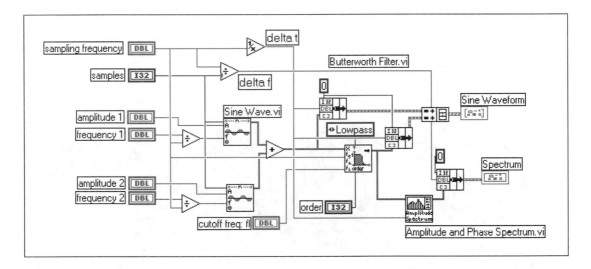

  Sine Wave VI generates the two sine waves.

  Amplitude and Phase Spectrum VI determines the amplitude and phase spectrum of the output of the filtered signal. In this activity, you are interested in only the amplitude spectrum.

The reciprocal of the sampling frequency gives the time interval, *dt*, between samples.

Remember that the frequency spacing, Δf, is obtained by dividing the sampling frequency by the number of samples.

The two sine waves being combined have frequencies of 10 Hz and 2 Hz. To separate them, you should set the cutoff frequency of the lowpass filter to somewhere between these two values.

3. Switch back to the front panel. Keeping the **cutoff freq: fl** control at 7 Hz, run the VI. Observe that in the **Spectrum** plot the amplitude of the 2-Hz signal is much larger than that of the 10-Hz signal.

4. Reduce the filter order to 5 and run the VI. Repeat with an order of 3. What do you notice about the spectrum amplitudes?

5. Increase the order to 12 and run the VI. Observe the spectrum amplitudes. Explain what happens.

6. When you finish, close the VI. Do not save any changes.

■ End of Activity 6-3

The Transient Response of IIR Filters

You have seen that the output of a general IIR filter is given by

$$y[i] = -a_1 y[i-1] - a_2 y[i-2] - ... - a_{N_y - 1} y[i-(N_y - 1)] + b_0 x[i] \qquad (6.4)$$
$$+ b_1 x[i-1] + b_2[i-2] + ... + b_{N_x - 1} x[i-(N_x - 1)]$$

where N_x is the number of forward coefficients, N_y is the number of reverse coefficients, and a_0 is assumed to be equal to 1. Consider a second-order filter where $N_x = N_y = 3$. The corresponding difference equation is:

$$y[i] = -a_1 y[i-1] - a_2 y[i-2] + b_0 x[i] + b_1 x[i-1] + b_2 x[i-2] \qquad (6.5)$$

To calculate the current output (at the *i*th instant) of the filter, you need to know the past two outputs (at the (*i*-1)st and the (*i*-2)nd time instants) as well as the current input (at the *i*th time instant) and past two inputs (at the (*i*-1) and (*i*-2) time instants).

Now suppose that you have just started the filtering process by taking the first sample of the input data. However, at this time instant you do not yet have the previous inputs ($x[i\text{-}1]$ and $x[i\text{-}2]$) or previous outputs ($y[i\text{-}1]$ and $y[i\text{-}2]$), so, by default, these values are assumed to be zero. When you get the second data sample, you already have the previous input ($x[i\text{-}1]$) and the previous output ($y[i\text{-}1]$) that you calculated from the first sample, but do not yet have $x[i\text{-}2]$ and $y[i\text{-}2]$. Again, by default, these are assumed to be zero. It is only after we start processing the third input data sample that all the terms on the right-hand side of Equation 6.5 now have the previously calculated values. Thus, there is a certain amount of delay before there are calculated values for all the terms on the right hand side (RHS) of the difference equation describing the filter. The output of the filter during this time interval is a transient and is known as the *transient response*. For lowpass and highpass filters implemented in the LabVIEW/BridgeVIEW Analysis Library, the duration of the transient response is equal to the order of the filter. For bandpass and bandstop filters, this duration is twice the order.

IIR filters in the analysis library contain the following properties.

- Samples with negative indices resulting from Equation 6.4 are assumed to be zero the first time you call the VI.

- Because the initial filter state is assumed to be zero, a transient occurs before the filter reaches a steady state.

- The filtered signal will be shifted, or delayed, by an amount proportional to the order of the filter.

So, each time one of the filter VIs is called, this transient appears at the output. This is seen in Figure 6–13. You can eliminate this transient response on successive calls by enabling the state memory of the VI. To enable state memory, set the **init/cont** con-

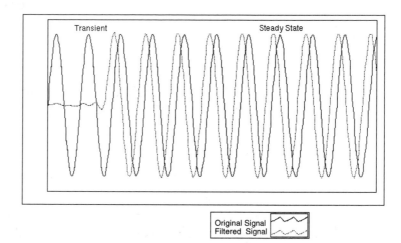

Figure 6–13
The transient response of a filter

trol of the VI to TRUE (continuous filtering). You will see how to do this in a later activity. Note that

- The number of elements in the filtered sequence equals the number of elements in the input sequence.
- The filter retains the internal filter state values when the filtering completes.

Activity 6-4

*Objective: To see
the difference in the
filter response with
and without
enabling state
memory*

1. Open the **Low Pass Filter** VI from the library `Activities.llb`.

■ Front Panel

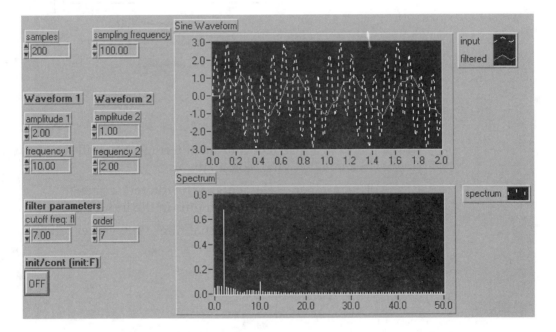

2. Run the VI with the values shown on the front panel above, but with **order** = 7. [Do not worry about the **init/cont (init: F)** control for now.]

   Observe the upper **Sine Waveform** graph, which shows two plots. The white dashed plot is the combined signal, whereas the green full-line plot is the filtered signal.

3. Change the filter order to 10 and run the VI. Observe the first few values of the plot corresponding to the filtered signal.

4. Change the filter order to 15, 20, and 25, and run the VI. Each time, observe the transient that occurs at the beginning of the plot corresponding to the filtered signal. This transient can be removed after the first call to the VI by enabling its state memory. This is done by setting the **init/cont** control of the VI to TRUE. Setting this control to TRUE is equivalent to continuous filtering, and except for the first call to the VI, each successive call will not have the transient.

5. Now connect a Boolean control, as shown in the diagram, to the **init/cont** input of the **Butterworth Filter VI.**

■ Block Diagram

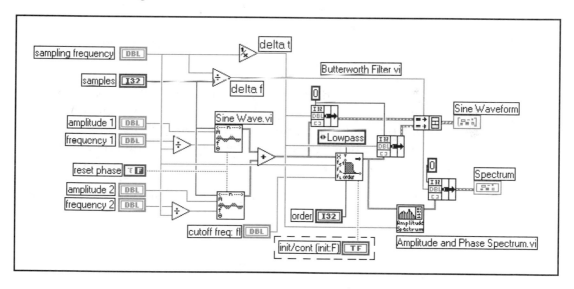

6. With the **init/cont(init:F)** control set to OFF, and **order** = 15, run the VI several times. Note the presence of the transient at the beginning of the filtered signal each time the VI runs.

7. Now set the **init/cont(init:F)** control to ON, and run the VI several times. Observe that the transient is present only the first time that the VI is run. On successive calls to the VI, the transient no longer exists.

8. When you finish, save the VI and close it.

■ End of Activity 6-4

Finite Impulse Response Filters

You have seen that the output of an IIR filter depends on the previous outputs as well as the current and previous inputs. Because of this dependency on the previous outputs, the IIR filter has infinite memory, resulting in an impulse response of infinite length.

On the other hand, the output of an FIR filter depends only on the current and past inputs. Because it does not depend on the past outputs, its impulse response decays to zero in a finite amount of time. The output of a general FIR filter is given by

$$y[i] = b_0 x[i] + b_1 x[i\text{-}1] + b_2 x[i\text{-}2] + \dots + b_{M\text{-}1} x[i - (M - 1)]$$

where M is the number of taps of the filter and $b_0, b_1, \dots, b_{M\text{-}1}$, are its coefficients. FIR filters have some important characteristics:

1. They can achieve linear phase response, and hence they can pass a signal without phase distortion.

2. They are always stable. During filter design or development, you do not need to worry about stability concerns.

3. FIR filters are simpler and easier to implement.

The graphs in Figure 6–14 plot a typical magnitude and phase response of FIR filters versus normalized frequency. The discontinuities in the phase response in the stopband arise from the discontinuities introduced when you compute the magnitude response using the absolute value. Notice that the discontinuities in phase are on the order of π, which correspond to a change in sign in the stopband response. The phase, however, is clearly linear.

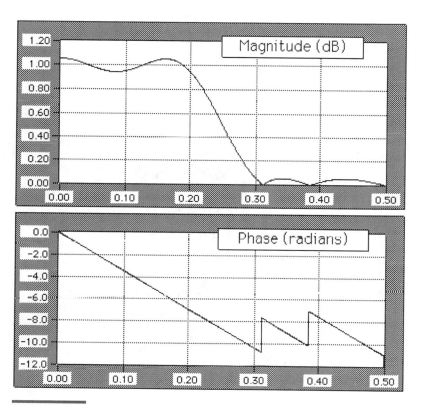

Figure 6–14
Magnitude and phase response of an FIR filter

Activity 6-5

Objective: To see the frequency response characteristics of FIR filters

In this activity, you will see the magnitude and phase response characteristics of FIR filters. You will also see the effect of using different windows on the filter response characteristics.

1. Open the **FIR Windowed Filter Design** VI from the library `Activities.llb`.

■ Front Panel

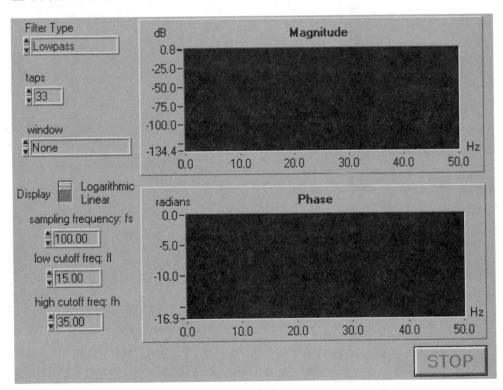

The **Filter Type** control specifies the type of FIR filter—lowpass, highpass, bandpass, or bandstop.

The **taps** control specifies the number of filter coefficients. Note that taps must be greater than 0. The higher the number of taps, the steeper the transition between the passband and the stopband.

The **window** control selects among nine different types of windows to be applied in truncating the infinite impulse response of an ideal filter. The result is a Finite Impulse Response (FIR) filter.

The **Display** control selects the display units on the **Magnitude** plot to be linear or logarithmic.

The **sampling frequency: fs**, **low cutoff frequency: fl**, and **high cutoff frequency: fh** controls specify the desired response characteristics of the filter. The **high cutoff frequency: fh** is required only when **Filter Type** is chosen as bandpass or bandstop.

The **Magnitude** and **Phase** plots show you the magnitude and phase response of the filter. Note that because the filter under consideration is an FIR filter, it is expected to have a linear phase response.

2. Switch to the block diagram and examine it.

■ Block Diagram

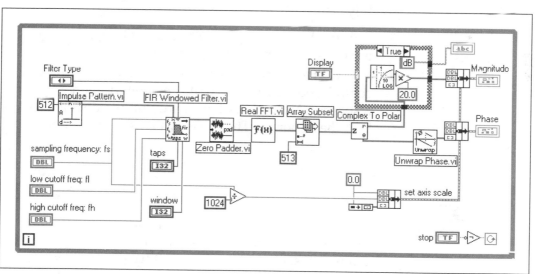

Impulse Pattern VI (**Analysis » Signal Generation** subpalette) generates an impulse to be applied to the input of the FIR filter. The response of the filter to the impulse will give us its impulse response. Finding the FFT of the impulse response gives the magnitude and phase response.

FIR Windowed Filter VI (**Analysis » Filters** subpalette) is an FIR filter that filters the input data sequence (in this case an impulse) using the set of windowed FIR filter coefficients specified by the **sampling frequency: fs, low cutoff frequency: fl, high cutoff frequency: fh**, and **taps** controls.

Zero Padder VI (**Analysis » Digital Signal Processing** subpalette) resizes the input sequence to the next higher power of two by adding zeros to the end of the sequence. By converting the total number of samples to a power of two, the calculation of the Fourier transform of the impulse response of the FIR filter can be done faster by the **Real FFT** VI.

The default values have been chosen so that you will see the response characteristics of a lowpass FIR filter with cutoff frequency of 15 Hz.

3. Run the VI with the default values. Observe the linear phase response in the passband.

4. Select different types of windows using the **window** control and observe both the magnitude and phase responses. Notice how the choice of a window affects both the responses. The phase in all cases will be linear.

The taps control affects the width of the transition region.

5. Observe the Magnitude response plot with the **Filter Type** control set to lowpass filter, the **low cutoff frequency: fl** set to 15 Hz, **sampling frequency: fs** set to 100 Hz, **window** set to Kaiser-Bessel, and **taps** set to 33.

6. Change **taps** to 55 and observe how the transition region becomes narrower.

7. Change **taps** to 10 and observe the increase in the width of the transition region.

8. You can experiment with different values of the other controls. When you are done, stop the VI by clicking on the **STOP** button in the lower-right corner.

Taps should be odd for highpass and bandstop filters because when taps is even, the magnitude response approaches zero as the frequency approaches half the sampling rate.

9. Close the VI without saving any changes.

■ End of Activity 6-5

Digital Filter Design Toolkit

While it is quite easy and straightforward to design a digital filter in LabVIEW or BridgeVIEW using the analysis VIs, it does require a basic understanding of digital signal theory. For the best tool to learn more about digital filter design, National Instruments offers an add-on toolkit called the Digital Filter Design (DFD) Toolkit. This toolkit was developed using G and has an excellent interactive graphical user interface. By adjusting the response in the frequency domain by using cursors, many different types of digital filters can easily be designed. The designed filter coefficients can then be integrated into Lab-VIEW, BridgeVIEW, LabWindows/CVI, and other programming environments. You will learn more about the DFD Toolkit in Chapter 11.

Wrap It Up!

- You have seen from the frequency response characteristics that practical filters differ from ideal filters.

- For practical filters, the gain in the passband may not always be equal to 1, the attenuation in the stopband may not always be zero, and there exists a transition region of finite width.

- The width of the transition region depends on the filter order, and the width decreases with increasing order.

- The output of FIR filters depends only on the current and past input values, whereas the output of IIR filters depends on the current and past input values as well as the past output values.

- IIR filters can be classified according to the presence of ripples in the passband and/or the stopband.

- Because of the dependence of an IIR filter's output on past outputs, a transient appears at the output of an IIR filter each time one of the filter VIs discussed in this chapter is called. This transient can be eliminated after the first call to the VI by setting its **init/cont** control to a TRUE value.

■ Review Questions

1. Name three practical examples where filtering is used.

2. If the stopband attenuation of a filter is –80 dB, what is the output of the filter to an input signal of 5 V_{peak} if the frequency of the signal lies in the stopband of the filter?

3. Which filter would you use for the following applications?

a) When you want the narrowest possible transition region with the smallest order.

b) When you cannot tolerate any ripples in either the passband or the stopband.

c) When linear phase is important.

4. Why does a transient initially appear at the output of a filter? How can it be eliminated on successive calls to a VI?

■ Additional Activities

1. Filtering is very common in communication systems where the received signal is corrupted with noise picked up from the channel (cable, atmosphere, etc) through which it has traveled. The **Pulse Demo VI (LabVIEW >> Examples >> Analysis >> Plsexmpl.llb)** simulates such a transmitter and receiver system. Open this VI and set the amount of noise to 0.2 by adjusting the **Additive Noise** control. Run the VI for several values of the **Filter Order** from 5 to 20. As you increase the filter order, observe the reduction in the higher-frequency noise at the output of the filter.

2. A cable running very close to a power source is picking up interference at the power line frequency (60 Hz). You are asked to build a VI that will remove this interference but will not affect the desired frequencies of interest which lie between 0 – 40 Hz and 80 – 99 Hz. The level of the interference at the output is to be less and 1/100th (40 dB) the level at the input. It is desired that

   i. there be no ripple in either the passband or the stopband of the filter.

   ii. the filter order be as small as possible.

      a) Build the VI and test it with a sine wave whose frequency can be controlled on the front panel.

b) What is the minimum order of the filter that meets the desired specifications?

3. Open and run the **Signal Generation and Processing** VI from **the Examples >> Apps >> Demos.llb** library in your LabVIEW folder. This VI generates two signals to each of which noise is added. Each of the noisy signals is then filtered and windowed. The plots on the front panel show the original noisy signal, the filtered and windowed signal, and the power spectrum of the filtered and windowed signal. You can specify the type of signal generated (sine, square, triangle, or none), the type of window (Hanning, Hamming, or none), and the type of filter (Butterworth, Chebyshev, or none). Experiment with different signals, of different frequencies, and see the results of the various types of windowing and filtering on the power spectrum.

This activity can only be done with the LabVIEW Full Development System.

4. In this activity, you will learn the use of Digital Signal Processing in the area of telecommunications. The process of modulation allows many individual signals to be transmitted over a single communication channel. This can be done by translating each signal to a different position in the frequency spectrum. Amplitude modulation is one form of modulation where the original signal is multiplied by a carrier signal to form a modulated signal. This modulated signal can then be transmitted over a channel that has limited bandwidth. While doing so, the information signal is not matched to the channel but has a bandwidth less than that of the channel. At the receiver end, the modulated signal can be demodulated to recover the original signal. We will build a very simple modulation-demodulation system, as

shown below, using some of the Analysis library VIs that you have learned about so far.

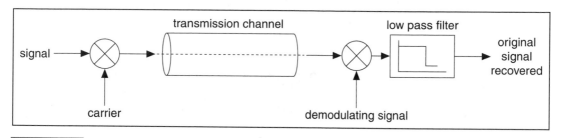

A modulation-demodulation system in a communication channel

Generate the information signal: Use the **Sine Wave** VI (**Analysis >> Signal Generation** subpalette) to generate a cosine wave with any frequency f_m.

A cosine wave is a sine wave with phase = 90 degrees.

Generate the carrier signal: Use the same VI to generate another cosine wave with a different frequency f_c, where $f_c > f_m$.

Generate the modulated signal: Multiply the information signal and the carrier signal to obtain the modulated signal. If you are interested, observe the spectral representation of the modulated signal by plotting the amplitude spectrum output of the **Amplitude and Phase Spectrum** VI (**Analysis >> Measurement** subpalette) on a waveform graph.

Process the received signal: Multiply the modulated signal by a demodulating signal. The demodulated signal is similar to the carrier signal (same frequency) with a different initial phase.

Filter the signal: Pass the processed signal through a low-pass filter to recover the original signal. Use the **Inverse Chebyshev** Filter VI (**Analysis >> Filters** subpalette).

Save the VI as **Modulation.vi** in the library `Activities.llb` and close it.

This is a challenging activity. You will have to play with different parameters such as frequencies f_m and f_c, initial phase of the demodulating signal, and parameters for the low pass filter to recover the original signal.

LabVIEW Exercise to Model and Analyze ECG/EKG Data

Richard Bayford
Middlesex University
London, UK

The Challenge To remove noise contamination from ECG data by the use of digital filter techniques.

The Solution The solution is implemented using LabVIEW and a standard data acquisition card in two stages: First, a simulation of the ECG data was created to test the performance of the filter system (this is the design phase) using simple pole-zero placement DSP techniques. Second, an acquisition card and isolation amplifier to obtain real ECG data recorded from a human subject is used to validate the filter system.

Introduction

In the area of biomedical signal processing a common problem is the requirement for signals which are free of contamination due to a number of sources. These signals are commonly used, for example, for foetal monitoring to assess the effects of the measurement system on the electrical activity of the foetal heart (Westgate 1994).

An ECG is a small electrical signal, which is produced due to the activity of the heart. Its source can be considered as a dipole located in the partially conducting medium of the thorax. This dipole induces a body-surface potential, which can be measured and used for diagnostic purposes. The signal ECG is characterised by five peacks and valleys labelled with successive letters of the alphabet P, Q, R, S, and T (Greene 1987). The ECG is said to consist of the P wave, QRS complex and T wave (Greene 1987). The reciprocal of the heart period, as shown in Figure 6–15, is the time interval between the R-to-R peaks (in milliseconds), multiplied by 60 000 gives the instantaneous heart rate.

The example shown in Figure 6–15 is termed the normal Sinus Rhythm. Impulses originate in the SA node regularly at a rate of 60-100 per minute in adults and at faster rates in older children

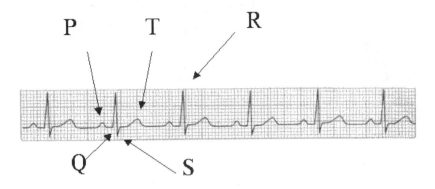

Figure 6–15
The electrocardiogram

(90-110), small children (100-120) and infants (120-160). The P waves are upright in L2 and negative in AVR and of uniform size and contour from beat to beat. The PR interval is 0.12-0.20 sec and constant when A-V conduction is normal; PR is prolonged and/or variable when A-V block is present. Each P is followed by a QRS with the resulting P:QRS ratio 1:1. The QRS may be less than 0.11 sec or QRS may be wide and bizarre when bundle branch block is present. The RR intervals may be slightly irregular, especially in the young and elderly.

The basic heart rate can be calculated from:

$$Heart\ rate\ (beats\ min^{-1}) = \frac{1}{heart\ period(ms)} \times 60000$$

There are many factors that must be considered in the design of a system that is capable of measuring these signals without introducing contamination into the signal. Patients who are having their ECGs taken on either a clinical electrocardiograph, or continuously on a cardiac monitor, are often connected to the pieces of electric apparatus. Each electrical device has its own ground connection either through the power line or, in some cases, through a heavy ground wire attached to some point in the room. A ground loop can exist when two or more electrical monitoring devices are connected to the patient. Another problem caused by the ground currents is related to the fact that, because the ground lead of the electrocardiograph usually runs alongside the signal leads, magnetic fields caused by the current in the grounding circuit can induce small voltages in the signal lead wires. This can produce interference on the recorded data.

A major source of interference is the electric-power system. Besides providing power to the ECG system itself, power lines are connected to other pieces of equipment in a typical hospital or physician's office. Such interferences can appear on the recorded data as a result of two mechanisms, each operating singly or, in some cases, both together. The first is electric-field coupling between the power lines and the electrocardiograph and/ or patient and is the result of the electric field surrounding the main power lines. The other source of interference from power lines is magnetic induction. Current in the power lines establishes a magnetic filed in the vicinity of the line. If basic precautions are taken a great deal of this type of contamination can be minimised (Webster 1995).

To reduce the above contamination it is possible to employ simple signal processing techniques. The example detailed in this text shows how LabVIEW can be used to achieve this. We begin by modelling the ECG signal, then applying a digital filter to remove a selected component of the signal. The system is also capable of filtering real ECG data. This is demonstrated at the end of the example. The basic aims and objectives of this example are detailed below:

1. To create a VI that will simulate a heartbeat signal that is contaminated by a signal at the mains frequency (50Hz)
2. To develop an IIR filter to reduce the contamination
3. To test the VI's using real ECG data obtained using a data acquisition card with a medical safe isolation amplifier.
4. To generate a LabVIEW program to simulate an ECG signal plus 50Hz contamination.

5. To use a standard LabVIEW function to create an IIR filter that will notch out the 50HZ contamination and create an appropriate front panel display to show the raw and filtered ECG data.
6. To devise a means of counting the beats per minute.

The VI corresponding to this example can be found on the CD-ROM included with this book.

Procedure

This example is designed in two parts, the first assumes that the ECG signal is modelled and the second will use real ECG data obtained from a patient. The first requirement is to simulate the heat beat. To achieve this a simple method can be used which combines the sum of the sinusoidal signals that represent the basic components of an ECG signal. Careful selection of the frequencies can be made to accurately simulate the contamination due to a ground loop. These are 60, 40 and 20Hz respectively. (Note the assumption at this system is design for UK use where the mains frequency is 50Hz). The contamination can vary from this centre frequency of 50Hz. To construct a simulation of this a white noise source was chosen. This noise source was fed into a 5th order Butterworth bandpass filter with a bandwidth of 49 to 51Hz.

The amplitude of the white noise source is set to 150 and the gain of the **Sine Pattern** generators are unity. The sampling rate was set at 600 samples/second. The details of the VI are illustrated in Figure 6–16 and the output waveform it generates is shown in Figure 6–17.

To achieve the objective of removing the contamination from the simulated ECG signal an Infinite Impulse Response (IIR) filter was chosen. The chosen design procedure is the pole-zero

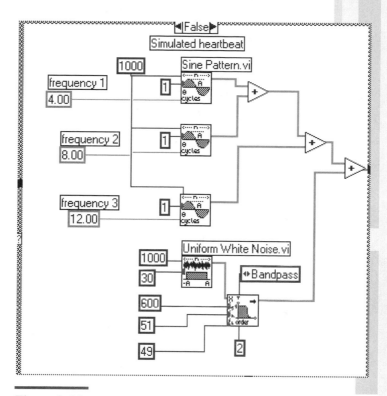

Figure 6–16
Simulation of heartbeat corrupted by mains frequency

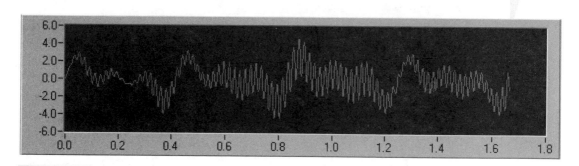

Figure 6–17
Simulated ECG display with 50Hz mains contamination

placement method (Emmanuel and Jervis 1993) with the following specification.

Notch frequency   50Hz
3dB width of Notch   5Hz
Sampling frequency   600Hz

The component in a signal may be rejected by placing a pair of zeros at points on the unit circle of the z-plane corresponding to 50Hz (Figure 6–18).

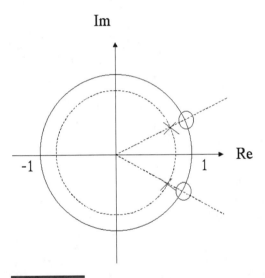

Figure 6–18
Pole-zero diagram

Hence pole placement:

$$2\pi \times \frac{50}{600} = 0.523 \, rad/s$$

The position of the poles in relation to the zeros determines the sharpness and amplitude response on either side of the notch.

Thus the poles are placed on the same radius as the zeros at a radius r<1 where

$$r \approx 1 - \frac{BW}{f_s}\pi$$

Where BW is the required notch width. Hence

$$r = 1 - \frac{10}{600}\pi = 0.9476$$

Thus the poles are positioned at the same radius as the zeros. The transfer function is given by:

$$H(z) = \frac{(z - e^{-j30})(z - e^{j30})}{(z - 0.9476e^{-j30})(z - 0.9476e^{j30})}$$

$$H(z^{-1}) = \frac{1 - 1.732z^{-1} + z^{-2}}{1 - 1.6413z^{-1} + 0.898z^2}$$

$$H(z) = \frac{z^2 - 1.732z + 1}{z^2 - 1.6413z + 0.898}$$

Therefore the difference equation is:

$$y(n) = x(n) - 1.732x(n-1) + x(n-2) + 1.6413y(n-1) - 0.898y(n-2)$$

Hence the coefficients of the notch filter are:

$$a_0 = 1 \quad b_0 = 1$$
$$a_1 = -1.732 \quad b_1 = -1.641$$
$$a_2 = 1 \quad b_2 = 0.898$$

The implementation of the filter is achieved using a standard digital filter VI and entering the above coefficient values. This is illustrated in Figure 6–19.

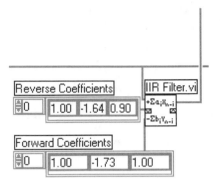

Figure 6–19
Basic VI program for filtering the 50Hz contamination

The next objective of the design example is to count the number of beats per minute. To achieve this there must be no d.c. off-set on the signal and the effect of artefact owing to muscle noise must be removed. The aim here is to count the number of peaks in the signal. Obviously, if the number of peaks within a given period were to be counted for the calculation of beats/min then it would be desired that the beat signal should remain at an average constant level and not drift. To remove the d.c. drift requires removal of the low frequency component of the signal. This can be achieved by placing a highpass filter is series with the output of the notch-filter. The beat rate can be as low as 30 beats per minute, which requires that the lowest frequency component of the ECG signal be at least 0.5Hz. This can occur when a patient is monitored in a coma. As a compromise, the cut-off frequency for the highpass filter is set to 0.25Hz.

A second source of contamination which can affect the estimation of the beat rate is muscle noise, this can occur at approximate 40Hz or more. This can simply be eliminated by lowpass filtering of the ECG signal. These extra filters can be reduced to a bandpass filter with cut-off frequencies at:

Lower cut-off = 0.25Hz
Upper cut-off = 40.00Hz

By placing the bandpass filter in series with the IIR sub-filter, and connecting the output of the bandpass to the waveform display, we eliminate any d.c. off-set and muscle artefact in the beat signal. The full implementation is given in Figure 6–20.

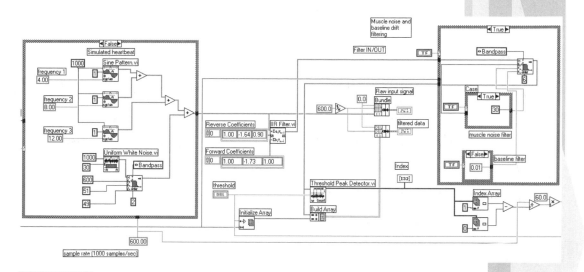

Figure 6–20
Full system, showing all the parts of the ECG diagram

As a further exercise the system can be used to test real ECG data. A data acquisition card can be placed in a PC with an isolation amplifier. Note this isolation amplifier must be medically safe! For optimum placement of the electrodes see Webster (1995). The **AI Acquire Waveform** VI enables the retrieval of data from the DAQ card. This VI can be found in the **Data Acquistion** menu in the **Functions** palette. The sampling rate and the number of samples should be set at 600 units. Figure 6–21 shows a diagram to illustrate the connection of the data acquisition system to the patient.

Two conductive pads are placed on the subject from whom the real ECG signal is to be obtained. A medically safe optical isolated differential amplifier is connected to the two conductive pads. This signal is then fed into an antialising filter and the signal is monitored using an oscilloscope and fed into a data acquisition card. Care must be taken to avoid bypassing the isolation when using the oscilloscope. Real ECG data is shown in the following figures.

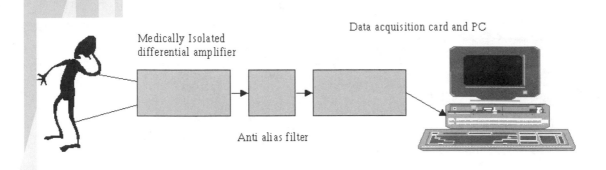

Patient

Figure 6–21
Connection of data acquisition to patient

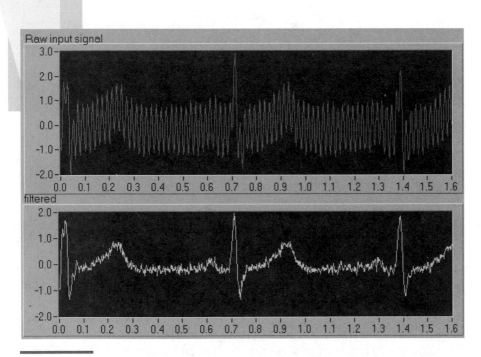

Figure 6–22
Real ECG displayed before and after mains filtering

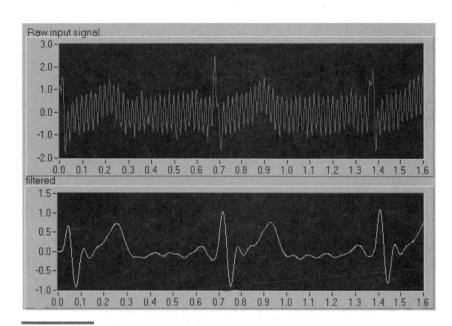

Figure 6–23
Real ECG data with mains and muscle filtering

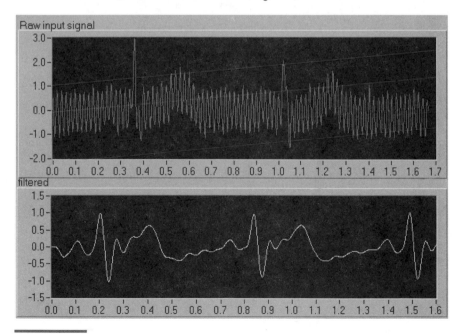

Figure 6–24
Real ECG data with all filters operating

References

Westgate, J., and Greene, K. (1994). The use of the fetal electro-cardiogram in labour, *Br. J. Obstet Gynaecol*.

Greene, K. R. (1987). The ECG waveform. In *Balliere's Clinical Obstetrics and Gynaecology* (M. Whittle, ed.), Vol 1 pp. 131–155.

Ifeachor, I. C. and Jervis B. W. (1993). *Digital Signal Processing: A Practical Approach*, Addison-Wesley.

■ **Contact Information**

Richard Bayford
Middlesex University
London, UK
E-mail: r.bayford@ucl.ac.uk
E-mail: r.bayford@mdx.ac.uk

Overview

In this chapter, you will learn about the Analysis VIs that are used to fit curves to data points.

GOALS

- Understand the concept of curve fitting and its applications.
- Learn how to use the **General Least Squares Linear Fit** VI.
- Learn how to use the **Nonlinear Levenberg-Marquardt Fit** VI.
- Be able to fit a curve to Gaussian (Normal) data points.

KEY TERMS

- linear
- exponential
- polynomial
- fit coefficients

- mean squared error
- least square fit
- applications
- regression

Curve Fitting

7

About Curve Fitting

In the digital domain, a data set can be represented by two input sequences: *Y Values* and *X Values*. A sample or point in the data set can be written as

$$(x[i], y[i])$$

where $x[i]$ is the ith element of the sequence *X Values*, and $y[i]$ is the ith element of the sequence *Y Values*. Each $y[i]$ is related to the corresponding $x[i]$. You are interested in finding the relationship between $y[i]$ and $x[i]$ in the digital domain and expressing it in the form of an equation in the analog domain.

Curve fitting acts as a bridge between the digital and analog worlds. Using curve fitting, digital data can be represented by a continuous model having a certain set of parameters. The basic idea is to extract a set of curve parameters or coefficients from the data set to obtain a functional description of the data set. This functional description consists of the set of parameters a_0, a_1, ..., a_k that best matches the experimental model from which the data samples $x[i]$ and $y[i]$ were obtained. Once you obtain the functional model, you can use it to estimate missing data points, interpolate data, or extrapolate data.

Note that $y[i]$ is a function of both the parameters a_k, as well as the data $x[i]$. The following discussion refers to the terms "linear" and "nonlinear," which apply to the relationship between y and a, and not y and x.

The Analysis Library offers both linear and nonlinear curve fitting algorithms. The different types of curve fitting algorithms are outlined below.

- *Linear Fit*—fits experimental data to a straight line of the form $y = mx + c$

$$y[i] = a_0 + a_1 x[i]$$

- *Exponential Fit*—fits data to an exponential curve of the form $y = a\exp(bx)$

$$y[i] = a_0\exp(a_1 x[i])$$

- *General Polynomial Fit*—fits data to a polynomial function of the form $y = a + bx + cx^2 + \ldots$

$$y[i] = a_0 + a_1 x[i] + a_2 x[i]^2 \ldots$$

but with selectable algorithms for better precision and accuracy.

- *General Linear Fit*—fits data to

$$y[i] = a_0 + a_1 f_1(x[i]) + a_2 f_2(x[i]) + \ldots$$

where $y[i]$ is a linear combination of the parameters $a_0, a_1, a_2 \ldots$ For example, $y = a_0 + a_1\sin(x)$ is a linear fit because y has a linear relationship with parameters a_0 and a_1. Polynomial fits are always linear fits for the same reason, but special algorithms can be designed for the polynomial fit to speed up the fitting processing and improve accuracy. The general linear fit also features selectable algorithms for better precision and accuracy.

- *Nonlinear Levenberg-Marquardt Fit*—fits data to

$$y[i] = f(x[i], a_0, a_1, a_2 \ldots)$$

where $a_0, a_1, a_2\ldots$ are the parameters. This method is the most general method and does not require y to have a linear relationship with $a_0, a_1, a_2....$ It can be used to fit linear or nonlinear curves but is almost always used to fit a nonlinear curve, because the general linear fit method is better suited to linear curve fitting. The Levenberg-Marquardt method does not always guarantee a correct result, so it is absolutely necessary to verify the results.

■ Mean Squared Error

The algorithm used to fit a curve to a particular data set is known as the *least squares* method. Let the observed data set be denoted by $y(x)$, and let $f(x,a)$ be the functional description of the data set where a is the set of curve coefficients that best describes the curve. The error $e(a)$ between the observed values, and its functional description, is defined as

$$e(a) = [f(x,a) - y(x)]$$

For example, let **a** be the vector $\mathbf{a} = [a_0, a_1]$. The functional description of a line is

$$f(x,a) = a_0 + a_1 x$$

The least squares algorithm estimates the values for **a** from the values of $y[i]$ and $x[i]$. After you have the values for **a**, you can obtain an estimate of the observed data set for any value of x using the functional description $f(x,a)$. The curve fitting VIs automatically set up and solve the necessary equations and return the set of coefficients that best describes your data set. You can thus concentrate on the functional description of your data and not worry about the methods used for solving for **a**.

For each of the observed data points $x[i]$, the differences between the polynomial value $f(x[i],a)$ and the original data $y(x[i])$ are called the *residuals* and are given by

$$e_i(a) = f_i(x[i],a) - y_i(x[i])$$

The Mean Squared Error (MSE) is a relative measure of these residuals between the expected curve values calculated by the functional description $f(x[i], a)$, and the actual observed values of $y[i]$, and is given by

$$MSE = \frac{1}{N} \sum_{i=0}^{N-1} (f_i - y_i)^2 = \frac{1}{N} \sum_{i=0}^{N-1} (e_i(a))^2$$

where **f** is the sequence representing the fitted values, **y** is the sequence representing the observed values, and N is the number of sample points observed. The smaller the MSE, the better is the fit between the functional description and the observed $y(x)$.

In general, for each predefined type of curve fit, there are two types of VIs, unless otherwise specified. One type returns only the coefficients. The other type returns the coefficients, the corresponding expected or fitted curve, and the MSE.

■ Applications of Curve Fitting

The practical applications of curve fitting are numerous. Some of them are listed below.

- Removal of measurement noise.
- Filling in missing data points (for example, if one or more measurements were missed or improperly recorded).
- Interpolation (estimation of data between data points) (for example, if the time between measurements is not small enough).
- Extrapolation (estimation of data beyond data points) (for example, if you are looking for data values before or after the measurements were taken).
- Differentiation of digital data. (For example, if you need to find the derivative of the data points. The discrete data can be modeled by a polynomial, and the resulting polynomial equation can be differentiated.)
- Integration of digital data (for example, to find the area under a curve when you have only the discrete points of the curve).
- To obtain the trajectory of an object based on discrete measurements of its velocity (first derivative) or acceleration (second derivative).

Activity 7-1

*Objective: To
perform a linear
curve fit on
experimental data*

■ Front Panel

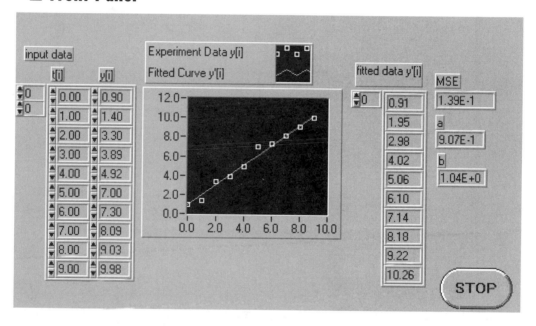

1. Open the **Linear Curve Fit** VI from the library `Activities.llb`. This example assumes that you have collected 10 pairs of experimental data **t** and **y** and have reason to believe there may be a linear relationship between each pair.

2. On the front panel, the **input data** control on the left shows the values of the data points $t[i]$ and $y[i]$. After calculating the equation for the linear fit, the calculated values of $y[i]$ for the measured values of $t[i]$ are shown on the **fitted data y[i]** indicator on the right. The VI also gives the values of the parameters a_0 and

a_1 (**a** and **b** indicators on the front panel) and the resulting *MSE*.

3. Switch to the block diagram.

■ Block Diagram

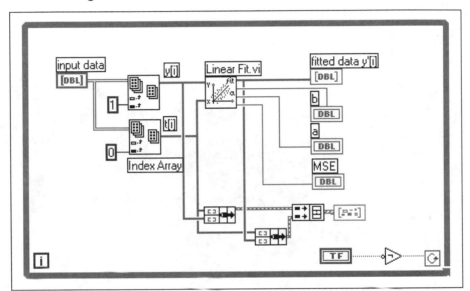

4. Examine the block diagram.
    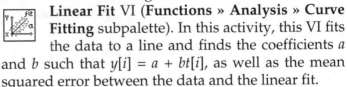 **Linear Fit** VI (**Functions » Analysis » Curve Fitting** subpalette). In this activity, this VI fits the data to a line and finds the coefficients *a* and *b* such that $y[i] = a + bt[i]$, as well as the mean squared error between the data and the linear fit.

5. The input data is in a two-dimensional array, which is a common format when the data sequence is collected from DAQ hardware. You use the **Index Array** VI to obtain two one-dimensional arrays, $y[i]$ and $t[i]$.

6. The **MSE** is the mean squared error. A smaller error indicates a better fit.

7. Switch back to the front panel and run the VI. The graph should display the experiment data, as well as the linear fitted curve.

8. Change the values of $y[i]$ in the **input data** control. Observe the corresponding plots and the effect on the MSE and slope when there are points that don't fit well.

9. When you finish, stop the VI by clicking on the **STOP** button.

10. Close the VI. Do not save any changes.

■ End of Activity 7-1

Activity 7-2

Objective: To perform a polynomial curve fit on experimental data

■ Front Panel

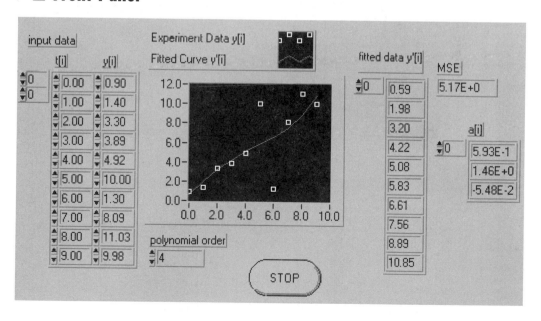

1. Open the **Polynomial Fit** VI from the library `Activities.llb`. This example assumes that the experimental input data sequence has a polynomial relationship where

   $$y[i] = a_0 + a_1t[i] + a_2t[i]^2 \ldots$$

2. When the polynomial order is 1, there are two coefficients (a_0 and a_1) and the result is a linear fit as in Activity 7-1. However, when the order is 2, it is a second-order polynomial fit with three coefficients. The polynomial coefficients are stored in an array indicator **a[i]**. You can use the **polynomial order** control to choose the order of the polynomial.

3. Switch to the block diagram.

■ Block Diagram

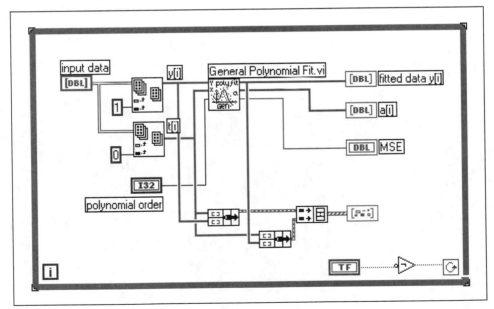

4. Examine the block diagram.

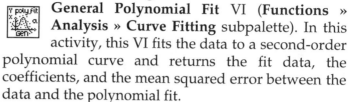 **General Polynomial Fit** VI (**Functions »
Analysis » Curve Fitting** subpalette). In this
activity, this VI fits the data to a second-order
polynomial curve and returns the fit data, the
coefficients, and the mean squared error between the
data and the polynomial fit.

5. You use the polynomial fit to obtain the fitting coefficients a_0, a_1, a_2, and so on. In general, you want to use
the lowest order possible to fit the polynomial.

6. Switch to the front panel and run the VI. The experiment data as well as the fitted data should appear on
the graph.

7. Change the values of $y[i]$ in the **input data** control
and the order of the polynomial in the **polynomial
order** control and observe the changes in the plots
and the MSE.

8. When you finish, stop the VI by clicking on the
STOP button.

9. Close the VI. Do not save any changes.

■ End of Activity 7-2

Activity 7-3

Objective: To use and compare the Linear, Exponential and Polynomial Curve Fit VIs to obtain the set of least square coefficients that best represent a set of data points

1. Open the **Regressions Demo** VI from the library `Activities.llb`. The front panel and block diagram are already built for you.

■ Front Panel

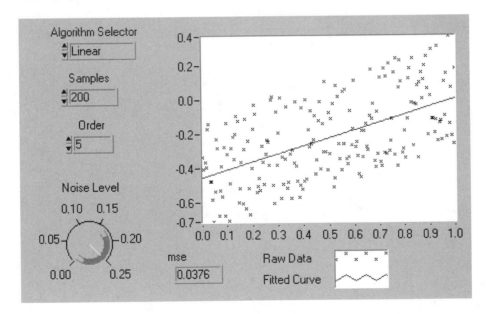

This VI generates "noisy" data samples that are approximately linear, exponential, or polynomial. It then uses the corresponding curve fitting VIs to determine the parameters of the curve that best fits those data points. (At this stage, you do not need to worry about how the noisy data samples are generated.) You can control the noise amplitude with the **Noise Level** control on the front panel.

■ Block Diagram

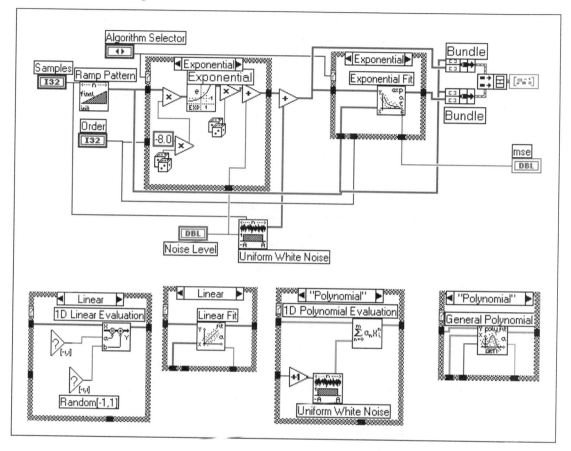

2. Select *Linear* in the **Algorithm Selector** control, and set the **Noise Level** control to about 0.1. Run the VI. Note the spread of the data points and the fitted curve (straight line).

3. Experiment with different values of **Order** and **Noise Level**, keeping one constant while changing the other. What do you notice? How does the MSE change?

4. Change the **Algorithm Selector** to *Exponential* and run the VI. Experiment with different values of **Order** and **Noise Level**. What do you notice?

5. Change the **Algorithm Selector** to *Polynomial* and run the VI. Experiment with different values of **Order** and **Noise Level**. What do you notice?

6. In particular, with the **Algorithm Selector** control set to *Polynomial*, change the **Order** to 0 and run the VI. Then change it to 1 and run the VI. Explain your observations.

7. Depending on your observations in steps 2, 3, 4, and 5, for which of the algorithms (Linear, Exponential, Polynomial) is the **Order** control the most effective? Why?

8. Close the VI. Do not save any changes.

■ End of Activity 7-3

General Least Square Linear Fit

The **Linear Fit** VI calculates the coefficients a_0 and a_1 that best fits the experimental data ($x[i]$ and $y[i]$) to a straight line model given by

$$y[i] = a_0 + a_1 x[i]$$

Here, $y[i]$ is a linear combination of the coefficients a_0 and a_1. You can extend this concept further so that the multiplier for a_1 is some function of x. For example:

$$y[i] = a_0 + a_1 \sin(\omega x[i]) \text{ or}$$
$$y[i] = a_0 + a_1 x[i]^2 \text{ or}$$
$$y[i] = a_0 + a_1 \cos(\omega x[i]^2)$$

where ω is the angular frequency. In each of these cases, $y[i]$ is a linear combination of the coefficients a_0 and a_1. This is the basic idea behind the **General LS Linear Fit** VI, where $y[i]$ can be a linear combination of several coefficients, each of which may be multiplied by some function of $x[i]$. Therefore, you can use it to calculate coefficients of the functional models that can be represented as linear combinations of the coefficients. Some such examples are:

$$y = a_0 + a_1 \sin(\omega x) \text{ or}$$
$$y = a_0 + a_1 x^2 + a_2 \cos(\omega x^2)$$
$$y = a_0 + a_1(3 \sin(\omega x)) + a_2 x^3 + a_3 / x + \dots$$

In each case, note that y is a *linear* function of the coefficients (although it may be a nonlinear function of x).

You will now see how to use the **General LS Linear Fit** VI to find the best linear fit to a set of data points. The inputs and outputs of the **General LS Linear Fit** VI are shown in Figure 7–1.

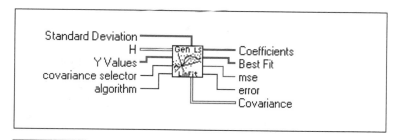

Figure 7–1
Terminal connections for the General LS Linear Fit VI

A matrix **H**, known as the observation matrix, is connected to the input **H** of the VI. The set of observed data points $y[i]$ is connected to the input **Y Values**. The **Covariance** output is the matrix of covariances between the coefficients a_k, where c_{ij} is the covariance between a_i and a_j, and c_{kk} is the variance of a_k. At this stage, you need not be concerned about the inputs **Standard Deviation**, **covariance selector**, and **algorithm**. For now, you will just use their default values. You can refer to the LabVIEW Online Help for more details on these inputs.

For example, suppose you have collected samples (**Y Values**) from a transducer and you want to solve for the coefficients of the model:

$$y = a_o + a_1 \sin(\omega x) + a_2 \cos(\omega x) + a_3 x^2$$

You see that the multiplier for each a_j $(0 \le j \le 3)$ is a different function. For example, a_0 is multiplied by 1, a_1 is multiplied by $\sin(\omega x)$, a_2 is multiplied by $\cos(\omega x)$, and so on. To build **H**, you set each column of **H** to the independent functions evaluated at each x value, $x[i]$. Assuming there are 100 "x" values, **H** would be

$$
\mathbf{H} = \begin{bmatrix}
1 & \sin(\omega x_0) & \cos(\omega x_0) & x_0^2 \\
1 & \sin(\omega x_1) & \cos(\omega x_1) & x_1^2 \\
1 & \sin(\omega x_2) & \cos(\omega x_2) & x_2^2 \\
\cdots & \cdots & \cdots & \cdots \\
1 & \sin(\omega x_{99}) & \cos(\omega x_{99}) & x_{99}^2
\end{bmatrix}
$$

If you have N data points and k coefficients $(a_0, a_1, ..., a_{k-1})$ for which to solve, **H** will be an N-by-k matrix with N rows and k columns. Thus, the number of rows of **H** is equal to the number of elements in **Y Values**, whereas the number of columns of **H** is equal to the number of coefficients for which you are trying to solve.

In practice, **H** is not available and must be built. Given that you have the N independent **X Values** and observed **Y Values**, the block diagram in Figure 7–2 demonstrates how to build **H** and use the **General LS Linear Fit** VI.

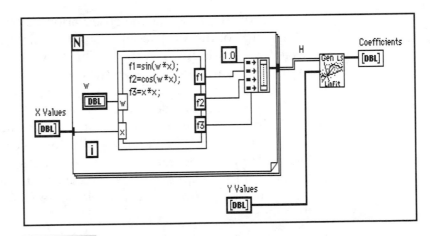

Figure 7–2
Building the observation matrix using the formula node and a for loop

Activity 7-4

Objective: To learn how to set up the input parameters and use the General LS Linear Fit VI

This activity demonstrates how to use the **General LS Linear Fit** VI to obtain the set of least square coefficients **a** and the fitted values, and also how to set up the input parameters to the VI.

The purpose is to find the set of least square coefficients **a** that best represent the set of data points ($x[i]$, $y[i]$). As an example, suppose you have a physical process that generates data using the relationship

$$y = 2h_0(x) + 3h_1(x) + 4h_2(x) + \text{noise} \tag{7.1}$$

where

$$h_0(x) = \sin(x^2)$$

$$h_1(x) = \cos(x)$$

$$h_2(x) = \frac{1}{x+1}$$

and *noise* is a random value. Also, assume you have some idea of the general form of the relationship between x and y but are not quite sure of the coefficient values. So, you may think that the relationship between x and y is of the form

$$y = a_0 f_0(x) + a_1 f_1(x) + a_2 f_2(x) + a_3 f_3(x) + a_4 f_4(x) \tag{7.2}$$

where

$$f_0(x) = 1.0$$

$$f_1(x) = \sin(x^2)$$

$$f_2(x) = 3\cos(x)$$

$$f_3(x) = \frac{1}{x+1}$$

$$f_4(x) = x^4$$

Equations 7.1 and 7.2 correspond, respectively, to the actual physical process and to your guess of this process. The coefficients you choose in your guess may be close to the actual values or may be far away from them. Your objective now is to accurately determine the coefficients **a**.

■ Building the Observation Matrix

To obtain the coefficients **a**, you must supply the set of $(x[i], y[i])$ points in the arrays **H** and **Y Values** (where the matrix **H** is a 2D array) to the **General LS Linear Fit** VI. The $x[i]$ and $y[i]$ points are the values observed in your experiment. A simple way to build the matrix **H** is to use the Formula Node as shown in the following block diagram:

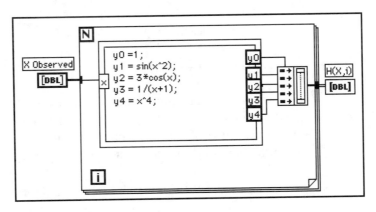

Building the observation matrix, **H**

You can easily edit the formula node to change, add, or delete functions. At this point, you have all the necessary inputs to use the **General LS Linear Fit** VI to solve for **a**. To obtain Equation 7.1 from Equation 7.2, you need to multiply $f_0(x)$ by 0.0, $f_1(x)$ by 2.0, $f_2(x)$ by 1.0, $f_3(x)$ by 4.0 and $f_4(x)$ by 0.0. Thus, looking at Equations 7.1 and 7.2, note that the expected set of coefficients are

$$\mathbf{a} = [0.0, 2.0, 1.0, 4.0, 0.0]$$

The following block diagram demonstrates how to set up the **General LS Linear Fit** VI to obtain the coefficients and a new set of **y** values:

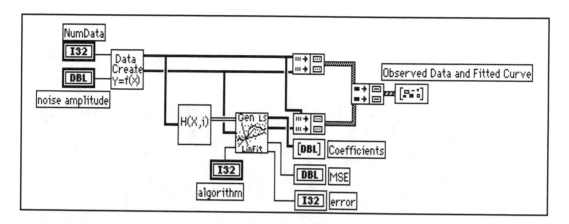

Using the General LS Linear Fit VI

 The subVI labeled **Data Create** generates the **X** and **Y** arrays. You can replace this icon with one that actually collects the data in your experiments. The icon labeled **H(X,i)** generates the 2D matrix **H**.

The last portion of the block diagram overlays the original and the estimated data points and produces a visual record of the General LS Linear Fit. Executing the **General LS Linear Fit** VI with the values of **X**, **Y**, and **H** returns the following set of coefficients:

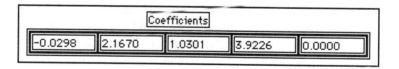

The resulting equation is thus

$$y = -0.0298(1) + 2.1670 \sin(x^2) + 1.0301(3 \cos(x)) +$$
$$3.9226/(x+1) + 0.00(x^4)$$
$$= -0.0298 + 2.1670 \sin(x^2) + 1.0301(3 \cos(x)) +$$
$$3.9226/(x+1)$$

The following graph displays the results:

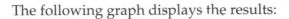

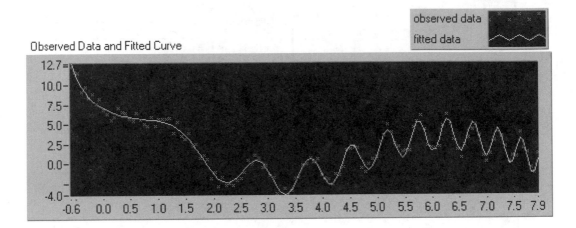

You will now see the VI in which this particular example has been implemented.

1. Open the **General LS Fit Example** VI from the library `Activities.llb`.

2. Examine the front panel.

    noise amplitude: Can change the amplitude of the noise added to the data points. The larger this value, the more the spread of the data points.

    NumData: The number of data points that you want to generate.

    algorithm: Provides a choice of six different algorithms to obtain the set of coefficients and the fitted values. In this particular example, there is no significant difference among different algorithms. You can select different algorithms from the front panel to see the results. In some cases, different algorithms may have significant differences, depending on your observed data set.

    MSE: Gives the mean squared error. The smaller the MSE, the better the fit.

    error: Gives the error code in case of any errors. If error code = 0, it indicates no error. For a list of error codes, see Appendix B.

Coefficients: The calculated values of the coefficients (a_0, a_1, a_2, a_3, and a_4) of the model.

3. Run the VI with progressively larger values of the **noise amplitude**. What happens to the observed data plotted on the graph? What about the MSE?

4. For a fixed value of **noise amplitude**, run the VI by choosing different algorithms from the **algorithm** control. Do you find that any one algorithm is better than the other? Which one gives you the lowest MSE?

5. When you finish, close the VI. Do not save any changes.

■ End of Activity 7-4

Activity 7-5

Objective: To predict production costs using the General LS Linear Fit VI

The VIs that you have seen so far have been used to fit a curve to a function of only one variable. In this activity, you will use the **General LS Linear Fit** VI to fit a curve to a multivariable function. In particular, the function will have two variables, X_1 and X_2. You can, however, generalize it to functions of three or more variables.

Suppose you are the manager of a bakery and want to estimate the total cost (in dollars) of a production of baked scones using the quantity produced, X_1, and the price of one pound of flour, X_2. To keep things simple, the following five data points form this sample data table (Table 7-1):

Table 7–1   Cost of production of baked scones

Cost (dollars) Y	Quantity X_1	Flour Price X_2
$150	295	3.00
$75	100	3.20
$120	200	3.10
$300	700	2.80
$50	60	2.50

You want to estimate the coefficients to the equation

$$Y = a_0 + a_1X_1 + a_2X_2$$

You can use the **General LS Linear Fit** VI. It must have the inputs **H**, the observation matrix, and **Y Values**, a vector of values of the left hand side of the above equation. Each column of **H** is the observed data for each of the independent variables associated with the coefficients a_0, a_1, and a_2. Note that the first column is one because the coefficient a_0 is not associated with any independent variable. Thus, **H** should be filled in as

$$\mathbf{H} = \begin{bmatrix} 1 & 295 & 3.00 \\ 1 & 100 & 3.20 \\ 1 & 200 & 3.10 \\ 1 & 700 & 2.80 \\ 1 & 60 & 2.50 \end{bmatrix}$$

In LabVIEW (or BridgeVIEW), the observed data would normally appear in three arrays (Y, X_1, and X_2). The following block diagram demonstrates how to build **H**.

■ Block Diagram

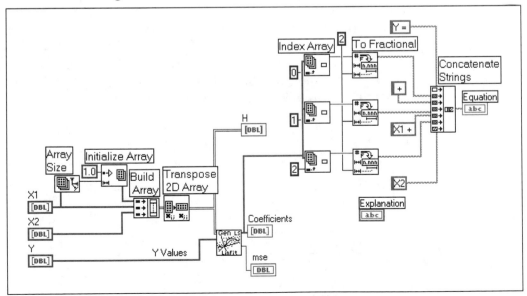

1. Open the **Predicting Cost** VI from the library `Activities.llb`. The values of Y, X_1, and X_2 have already been entered into the corresponding controls on the front panel.

2. Examine the block diagram. Be sure you understand how to build the matrix **H**. Most of the rest of the block diagram is used to build the string that displays the equation of the functional model. You do not need to worry about the rest of the diagram for now.

3. Switch to the front panel and run the VI. Check to see whether the matrix **H** was created correctly.

■ Front Panel

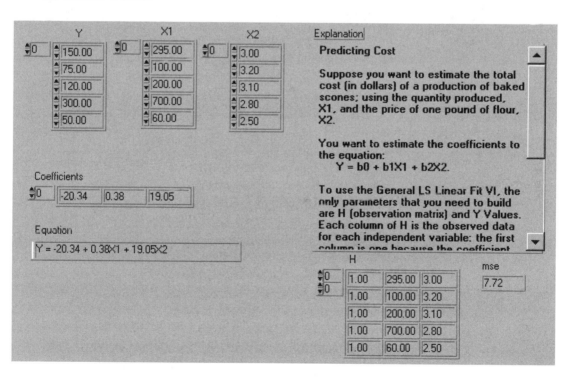

After running the **Predicting Cost VI**, the following coefficients are obtained:

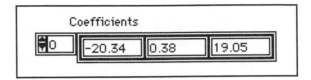

and the resulting equation for the total cost of scone production is therefore

$$Y = -20.34 + 0.38X_1 + 19.05X_2$$

4. Experiment with different values of X_2 (flour price).
5. When you finish, close the VI. Do not save any changes.

■ End of Activity 7-5

Nonlinear Levenberg-Marquardt Fit

So far, you have seen VIs that are used when there is a linear relationship between y and the coefficients a_0, a_1, a_2, …. However, when a nonlinear relationship exists, you can use the **Nonlinear Lev-Mar Fit** VI to determine the coefficients. This VI uses the Levenberg-Marquardt method, which is very robust, to find the coefficients $\mathbf{a} = \{a_0, a_1, a_2, ..., a_k\}$ of the nonlinear relationship between $\mathbf{a}$ and $y[i]$. The VI assumes that you have prior knowledge of the nonlinear relationship between the x and y coordinates.

As a preliminary step, you need to specify the nonlinear function in the **Formula Node** on the block diagram of one of the sub-VIs of the **Nonlinear Lev-Mar Fit** VI. This particular sub-VI is the **Target Fnc and Deriv NonLin** VI. You can access the **Target Fnc and Deriv NonLin** VI by selecting it from the menu that appears when you select **Project » This VI's SubVIs**.

When using the Nonlinear Lev-Mar Fit VI, you also need to specify the nonlinear function in the Formula Node on the block diagram of the Target Fnc and Deriv NonLin VI.

The connections to the **Nonlinear Lev-Mar Fit** VI are shown in Figure 7–3.

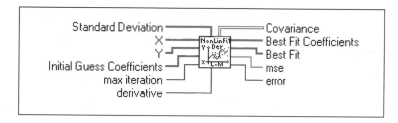

Figure 7–3
Terminal connections for the Nonlinear Lev-Mar Fit VI

X and **Y** are the input data points $x[i]$ and $y[i]$.

Initial Guess Coefficients is your initial guess as to what the coefficient values are. The coefficients are those used in the formula that you entered in the **Formula Node** of the **Target Fnc and Deriv NonLin** VI. Using the **Nonlinear Lev-Mar Fit** VI successfully sometimes depends on how close your initial guess coefficients are to the actual solution. Therefore, it is always worth taking the time and effort to obtain a good initial guess to the solution from any available resource.

For now, you can leave the other inputs to their default values. For more information on these inputs, see the LabVIEW Online Help.

Best Fit Coefficients: The values of the coefficients (a_0, a_1, ...) that best fit the model of the experimental data.

Activity 7-6

Objective: To create a general exponential a exp(bx) + c + noise and then use the Nonlinear Lev-Mar Fit VI to fit the data and get the best guess coefficients a, b, and c of the general exponential signal

In this activity, you will see how to use the **Nonlinear Lev-Mar Fit** VI to determine the coefficients a, b, and c, of a nonlinear function given by $a \exp(bx) + c$.

1. Open the **Nonlinear Lev-Mar Exponential Fit** VI from the library `Activities.llb`. The front panel is shown below.

■ Front Panel

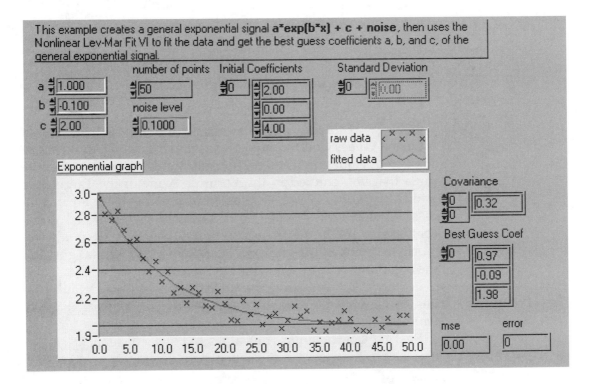

The **a**, **b**, and **c** controls determine the actual values of the coefficients a, b, and c. These controls are used to generate simulated data. To simulate a more practical example, you also add noise to this equation, thus making it of the form

$$a \exp(bx) + c + \text{noise}$$

The **Initial Coefficients** control is your educated guess as to the actual values of a, b, and c. Finally, the **Best Guess Coef** indicator gives you the values of a, b, and c calculated by the **Nonlinear Lev-Mar Fit** VI.

The **noise level** control adjusts the noise level. Note that the actual values of a, b, and c being chosen are +1.0, −0.1 and 2.0. In the **Initial Coefficients** control, the default guess for these is $a = 2.0$, $b = 0$, and $c = 4.0$.

2. Examine the block diagram.

■ Block Diagram

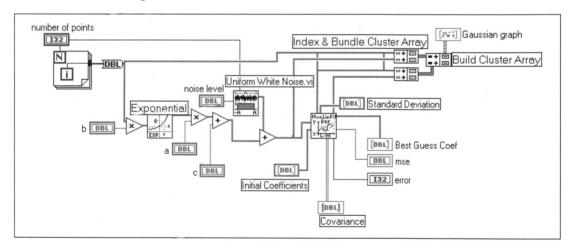

The data samples of the exponential function are simulated using the **Exponential** VI (**Functions » Numeric » Logarithmic** subpalette) and uniform white noise is added to the samples with the help of the **Uniform White Noise** VI (**Functions » Analysis » Signal Generation** subpalette).

3. From the **Project** menu, select **Unopened SubVIs »
   Target Fnc and Deriv NonLin** VI. The front panel of
   the **Target Fnc and Deriv NonLin** VI opens, as
   shown below.

■ Front Panel

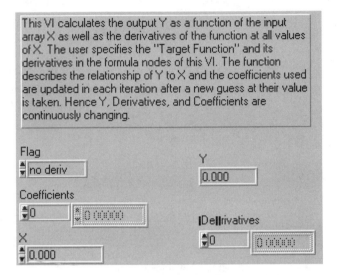

4. Switch to the block diagram of the VI that you
   opened in the previous step.

■ Block Diagram

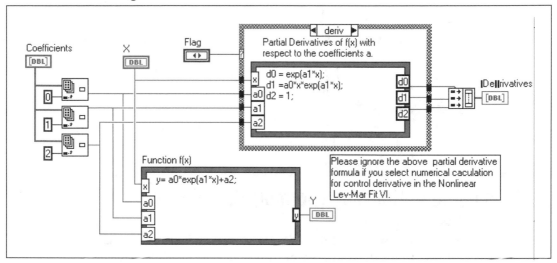

Observe the formula node in the block diagram. It has the form of the function whose parameters (a_0, a_1, and a_2) you are trying to evaluate.

5. Close the front panel and the block diagram of the **Target Fnc and Deriv NonLin** VI.

6. Run the **NonLinear Lev-Mar Exponential Fit** VI. Note that the values of the coefficients returned in **Best Guess Coef** are very close to the actual values entered in the **Initial Coefficients** control. Also note the value of the **mse**.

7. Increase the **noise level** from 0.1 to 0.5. What happens to the **mse** and the coefficient values in **Best Guess Coef**? Why?

8. Change the **noise level** back to 0.1 and the **Initial Coefficients** to 5.0, -2.0, and 10.0, and run the VI. Note the values returned in the **Best Guess Coef** and the **mse** indicators.

9. With the **noise level** still at 0.1, change your guess of the **Initial Coefficients** to 5.0, 8.0, and 10.0, and run the VI. This time, your guess is further away than the

one you chose in step 4. Note the error! This goes to show how important it is to have a reasonably educated guess for the coefficients.

10. When you finish, close the VI. Do not save any changes.

■ End of Activity 7-6

Fitting a Curve to Gaussian (Normal) Data Points

Real-world data are very often Gaussian distributed. That means that many of the data points lie close to a particular value, known as the mean, and the number of data points is smaller as you get further away from the mean. The mathematical description of a Gaussian (also known as a Normal) distribution is

$$f(x) = \frac{1}{\sigma\sqrt{2\pi}}\exp\left[-\frac{1}{2}\left(\frac{x-\mu}{\sigma}\right)^2\right] \tag{7.3}$$

where μ is the *mean* and σ is the *standard deviation*. Figure 7–4 shows the Gaussian distribution with $\mu = 0$ and $\sigma = 0.5$, 1.0, and 2.0.

As seen from the figure, the curve is bell shaped and is symmetric about the mean, μ. The peak of the curve occurs at μ. The standard deviation, σ, determines the "spread" of the curve around the mean. The smaller the value of σ, the more concentrated the curve around the mean, the higher the peak at the mean, and the steeper the descents on both sides.

If you have data that is Normally distributed, you will find that the standard deviation is an important parameter in determining the limits within which a certain percentage of your data values are expected to occur. For example, as shown in Figure 7–5,

1. About two-thirds of the values will lie between μ-σ and μ+σ.

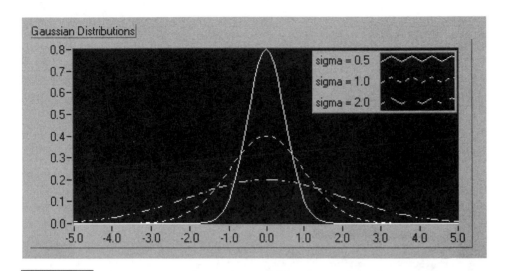

Figure 7–4
Gaussian distributions with zero mean and standard deviation - 0.5, 1.0, and 2.0

2. About 95 percent of the values will lie between μ-2σ and μ+2σ.

3. About 99.75 percent of the values will lie between μ-3σ and μ+3σ. Thus, you see that almost all the data values lie between μ-3σ and μ+3σ.

Notice that the two parameters that completely describe Gaussian data are the *mean* and the *standard deviation* of the data. If you believe your data has a Gaussian distribution, you could determine its mean and standard deviation. This has numerous applications, such as determining whether

- The dimensions of products being manufactured (for example, thickness of plates) are within specified limits.

- The values of components (for example, resistance of resistors) are within a specified tolerance.

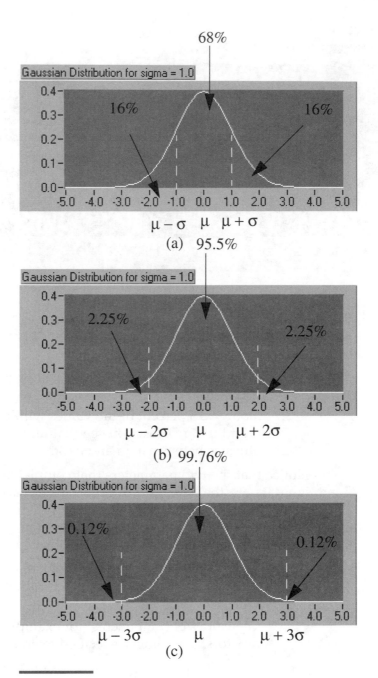

Figure 7–5
Percentage of data values between $\mu - n\sigma$, where n = ±1, ±2, and ±3

Activity 7-7

Objective: To fit a curve to noisy Gaussian data

1. Open the **Normal (Gaussian) Fit** VI from the library `Activities.llb`. The Front Panel is as shown.

■ Front Panel

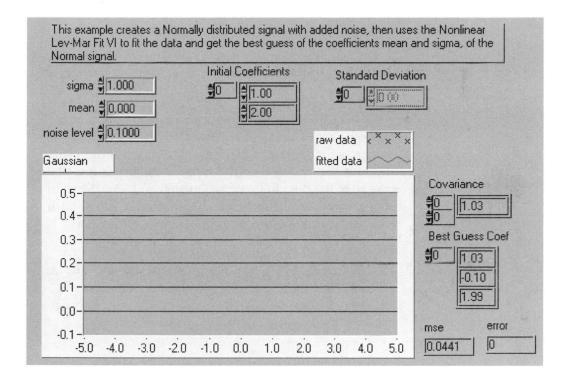

This example creates a Normally distributed signal with added noise, then uses the Nonlinear Lev-Mar Fit VI to fit the data and get the best guess of the coefficients mean and sigma, of the Normal signal.

sigma 1.000
mean 0.000
noise level 0.1000

Initial Coefficients
0 | 1.00
2.00

Standard Deviation
0 | 0.00

Gaussian

raw data
fitted data

Covariance
0
0 | 1.03

Best Guess Coef
0 | 1.03
-0.10
1.99

mse
0.0441

error
0

2. Switch to the block diagram.

■ Block Diagram

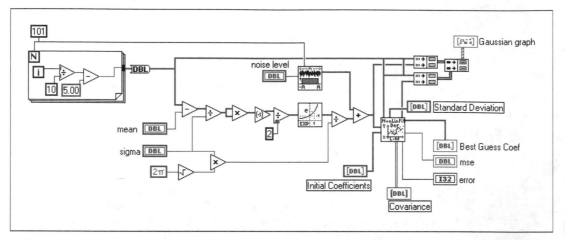

Observe that the parameters *a* and *b* in the **Nonlinear Lev-Mar Exponential Fit** VI of the previous exercise have been replaced by *mean* and *sigma*. Note that the controls for the parameter *c* and the **number of points** have been removed. The For Loop generates the range of the data to lie between –5.0 and +5.0. Most of the additions on the block diagram have to do with implementing Equation 7.3. Uniform white noise is then added to the Gaussian data generated.

3. Select **Project » Unopened SubVIs » Target Fnc & Deriv Nonlin** VI.

4. Modify the block diagram of the **Target Fnc & Deriv Nonlin** VI as shown below.

■ Block Diagram

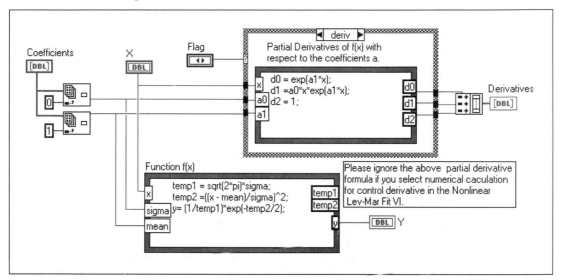

5. Go back to the **Normal (Gaussian) Fit** VI and enter the following values in the controls on the front panel:

sigma	1.0
mean	0.0
noise level	0.1

6. Clear the **Initial Coefficients** array by popping up on it with the right mouse button and selecting **Data Operations » Empty Array**. Set the values in **Initial Coefficients** to 2.0 and 2.0. These are your guesses of sigma and mean, the actual values of which you chose as 1.0 and 0.0 in the corresponding controls.

7. Run the VI several times and observe the values of sigma and mean in the **Best Guess Coeff** indicator. In the plot, you can also see the spread of the data and the fitted Gaussian curve.

8. Change the value of sigma to some value other than 1.0 and run the VI several times. Each time observe the fitted curve, the MSE, and the values of the **Best Guess Coeff**.

9. Change **Initial Coefficients** [0] (the guess for sigma) to 50.0 and run the VI several times. Does the VI obtain a good estimate of sigma in the array **Best Guess Coeff**?

10. Change **Initial Coefficients** [0] (the guess for sigma) to 500.0 and again run the VI several times. Now how accurate is the estimate?

11. When you finish, save and close both the VIs.

In this activity and the previous one, you had to open the **Target Fnc and Deriv NonLin** VI and enter equations in the formula node. This requires a certain amount of work, as well as an understanding of the formula node, on your part. A much simpler and easier method is to have the flexibility of entering the formulas directly on the front panel, without accessing the block diagram. As you will see in a later chapter, this is possible by using the G Math Toolkit, which is a mathematics package (written entirely in G) that is extremely useful for solving differential equations, optimization, and a wide variety of other common math problems.

■ End of Activity 7-7

Wrap It Up!

- Curve fitting is useful in fitting an equation to a set of data points. Some of the practical applications of curve fitting are to remove measurement noise, to fill in missing data points, interpolation, extrapolation, and integration and differentiation of digital data.

- In particular, you have learned how to use the **Linear, Exponential, Polynomial, General Linear LS Fit**, and **Nonlinear Lev-Mar Fit** VIs to perform several different types of linear and nonlinear fits.

- The MSE is a useful criterion in determining the fit accuracy.

■ Review Questions

1. Name five applications of curve fitting.
2. Which curve fitting VI would you use to determine the parameters (denoted by the a_i ... i integer) of the following models?

   a) $y = a_0 \exp(a_1 x)$
   b) $y = a_0 \exp(a_1 x) + a_2$
   c) $y = a_1 x + a_2 x^2$
   d) $y = a_1 \sin(x) + a_2 \cos(x) + \dfrac{a_3}{x+1}$
   e) $t = a_1 y + a_2 x^2$ where x and y are different variables

■ Additional Activities

1. Table 7-2 shows the time and distance measurements of a missile launched at an enemy aircraft from a battleship. (Eleven seconds after the launch, the aircraft is hit.)

Table 7–2   *Time and distance measurements of a missile launched from a battleship*

Time (secs)	Distance (100's of feet)
0	0
1	139
2	361
3	619
4	799
5	1080
6	1309
7	1430
8	1525
9	1675
10	1734
11	1772

Use the **Polynomial Fit VI** to fit a polynomial to this set of data points. What polynomial order gives you the smallest MSE?

2. In Activity 7-6 you saw how to use the **Nonlinear Lev-Mar Fit VI** to determine the coefficients of a general exponential signal. Suppose you had to determine the coefficients of a biexponential signal

$a \exp(-bx) + c + d \exp(-ex)$

How would you modify the **Target Fnc & Deriv Nonlin VI** to achieve this?

3. Consider the following data set which represents the position of a planet as it follows an elliptical orbit in space.

Table 7–3 Position of a planet in an elliptical orbit

X	Y
1.02	0.39
0.95	0.32
0.87	0.27
0.77	0.22
0.67	0.18
0.56	0.15
0.44	0.13
0.30	0.12
0.16	0.13
0.01	0.15

The elliptical orbit of the planet can be represented in a Cartesian coordinate system (x,y) by the equation $ay^2 + bxy + cx + dy + e = x^2$. Use the **General Polynomial Fit** VI to compute the orbital parameters a, b, c, d and e. Plot the resulting orbit (fitted data) and the given data points on a XY Waveform Graph. Try different values for the polynomial order such as 1, 3, 5, 9, 11, 15 and 20 to see which value gives the smallest mean square error. From this activity, you will learn that arbitrarily increasing the order of the polynomial does not necessarily give the best fit.

Save the VI as **Planet.vi** in the library `Activities.11b` and close it.

LabVIEW Signal Processing in Biomechanics

Rod Barrett
School of Physiotherapy and Exercise Science
Griffith University, Queensland, Australia

The Challenge In 1995 a Biomechanics Laboratory was established at Griffith University for the purpose of (i) teaching undergraduate students undertaking study in the fields of Exercise Science and Physiotherapy; and (ii) facilitating high quality Biomechanics research by staff and postgraduate students. In order to achieve these outcomes it was necessary to develop data acquisition and analysis systems that could be used to acquire data from devices such as force platforms, video motion capture systems, telemetric EMG systems, force, pressure and acceleration transducers and numerous other types of biomechanics equipment.

The Solution LabVIEW was chosen because of its user-friendliness, versatility, data processing capbilities as well as due to the way software and hardware are easily integrated via the use of National Instruments Data Acquisition Boards.

Biomechanics

According to the most accepted definition of the term, *biomechanics* is the science that examines the internal and external forces acting on biological systems and the effects produced by these forces. The area can be further subdivided into *kinematics*, the study of time and space factors of a system in motion, and *kinetics*, the study of the forces acting on a body that influence its movement. By virtue of this definition the field of biomechanics is vast and encompasses many diverse areas of research. At the last International Congress of Biomechanics (Japan, 1997), papers were organised into the categories of sports biomechanics, muscle mechanics, occupational biomechanics, instrumentation and methods, computer modeling, gait analysis, motor control, athropometry and orthopaedic biomechanics, tissue mechanics and electromyography.

Common to each of these research categories is an instrumentation based methodological approach to problem solving. The Biomechanics laboratory at Griffith University is equipped to perform both kinetic and kinematic analyses. The lab is designed around a walkway on which dual force platforms are mounted that are used to measure ground reaction forces exerted by the subject as they perform the activities for analysis. A video based motion analysis system is used to quantify the subject's motion and a telemetric elelctromyogram (EMG) is used to measure the electrical activity associated with the subject's muscular contractions. In addition, a variety of other transducers are used in various experiments to measure displacement, acceleration, force and pressure. We use the LabVIEW graphical programming language extensively in our laboratory to both acquire and process biomechanical data by interfacing our data acquisition devices (ie. transducers) to PCs via National Instruments data acquisition cards. A few simple examples of how we use LabVIEW to acquire and process data are as follows:

EMG Data Acquisition and Analysis

EMG refers to the spatial and temporal summation of motor unit action potentials during muscular contraction and can be measured with the aid of a suitable set of electrodes and a differential amplifier. Once the signal has been acquired, various signal processing techniques can be employed to estimate (a) if and when a muscle is active; (b) the force of muscular contraction; (c) the extent to which a muscle is experiencing fatigue.

Time domain processing of the EMG signal is routinely used to determine the time of onset of muscular activity as well as to examine the EMG-force relation. We have implemented a number of signal processing techniques for determining the time of onset of muscular activity (Bogey et al., 1992; di Fabio, 1987; Walter, 1984). Typically these involve the use of linear envelopes obtained by full wave rectification and low pass filtering of the raw EMG waveform that are assessed using decision making algorithms that attempt distinguish the signal from baseline noise. An amalgam of these methods has been used to determine the time of impact in surface testing experiments where an instrumented drop head was released from set heights onto a variety of sports surfaces (see section below on mechanics of impact for a more detailed description).

EMG is known to be a course predictor of muscle tension for muscles whose length is not changing rapidly. Because the mean of the raw EMG signal is zero, rectification is performed as the first step in relating EMG amplitude to muscle force. To obtain a linear envelope the full wave rectified signal is smoothed using a low pass digital filter (eg. Butterworth) with a cut-off frequency of approximately 5 Hz resulting in a signal that is believed to closely resemble the shape of the muscle's tension curve. As a final step the integrated signal is calculated from the area under the linear envelope. Since the integrated waveform will always increase when EMG activity is present it is usually necessary to reset the integrator either with time or at a set voltage level. When re-setting on a voltage level, the integrator resets when a voltage threshold is reached so that the total amount of EMG can be determined from the product of the threshold voltage and the number of resets.

The front panel and a subset of the block diagram we use to process EMG in the time domain is displayed in Figures 7–6 and 7-7. The three Waveform Graphs on the Front Panel in Figure 7–6 show the raw EMG obtained during a maximum voluntary isometric contraction of the biceps brachii muscle (top), the full wave rectified signal and the linear envelope (middle), and the integrated EMG signal obtained using resets on a voltage threshold. LabVIEW's built in **Butterworth Filter.vi** is used to low pass filter the raw EMG data and a For Loop with Shift Registers and the Select node is used to obtain the "Reset" integrated EMG.

We have also used frequency domain analysis of the EMG signal to quantify muscular fatigue. A power density spectra of the EMG signal is obtained using a FFT and the mean and or median frequency computed and used as a spectral compression index since the high frequency content of the signal is known to decrease with increasing muscular fatigue.

Our LabVIEW implementation of the frequency domain analysis of raw EMG data is shown in Figures 7–8 and 7–9. Raw EMG

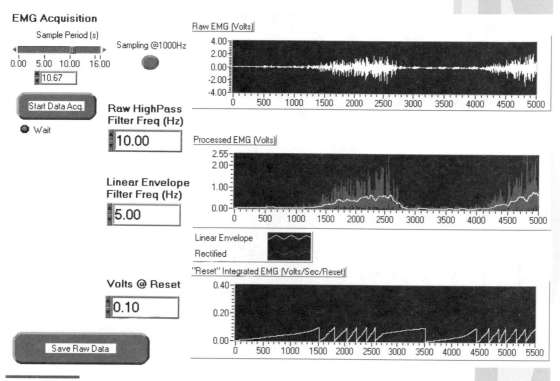

Figure 7–6

Front panel of a VI used to process raw EMG data in the time domain. Raw EMG data is plotted on the waveform graph at the top, full wave rectified EMG and a linear envelope is displayed on the middle waveform graph, and integrated EMG using a voltage threshold reset is shown on the bottom waveform graph. Other controls are used to acquire and save the data.

data obtained from the gastrocnemius muscle during a vertical jump is plotted on a Waveform Graph in Figure 7–8 together with the corresponding power spectrum obtained using LabVIEW's **Auto Power Spectrum.vi**. The power spectrum is normalised to the maximum power present following low pass filtering of the power spectrum via application of a dual pass zero phase lag Butterworth filter and is plotted against frequency (measured in Hertz) on an X-Y Graph. Based on the normalised power spectrum, the lower and upper limits of the bandwidth, bandwidth, and the mean, median and peak power frequencies were calculated and displayed on Digital Indicators. The upper and lower limits of the bandwidth were defined by the power cutoff speci-

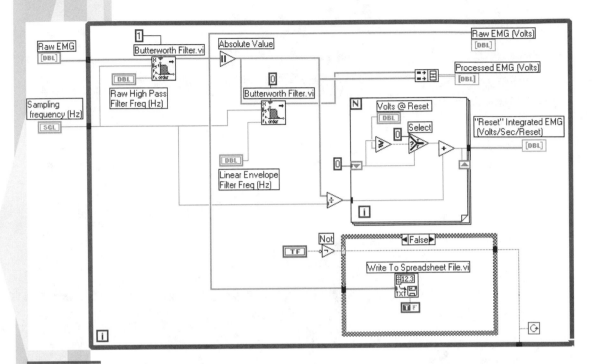

Figure 7–7
Subset of the block diagram used to process the raw EMG signal. Filtering (high pass and low pass) was performed using the Butterworth.vi . A For Loop with a Shift Register was used to obtain the "Reset" integrated EMG and a Case Structure was used to give the user control over whether the data was saved or not. Execution terminates once the data is saved.

fied by the user. Once the bandwidth was known it was then possible to calculate the mean, median and peak power frequencies. These were calculated as follows: The mean frequency was defined as the sum of the first moments of frequency divided by the area under the power-frequency curve; the median frequency was defined as the frequency that divides the power of the EMG into two equal parts; and the peak power frequency was defined as the frequency at which the power of the signal is greatest (Nigg & Herzog, 1994). The Block Diagram for this VI is displayed in Figure 7–9 and makes use of a Sequence Structure to maintain control flow on account of the use of a number of Local Variables.

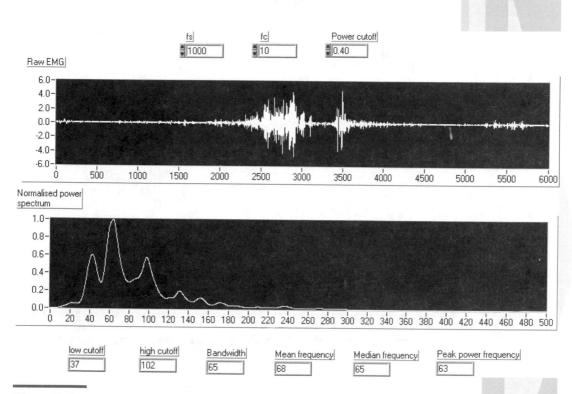

Figure 7–8

Front panel of a VI used to process raw EMG data in the frequency domain. The user is required to specify the sampling frequency (fs), the cutoff frequency for the low pass filter used to smooth the power spectrum (fc) and the power threshold used to determine the lower and upper limits of the bandwidth of the power spectrum (Power cutoff). The raw EMG is displayed on a waveform graph and the normalised power spectrum is displayed on an X-Y graph up to the Nyquist limit (ie. half the sampling frequency). Digital indicators are used to the display the lower and upper limits of the bandwidth, bandwidth, and the mean, median and peak power frequency.

In the first frame of the Sequence Structure a For Loop is used to generate the array of frequencies that represent the horizontal axis of the XY Graph and pair of While Loops are used to determine the array indexes at which the normalised power exceeds power cutoff value in order to determine the lower and upper limits of the bandwidth. A further While Loop is used in the second frame of the Sequence Structure to ascertain the index of the power corresponding to the median frequency.

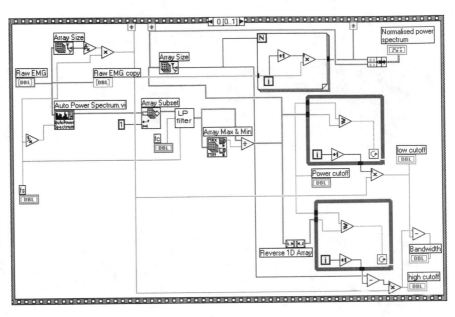

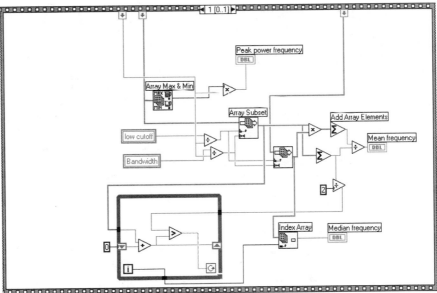

Figure 7–9
Front panel of a VI used to process raw EMG data in the frequency domain. The normalised power spectrum of a raw EMG trace is determined using the Auto Power Spectrum.vi which is then used to calculate bandwidth, mean frequency, median frequency, and peak power frequency.

Force Platform Data Acquisition and Analysis

The force platforms used in our laboratory have 8 voltage outputs that are proportional to the forces acting on 12 separate piezoelectric sensors (the outputs from two sets of sensors are summed in the hardware). The LabVIEW graphical programming language is used to acquire and then to post-process the 8 channels of data so that the 3 orthogonal components (vertical, front-to-back and side-to-side) of the ground reaction force and its point of application (centre of pressure) are known. Fast Fourier Transforms (FFTs) are used to determine the frequency content of the signal and digital filters are used, where necessary to smooth the data.

Measurement of Segmental Moments of Inertia Using the Quick Release Method

The quick release method for measuring moment of inertia requires the simultaneous measurement of force and acceleration and serves as a simple example of how we use LabVIEW to acquire and process and display data obtained from transducers. To measure the moment of inertia of the shank and foot an accelerometer is placed on the distal portion of the tibia, which itself is fixed in position via a cable attached to a force transducer and then via an electromagnet to the back of the chair on which the subject is seated. The subject attempts to extend the knee against the fixed resistance at which time the electromagnet is switched off causing the leg to extend and an acceleration to be registered. With the use of some simple mechanical relations it is then possible to calculate the moment of inertia of the combined shank and foot.

Inverse Dynamic Analysis of Lower Extremity Movement Patterns

Inverse dynamic analysis is performed in order to predict the net muscle moments acting about individual joint centres during movement. Knowledge of the net muscle moment is important because it tells us about the net effect of all muscle crossing the joint and provides an insight into the causes of the observed motion. To perform an inverse dynamic analysis anthropometric data, kinematic data and data describing the external forces acting on the body of interest are used an inputs to the equations of motion of Newton and Euler. The process is called inverse dynamics since kinematic measurements are used to calculate the torques responsible for them.

To calculate the net moments in the ankle, knee and hip during activities such as walking, running, jumping, squatting and the sit-to-stand motion, we have developed a full 2D inverse dynamic model of the lower extremity in LabVIEW.

Mechanics of Impact

We recently investigated the shock absorbing characteristics of a number of sports surfaces using an accelerometer attached to a drop mass that was dropped from height that yielded impact energies consistent with those associated with running. In order to accurately measure the temporal characteristics of each impact it was important to be able to accurately determine both the occurrence and cessation of impact. The accelerometer signals were processed using LabVIEW as follows:

The accelerations measured in the experiment were optimally filtered with a Butterworth low pass zero lag filter using a cut-off frequency (Hz, Mean ± SD) determined from an analysis of residuals (Winter, 1990) An algorithm was developed that compared the value of the *n-th* element in the array with j (the mean of the previous k elements plus m standard deviations). If the value of the *n-th* element exceeded the value of *j* on *l* consecutive occasions, the *n-lth* element was judged to be the point at which impact occurred (ie. the front edge). The acceleration data array was reversed to determine the rear edge so that the impact time could be calculated. To ensure that the algorithm did not detect a false positive (ie. detect an edge that was not associated with impact), the algorithm was used in the framework of an interactive graphics approach whereby the index of the proposed edge was plotted on a graphical display of the acceleration data which was then be either accepted or rejected by the operator. If rejected, the algorithm was repeated to find the next edge. An additional feature of the program for detecting the onset and cessation of impact was that the operator was able to specify the index at which the algorithm would commence searching for the edge. This feature was also used in situations where the baseline acceleration signal was unstable and likely to result in a false positive.

In addition to using the LabVIEW graphical programming language as a way of acquiring and analysing experimental data in the Biomechanics laboratory, we also use LabVIEW extensively in the undergraduate teaching programs within the School of Physiotherapy and Exercise Science. All students in our school must complete a full semester subject (PES2005 Bioinstrumentation) where principles of LabVIEW programming are presented in lectures and then reinforced in a self-paced tutorial program. Students are required to develop a Virtual Instrument (VI) that is relevant to a an area of interest within the field of Exercise Science as part of their assessment. As part of this subject students are also introduced to basic electronic principles underlying the operation of electronic instrumentation which includes sections on AC and DC circuits, test equipment, transducers, basic laboratory skills, noise reduction techniques, safety, digital and analog I/O. In our opinion this subject provides students with an excellent introduction to laboratory data acquisition and analysis from both a hardware and software perspective.

References

Barrett, R.S., Neal, R.J., and Roberts, L.J. (1998). The dynamic loading response of surfaces encountered in beach running. *Journal of science and medicine in sport, 1*(1), 3–13.

Bogey, R.A., Bartners, L.A., and Perry, J. (1992). Computer algorithms to characterise individual subject EMG profiles during gait. *Archives of physical medicine and rehabilitation, 73*, 835–841.

Di Fabio, R.P. (1987). Reliability of computerised surface electromyography for determining onset of muscular activity. *Physical therapy, 67*(1), 43–48.

Nigg, B.M., and Herzog, W. (1994). *Biomechanics of musculoskeletal system.* New York: John Wiley and Sons.

Walter, C.B. (1984). Temporal quantification of electromyography with reference to motor control research. *Human movement science, 3*, 155–162.

■ **Contact information**

Rod Barrett
School of Physiotherapy and Exercise Science
Griffith University, Gold Coast Campus
PMB50 Gold Coast Mail Centre 9726
Queensland AUSTRALIA
Tel: +617 55948 934
Fax: +617 55948 674
email: r.barrett@nhs.gu.edu.au

Overview

In this chapter, you will learn the basic theory behind the Linear Algebra VIs in the Analysis Library and how these VIs can be used in different applications. Matrix computations, such as matrix-matrix multiplication and many others discussed throughout this chapter, form a significant component of linear algebra and are very important in analysis. It is essential that you completely understand the theory and different VIs discussed in this chapter because they form the basis of different algorithms used in many digital signal processing, control, and measurement applications. Some of the algorithms discussed in this chapter are used in solving the curve fitting problems that you have encountered previously. Some of these matrix operations will also be used in Chapter 10 on control systems.

GOALS

- Familiarize yourself about linear systems and matrix analysis.
- Master basic matrix operations and eigenvalue-eigenvector problems.
- Learn about the inverse of a matrix and solving systems of linear equations.
- Understand the concept behind matrix factorization.

KEY TERMS

- matrix
- diagonal
- transpose
- determinant
- singular
- condition number
- matrix norm
- eigenvalues
- eigenvectors
- inverse
- factorization
- matrix multiplication
- dot product
- outer product
- linear system

Linear Algebra

Linear Systems and Matrix Analysis

Systems of linear algebraic equations arise in many applications that involve scientific computations such as signal processing, computational fluid dynamics, and others. Such systems may occur naturally or may be the result of approximating differential equations by algebraic equations.

■ Types of Matrices

Whatever the application, it is always necessary to find an accurate solution for the system of equations in a very efficient way. In matrix-vector notation, a system of linear algebraic equations has the form $Ax = b$, where A is an $m \times n$ matrix, b is a given vector consisting of m elements, and x is the unknown solution vector to be determined. A matrix is represented by a 2D array of elements. These elements may be real numbers, complex numbers, functions, or operators. The matrix **A** shown below is an array of m rows and n columns with $m \times n$ elements.

$$\mathbf{A} = \begin{bmatrix} a_{0,0} & a_{0,1} & \cdots & a_{0,n-1} \\ a_{1,0} & a_{1,1} & \cdots & a_{1,n-1} \\ \cdots & \cdots & \cdots & \cdots \\ a_{m-1,0} & a_{m-1,1} & \cdots & a_{m-1,n-1} \end{bmatrix}$$

Here, $a_{i,j}$ denotes the (i,j)th element located in the ith row and the jth column. In general, such a matrix is called a *rectangular matrix*. When $m = n$, so that the number of rows is equal to the number of columns, it is called a *square matrix*. An $m \times 1$ matrix (m rows and one column) is called a *column vector*. A *row vector* is a $1 \times n$ matrix (1 row and n columns). If all the elements other than the diagonal elements are zero (that is, $a_{i,j} = 0$, $i \neq j$), such a matrix is called a *diagonal matrix*. For example,

$$\mathbf{A} = \begin{bmatrix} 4 & 0 & 0 \\ 0 & 5 & 0 \\ 0 & 0 & 9 \end{bmatrix}$$

is a diagonal matrix. A diagonal matrix with all the diagonal elements equal to one is called an *identity matrix*, also known as a *unit matrix*. If all the elements below the main diagonal are zero, then the matrix is known as an *upper triangular matrix*. On the other hand, if all the elements above the main diagonal are zero, then the matrix is known as a *lower triangular matrix*. When all the elements are real numbers, the matrix is referred to as a *real matrix*. On the other hand, when at least one of the elements of the matrix is a complex number, the matrix is referred to as a *complex matrix*. To make things simpler to understand, you will work mainly with real matrices in this lesson. However, for the adventurous, there are also some activities involving complex matrices.

■ Determinant of a Matrix

One of the most important attributes of a matrix is its *determinant*. In the simplest case, the determinant of a 2 x 2 matrix

$\mathbf{A} = \begin{bmatrix} a & b \\ c & d \end{bmatrix}$ is given by $ad - bc$. The determinant of a square

matrix is formed by taking the determinant of its elements. For example, if

$$\mathbf{A} = \begin{bmatrix} 2 & 5 & 3 \\ 6 & 1 & 7 \\ 1 & 6 & 9 \end{bmatrix}$$

the determinant of $\mathbf{A}$, denoted by $|\mathbf{A}|$, is

$$|\mathbf{A}| = \begin{vmatrix} 2 & 5 & 3 \\ 6 & 1 & 7 \\ 1 & 6 & 9 \end{vmatrix} = \left(2 \begin{vmatrix} 1 & 7 \\ 6 & 9 \end{vmatrix} - 5 \begin{vmatrix} 6 & 7 \\ 1 & 9 \end{vmatrix} + 3 \begin{vmatrix} 6 & 1 \\ 1 & 6 \end{vmatrix} \right)$$

$$= 2(-33) - 5(47) + 3(35)$$

$$= -196$$

The determinant tells many important properties of the matrix. For example, if the determinant of the matrix is zero, then the matrix is *singular*. In other words, the above matrix (with non-zero determinant) is *nonsingular*. You will revisit the concept of singularity later on in this chapter, when the solution of linear equations and matrix inverses is discussed.

■ Transpose of a Matrix

The *transpose* of a real matrix is formed by interchanging its rows and columns. If the matrix $\mathbf{B}$ represents the transpose of $\mathbf{A}$, denoted by $\mathbf{A}^\mathbf{T}$, then $b_{j,i} = a_{i,j}$. For the matrix $\mathbf{A}$ defined above,

$$\mathbf{B} = \mathbf{A}^{\mathbf{T}} = \begin{bmatrix} 2 & 6 & 1 \\ 5 & 1 & 6 \\ 3 & 7 & 9 \end{bmatrix}$$

In case of complex matrices, complex conjugate transposition is defined. If the matrix $\mathbf{D}$ represents the *complex conjugate transpose*[1] of a complex matrix $\mathbf{C}$, then

$$\mathbf{D} = \mathbf{C}^{\mathbf{H}} \Rightarrow d_{i,j} = c^*_{j,i}$$

That is, the matrix $\mathbf{D}$ is obtained by replacing every element in $\mathbf{C}$ by its complex conjugate and then interchanging the rows and columns of the resulting matrix.

A real matrix is called a *symmetric matrix* if the transpose of the matrix is equal to the matrix itself. The example matrix $\mathbf{A}$ is not a symmetric matrix. If a complex matrix $\mathbf{C}$ satisfies the relation $\mathbf{C} = \mathbf{C}^{\mathbf{H}}$, then $\mathbf{C}$ is called a *Hermitian matrix*.

■ Obtaining One Vector as a Linear Combination of Other Vectors (Linear Dependence)

A set of vectors $x_1, x_2, ..., x_n$ is said to be *linearly dependent* if and only if there exist scalars $\alpha_1, \alpha_2, ..., \alpha_n$, not all zero, such that

$$\alpha_1 x_1 + \alpha_2 x_2 + ... + \alpha_n x_n = 0$$

In simpler terms, if one of the vectors can be written in terms of a linear combination of the others, then the vectors are said to be linearly dependent.

If the only set of α_i for which the above equation holds is $\alpha_1 = 0, \alpha_2 = 0, ..., \alpha_n = 0$, the set of vectors $x_1, x_2, ..., x_n$ is said to be *linearly independent*. So, in this case, none of the vectors can be written in terms of a linear combination of the others. Given

---

1. Complex conjugate: If $a = x + jy$, complex conjugate $a^* = x - jy$.

any set of vectors, the above equation always holds for $\alpha_1 = 0$, $\alpha_2 = 0$, ..., $\alpha_n = 0$. Therefore, to show the linear independence of the set, you must show that $\alpha_1 = 0$, $\alpha_2 = 0$, ..., $\alpha_n = 0$ is the only set of α_i for which the above equation holds.

For example, first consider the vectors

$$\mathbf{x} = \begin{bmatrix} 1 \\ 2 \end{bmatrix} \qquad \mathbf{y} = \begin{bmatrix} 3 \\ 4 \end{bmatrix}$$

Notice that $\alpha_1 = 0$ and $\alpha_2 = 0$ are the only values for which the relation $\alpha_1 \mathbf{x} + \alpha_2 \mathbf{y} = 0$ holds true. Hence, these two vectors are linearly independent of each other. Now, consider vectors

$$\mathbf{x} = \begin{bmatrix} 1 \\ 2 \end{bmatrix} \qquad \mathbf{y} = \begin{bmatrix} 2 \\ 4 \end{bmatrix}$$

Notice that, if $\alpha_1 = -2$ and $\alpha_2 = 1$, then $\alpha_1 \mathbf{x} + \alpha_2 \mathbf{y} = 0$. Therefore, these two vectors are linearly dependent on each other. You must completely understand this definition of linear independence of vectors to fully appreciate the concept of the *rank* of the matrix as discussed next.

How Can You Determine Linear Independence? (Matrix Rank)

The *rank* of a matrix $\mathbf{A}$, denoted by $\rho(\mathbf{A})$, is the maximum number of linearly independent columns in $\mathbf{A}$. If you look at the example matrix $\mathbf{A}$, you will find that all the columns of $\mathbf{A}$ are linearly independent of each other. That is, none of the columns can be obtained by forming a linear combination of the other columns. Hence, the rank of the matrix is 3. Consider one more example matrix, $\mathbf{B}$, where

$$\mathbf{B} = \begin{bmatrix} 0 & 1 & 1 \\ 1 & 2 & 3 \\ 2 & 0 & 2 \end{bmatrix}$$

This matrix has only two linearly independent columns, because the third column of $\mathbf{B}$ is linearly dependent on the first two columns. Hence, the rank of this matrix is 2. It can be shown that the number of linearly independent columns of a matrix is equal to the number of independent rows. So, the rank can never be greater than the smaller dimension of the matrix. Consequently, if $\mathbf{A}$ is an $n \times m$ matrix, then

$$\rho(\mathbf{A}) \leq \min(n, m)$$

where *min* denotes the minimum of the two numbers. In matrix theory, the rank of a square matrix pertains to the highest order nonsingular matrix that can be formed from it. Remember from the earlier discussion that a matrix is singular if its determinant is zero. So, the rank pertains to the highest order matrix that you can obtain whose determinant is not zero. For example, consider a 4 x 4 matrix

$$\mathbf{B} = \begin{bmatrix} 1 & 2 & 3 & 4 \\ 0 & 1 & -1 & 0 \\ 1 & 0 & 1 & 2 \\ 1 & 1 & 0 & 2 \end{bmatrix}$$

For this matrix, $det(\mathbf{B}) = 0$, but

$$\left\| \begin{bmatrix} 1 & 2 & 3 \\ 0 & 1 & -1 \\ 1 & 0 & 1 \end{bmatrix} \right\| = -1$$

Hence, the rank of $\mathbf{B}$ is 3. A square matrix has full rank if and only if its determinant is different from zero. Matrix $\mathbf{B}$ is not a full-rank matrix.

■ "Magnitude" (Norms) of Matrices

You must develop a notion of the "magnitude" of vectors and matrices to measure errors and sensitivity in solving a linear system of equations. As an example, these linear systems can be obtained from applications in control systems and computational fluid dynamics. In two dimensions, for example, you cannot compare two vectors $\mathbf{x} = \begin{bmatrix} x_1 & x_2 \end{bmatrix}$ and $\mathbf{y} = \begin{bmatrix} y_1 & y_2 \end{bmatrix}$, because you might have $x_1 > y_1$ but $x_2 < y_2$. A vector norm is a way to assign a scalar quantity to these vectors so that they can be compared with each other. It is similar to the concept of magnitude, modulus, or absolute value for scalar numbers.

There are several ways to compute the norm of a matrix. These include the *2-norm* (Euclidean norm), the *1-norm*, the *Frobenius norm* (F-norm), and the *infinity norm* (inf-norm). Each norm has its own physical interpretation. Consider a unit ball containing the origin. The Euclidean norm of a vector is simply the factor by which the ball must be expanded or shrunk to encompass the given vector exactly. This is shown in Figure 8–1.

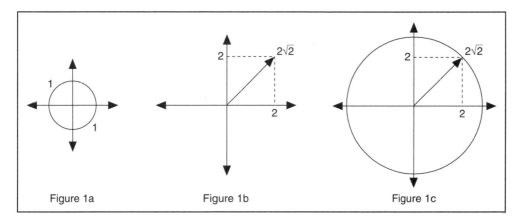

Figure 1a   Figure 1b   Figure 1c

Figure 8–1
Geometrical interpretation of the Euclidean norm

Figure 8.1a shows a unit ball of radius = 1 unit. Figure 8.1b shows a vector of length $\sqrt{2^2 + 2^2} = \sqrt{8} = 2\sqrt{2}$. As shown in Figure 8.1c, the unit ball must be expanded by a factor of $2\sqrt{2}$ before it can exactly encompass the given vector. Hence, the Euclidean norm of the vector is $2\sqrt{2}$.

The norm of a matrix is defined in terms of an underlying vector norm. It is the maximum relative stretching that the matrix does to any vector. With the vector *2-norm*, the unit ball expands by a factor equal to the norm. On the other hand, with the matrix *2-norm*, the unit ball may become an ellipsoidal (ellipse in 3D), with some axes longer than others. The longest axis determines the norm of the matrix.

Some matrix norms are much easier to compute than others. The *1-norm* is obtained by finding the sum of the absolute value of all the elements in each column of the matrix. The largest of these sums is called the 1-norm. In mathematical terms, the 1-norm is simply the maximum absolute column sum of the matrix.

$$\|\mathbf{A}\|_1 = \max_j \sum_{i=0}^{m-1} |a_{i,j}|$$

For example, $\mathbf{A} = \begin{bmatrix} 1 & 3 \\ 2 & 4 \end{bmatrix}$

then $\|\mathbf{A}\|_1 = \max(3, 7) = 7$. The *inf-norm* of a matrix is the maximum absolute row sum of the matrix

$$\|\mathbf{A}\|_\infty = \max_i \sum_{j=0}^{n-1} |a_{i,j}|$$

In this case, you add the magnitudes of all elements in each row of the matrix. The maximum value that you get is called the inf-norm. For the above example matrix, $\|\mathbf{A}\|_\infty = \max(4, 6) = 6$.

The *2-norm* is the most difficult to compute because it is given by the largest singular value of the matrix. Singular values are discussed later on in this chapter, and Activity 8-9 verifies the validity of the above statement.

■ Determining Singularity (Condition Number)

Whereas the norm of the matrix provides a way to measure the magnitude of the matrix, the *condition number* of a matrix is a measure of how close the matrix is to being singular. The condition number of a square nonsingular matrix is defined as

$$\mathrm{cond}(\mathbf{A}) = \|\mathbf{A}\|_p \cdot \|\mathbf{A}^{-1}\|_p$$

where p can be one of the four norm types already discussed. For example, to find the condition number of a matrix A, you can find the 2-norm of A, the 2-norm of the inverse[2] of the matrix A, denoted by A^{-1}, and then multiply them together. As mentioned earlier, the 2-norm is difficult to calculate on paper. You can use the **Matrix Norm** VI from the Analysis library to compute the 2-norm. For example,

$$A = \begin{bmatrix} 1 & 2 \\ 3 & 4 \end{bmatrix}, A^{-1} = \begin{bmatrix} -2 & 1 \\ 1.5 & -0.5 \end{bmatrix}, \|A\|_2 = 5.4650, \|A^{-1}\|_2$$

$$= 2.7325, \text{cond}(A) = 14.9331$$

The condition number can vary between 1 and infinity. A matrix with a large condition number is nearly singular, while a matrix with a condition number close to 1 is far from being singular. The matrix A above is nonsingular. However, consider the matrix

$$B = \begin{bmatrix} 1 & 0.99 \\ 1.99 & 2 \end{bmatrix}$$

The condition number of this matrix is 47168, and hence the matrix is close to being singular. As you might recall, a matrix is singular if its determinant is equal to zero. However, the determinant is not a good indicator for assessing how close a matrix is to being singular. For the matrix B above, the determinant (0.0299) is nonzero; however, the large condition number indicates that the matrix is close to being singular. Remember that the condition number of a matrix is always greater than or equal to one, the latter being true for identity and permutation matrices.[3] The condition number is a very useful quantity in assessing the accuracy of solutions to linear systems.

In this section, you have become familiar with some basic notation and fundamental matrix concepts such as determinant of a matrix and its rank. The following activity should help you

---

2. The inverse of a square matrix A is a square matrix B such that $AB = I$, where I is the identity matrix. Matrix inverses and their applications are described in more detail in a later section.

3. A *permutation matrix* is an identity matrix with some rows and columns exchanged.

further understand these terms, which will be used frequently throughout the rest of the chapter.

Activity 8–1

Objective: To construct matrices, compute condition number, rank, and determinant and observe how the condition number affects the accuracy of the solution

In this activity, you will complete a VI that constructs four different matrices of size 9×9. You will determine the condition number, rank, and determinant of these matrices. You will then solve a curve fitting problem and observe how the condition number of these matrices affects the accuracy of the final solution.

1. Open the **Construct Matrices** VI from the library `Activities.llb`.

■ Front Panel

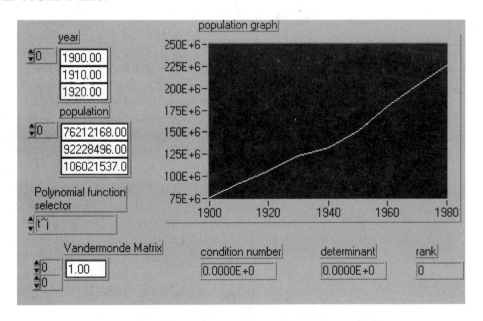

2. The array **population** (nine elements) contains the population data for the United States for the years 1900 to 1980, at intervals of 10 years. This data has

been plotted for you in the graph labeled **population graph**. To find the population in any one of the intermediate years, you need to interpolate between these nine data points. This interpolation can be achieved by first fitting a curve to these data points and then using the curve to obtain the intermediate values. This curve is represented by a unique polynomial of degree eight that interpolates these nine data points, but that polynomial can be represented in many different ways. Consider the following four ways to represent the individual terms in the polynomial:

(*i*) $b_j(t) = t^j$

(*ii*) $b_j(t) = (t - 1900)^j$

(*iii*) $b_j(t) = (t - 1940)^j$

(*iv*) $b_j(t) = (t - 1940)^j / 40^j$

For example, the actual polynomial in case (i) is given by

$$y = a_0 + a_1 t + a_2 t^2 + \ldots$$

where the a_i are the curve parameters to be determined.

To determine these parameters, it is necessary to solve a linear system of equations $\mathbf{V}\mathbf{a} = \mathbf{y}$, where $\mathbf{V}$ is a matrix, $\mathbf{a} = [a_0, a_1, \ldots]$, and $\mathbf{y}$ is the population vector.

For each of these four representations, you can generate the matrix $\mathbf{V}$, where the (i,j)th element of this matrix is given by

$$v_{i,j} = b_j(t[i])$$

Such a matrix $\mathbf{V}$ is referred to as the *Vandermonde matrix*. For example, if you are using the polynomial representation (i) above, the Vandermonde matrix will look like

$$
V = \begin{bmatrix}
1 & t_0 & t_0^2 & \cdots & t_0^{n-1} \\
1 & t_1 & t_1^2 & \cdots & t_1^{n-1} \\
\cdots & \cdots & \cdots & \cdots & \cdots \\
1 & t_{n-1} & t_{n-1}^2 & \cdots & t_{n-1}^{n-1}
\end{bmatrix}
$$

where t_i ($0 \leq i \leq n$-1) is the ith element of the vector "year."

■ Block Diagram

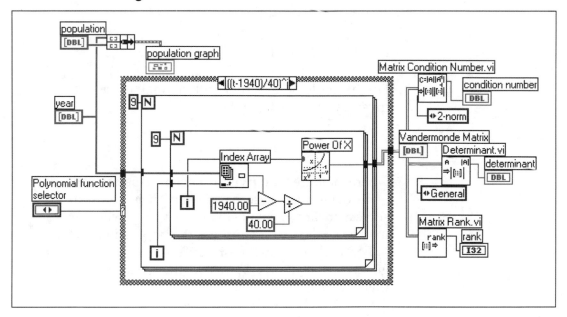

3. Open the block diagram for this VI and complete it as shown above, and discussed in the following steps.

Matrix Condition Number VI (**Functions » Analysis » Linear Algebra » Advanced Linear Algebra** subpalette). In this activity, this VI computes the matrix condition number.

 Determinant VI (**Functions » Analysis » Linear Algebra** subpalette). In this activity, this VI computes the matrix determinant.

 Matrix Rank VI (**Functions » Analysis » Linear Algebra » Advanced Linear Algebra** subpalette). In this activity, this VI computes the matrix rank.

 Bundle function (**Functions » Cluster** subpalette) In this activity, this function assembles the population array and the year array to plot the population array against the year array.

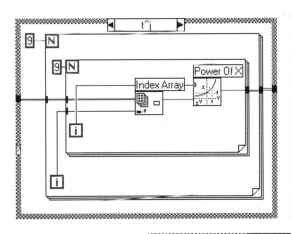
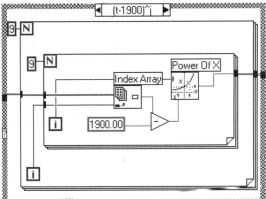
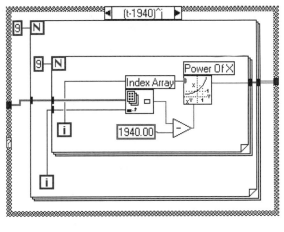

4. The subdiagrams in Case t^j, $(t-1900)^j$, $(t-1940)^j$, and $\left(\dfrac{t-1940}{40}\right)^j$ construct the matrices corresponding to the polynomial functions *(i)*, *(ii)*, *(iii)*, and *(iv)*, respectively. The subdiagrams in the first three cases, as shown above, are already built for you. You need only complete the subdiagram in Case $\left(\dfrac{t-1940}{40}\right)^j$.

5. The **Matrix Condition Number** VI has an input called **norm type**. Earlier in this chapter, you looked at different types of norms. This VI can compute the condition number using four different norm types. In this activity, **norm type** has been set to 2-norm. In case of a singular matrix, the **condition number** output may be set to NaN.

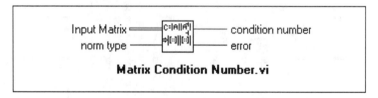

Matrix Condition Number.vi

6. The **Determinant** VI has an input called **matrix type**. In this activity, **matrix type** has been set to General.

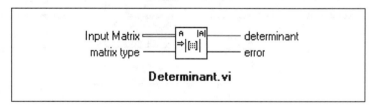

Determinant.vi

7. The **Matrix Rank** VI has an input called **tolerance**. Leave this terminal unconnected, using the default value for this activity.

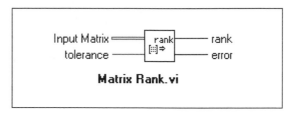

Matrix Rank.vi

8. Return to the front panel and run the VI. Choose different polynomial functions using the **polynomial function selector** control.

9. Look at the condition numbers of the four matrices. Which one is very close to being singular?

10. Save and close this VI.

11. You will now use the matrices (computed in the **Construct Matrices** VI) to calculate the population for any year between 1900 and 1980. To do so, open the **Compute Population** VI from the library `Activities.llb`. This VI computes a polynomial to interpolate the data values to the population data. It then computes the population for a specified year.

12. Set the **choose year** control to 1950. Using each of the four polynomial functions, run the VI to compute the population for this year. The red dot on the **population graph** shows the population value for the year chosen. Which of these values is closest to the true value of 151,325,798, according to the 1950 census?

13. Save the VI and close it.

■ End of Activity 8-1

Activity 8-2

Objective: To study special matrices

In this activity, you will learn to use the **Create Special Matrix** VI in the Analysis Library. Examine the different types of special matrices that this VI creates. Note that this VI also generates the Vandermonde matrix used in the previous activity.

■ Front Panel

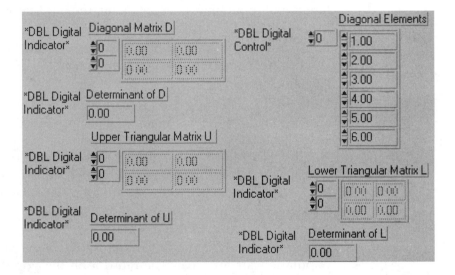

1. Build the front panel as shown above. You can resize the two-dimensional matrices **D, U,** and **L** to see all the elements in the matrix.

■ Block Diagram

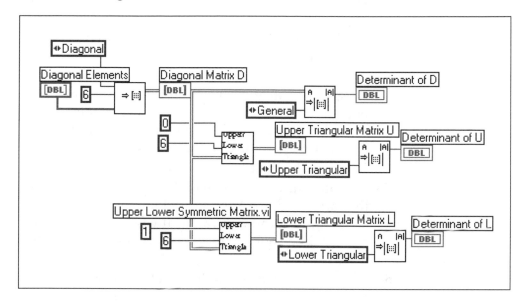

2. Build the block diagram as shown above. This block diagram is partially built for you. You will complete the remaining diagram as explained in steps 3, 4, and 5.

 Create Special Matrix VI (**Functions » Analysis » Linear Algebra » Advanced Linear Algebra** subpalette). In this activity, this function creates special matrices.

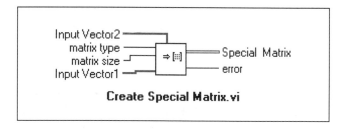

Create Special Matrix.vi

3. Construct **Diagonal Matrix D** using the **Create Special Matrix** VI. The **matrix size** input determines the

dimension size of the output **Special Matrix**. Choose **matrix size** = 6.

4. The **matrix type** input determines the type of matrix that is generated at the output **Special Matrix**. Choose **matrix type** = Diagonal.

5. Enter the diagonal elements 1, 2, 3, 4, 5, and 6 in the **Diagonal Elements** control on the front panel. Connect this control to the **Input Vector1** terminal. The **Input Vector1** terminal is the input to construct a special matrix depending on the **matrix type** chosen.

6. Return to the front panel and run the VI.

7. Compute the determinant of this matrix as you did in the earlier activity. Do you find anything interesting about this determinant value? If yes, what is it? If no, take a close look at the diagonal elements of the matrix **D**.

8. Select the **Upper Lower Symmetric Matrix** VI from the library `Activities.llb`. This subVI can be found below the top level VIs. Set **Select Matrix Type** = 0 (upper triangular matrix). Set **matrix Size** = 6. Wire the matrix **D** to the **input matrix** terminal. The **output matrix U** is an upper triangular matrix.

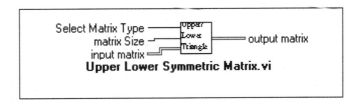

9. Compute the determinant of this matrix. Do you find anything interesting about this determinant value? If yes, what is it?

10. Once again, select the **Upper Lower Symmetric Matrix** VI from the library `Activities.llb`. This time, choose **Select Matrix Type** = 1 (lower triangular matrix). Set **matrix Size** equal to 6. Wire the matrix **D** to the **input matrix** terminal. The **output matrix L** is a lower triangular matrix. Repeat step 9.

11. Save the VI as **Special Matrix.vi** in the library `Activities.llb` and close it.

The determinant for all the three matrices is equal to the product of the diagonal elements of the matrices.

■ **End of Activity 8-2**

Basic Matrix Operations and Eigenvalue-Eigenvector Problems

In this section, we will discuss some very basic matrix operations. Two matrices, **A** and **B**, are said to be equal if they have the same number of rows and columns and their corresponding elements are all equal. Multiplication of a matrix **A** by a scalar α is equal to multiplication of all its elements by the scalar. That is,

$$\mathbf{C} = \alpha\mathbf{A} \Rightarrow c_{i,j} = \alpha a_{i,j}$$

For example,

$$2\begin{bmatrix} 1 & 2 \\ 3 & 4 \end{bmatrix} = \begin{bmatrix} 2 & 4 \\ 6 & 8 \end{bmatrix}$$

Two (or more) matrices can be added or subtracted if and only if they have the same number of rows and columns. If both matrices **A** and **B** have m rows and n columns, then their sum **C** is an m-by-n matrix defined as $\mathbf{C} = \mathbf{A} \pm \mathbf{B}$, where $c_{i,j} = a_{i,j} \pm b_{i,j}$. For example,

$$\begin{bmatrix} 1 & 2 \\ 3 & 4 \end{bmatrix} + \begin{bmatrix} 2 & 4 \\ 5 & 1 \end{bmatrix} = \begin{bmatrix} 3 & 6 \\ 8 & 5 \end{bmatrix}$$

For multiplication of two matrices, the number of columns of the first matrix must be equal to the number of rows of the second matrix. If matrix **A** has m rows and n columns and matrix **B** has n rows and p columns, then their product **C** is an m-by-p matrix defined as $\mathbf{C} = \mathbf{AB}$, where

$$c_{i,j} = \sum_{k=0}^{n-1} a_{i,k} b_{k,j}$$

for i = 0, 1, ... m-1
and j = 0, 1, ... p-1

For example,

$$\begin{bmatrix} 1 & 2 \\ 3 & 4 \end{bmatrix} \times \begin{bmatrix} 2 & 4 \\ 5 & 1 \end{bmatrix} = \begin{bmatrix} 12 & 6 \\ 26 & 16 \end{bmatrix}$$

So, you multiply the elements of the first row of **A** by the corresponding elements of the first column of **B** and add all the results to get the element in the first row and first column of **C**. Similarly, to calculate the element in the ith row and the jth column of **C**, multiply the elements in the ith row of **A** by the corresponding elements in the jth column of **B**, and then add them all. This is shown pictorially in Figure 8–2.

Matrix multiplication, in general, is not commutative. That is, **AB** ≠ **BA**. Also, remember that multiplication of a matrix by an identity matrix results in the original matrix.

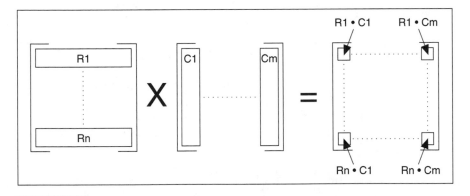

Figure 8–2
Pictorial representation of matrix–matrix multiplication

■ Dot Product and Outer Product

If **x** represents a vector and **y** represents another vector, then the *dot product* of these two vectors is obtained by multiplying the corresponding elements of each vector and adding the results. This is denoted by

$$\mathbf{x} \bullet \mathbf{y} = \sum_{i=0}^{n-1} x_i y_i$$

where n is the number of elements in **x** and **y**. Note that both vectors must have the same number of elements. The dot product is a scalar quantity and has many practical applications.

For example, consider the vectors $\mathbf{a} = 2i + 4j$ and $\mathbf{b} = 2i + j$ in a two-dimensional rectangular coordinate system as shown in Figure 8–3.

The dot product of these two vectors is given by

$$\mathbf{d} = \begin{bmatrix} 2 \\ 4 \end{bmatrix} \bullet \begin{bmatrix} 2 \\ 1 \end{bmatrix} = (2 \times 2) + (4 \times 1) = 8$$

The angle α between these two vectors is given by

$$\alpha = \text{inv } \cos\left(\frac{a \bullet b}{|a||b|}\right) = \text{inv } \cos\left(\frac{8}{10}\right) = 36.86^\circ$$

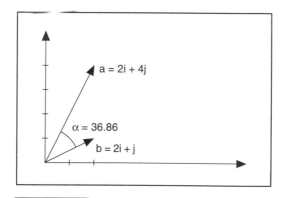

Figure 8–3
Two vectors in a rectangular coordinate system

where $|\mathbf{a}|$ denotes the magnitude of $\mathbf{a}$.

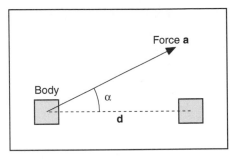

Figure 8–4   Force acting on a body causes displacement

As a second application, consider a body on which a constant force $\mathbf{a}$ acts. The work W done by $\mathbf{a}$ in displacing the body is defined as the product of $|\mathbf{d}|$ and the component of $\mathbf{a}$ in the direction of displacement $\mathbf{d}$. That is,

$$W = |\mathbf{a}||\mathbf{d}|\cos\alpha = \mathbf{a} \bullet \mathbf{d}$$

On the other hand, the *outer product* of two vectors is a matrix. The (i,j)th element of this matrix is obtained using the formula

$$z_{i,j} = x_i \times y_j$$

For example,

$$\begin{bmatrix} 1 \\ 2 \end{bmatrix} \times \begin{bmatrix} 3 \\ 4 \end{bmatrix} = \begin{bmatrix} 3 & 4 \\ 6 & 8 \end{bmatrix}$$

■ Eigenvalues and Eigenvectors

To understand eigenvalues and eigenvectors, let us start with the classical definition. Given an $n \times n$ matrix $\mathbf{A}$, the problem is to find a scalar λ and a nonzero vector $\mathbf{x}$ such that

$$\mathbf{A}\mathbf{x} = \lambda\mathbf{x}$$

Such a scalar λ is called an *eigenvalue*, and $\mathbf{x}$ is the corresponding *eigenvector*.

Calculating the eigenvalues and eigenvectors are fundamental principles of linear algebra. Completely understanding these principles allows you to solve many problems, such as systems of differential equations. Consider an eigenvector **x** of a matrix **A** as a nonzero vector that does not rotate when **x** is multiplied by **A** (except perhaps to point in precisely the opposite direction). **x** may change length or reverse its direction, but it will not turn sideways.

Consider the following example. One of the eigenvectors of the matrix **A**, where $\mathbf{A} = \begin{bmatrix} 2 & 3 \\ 3 & 5 \end{bmatrix}$, is $\mathbf{x} = \begin{bmatrix} 0.62 \\ 1.00 \end{bmatrix}$. Multiplying the matrix **A** and the vector **x** simply causes the vector **x** to be expanded by a factor of 6.85. Hence, the value 6.85 is one of the eigenvalues of the matrix **A**. For any constant α, the vector $\alpha\mathbf{x}$ is also an eigenvector with eigenvalue λ, because

$$\mathbf{A}(\alpha\mathbf{x}) = \alpha\mathbf{A}\mathbf{x} = \lambda\alpha\mathbf{x}$$

In other words, an eigenvector of a matrix determines a direction in which the matrix expands or shrinks any vector lying in that direction by a scalar multiple, and the expansion or contraction factor is given by the corresponding eigenvalue.

The following are some important properties of eigenvalues and eigenvectors:

- The eigenvalues of a matrix are not necessarily all distinct. In other words, a matrix can have multiple eigenvalues.

- All the eigenvalues of a real matrix need not be real. However, complex eigenvalues of a real matrix must occur in complex conjugate pairs.

- The eigenvalues of a diagonal matrix are its diagonal entries, and the eigenvectors are the corresponding columns of an identity matrix of the same dimension.

- A real symmetric matrix always has real eigenvalues and eigenvectors.

- As discussed earlier, eigenvectors can be scaled arbitrarily.

There are many practical applications in the field of science and engineering for an eigenvalue problem. For example, the stability of a structure and its natural modes and frequencies of vibration are determined by the eigenvalues and eigenvectors of an appropriate matrix. Eigenvalues are also very useful in analyzing numerical methods, such as convergence analysis of iterative methods for solving systems of algebraic equations, and the stability analysis of methods for solving systems of differential equations.

The LabVIEW/BridgeVIEW **EigenValues and Vectors** VI is shown in Figure 8.5. The **Input Matrix** is an n × n real square matrix. **Matrix type** determines the type of the input matrix. **Matrix type** could be 0, indicating a *general matrix*[4], or 1, indicating a *symmetric matrix*. A symmetric matrix always has real eigenvalues and eigenvectors.

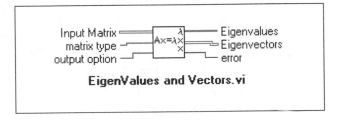

Figure 8–5
EigenValues and Vectors VI with Input–Output terminals

Output option determines what needs to be computed. Output option = 0 indicates that only the eigenvalues need to be computed. Output option = 1 indicates that both the eigenvalues and the eigenvectors should be computed. It is computationally very expensive to compute both the eigenvalues and the eigenvectors. So, it is important that you use the **output option** con-

---

4. General matrix: A matrix with no special property such as symmetry or triangular structure.

trol in the **EigenValues and Vectors** VI very carefully. Depending on your particular application, you might just want to compute the eigenvalues or both the eigenvalues and the eigenvectors. Also, a symmetric matrix needs less computation than an unsymmetric matrix. So, choose the **matrix type** control carefully.

In this section, you learned about some basic matrix operations and the eigenvalues-eigenvectors problem. The next activity introduces some VIs in the Analysis Library that perform these operations.

Activity 8-3

Objective: To learn basic matrix operations

You will build a VI that will help you further understand the basic matrix operations discussed in the first part of the previous section. You will also learn some very interesting matrix properties.

■ Front Panel

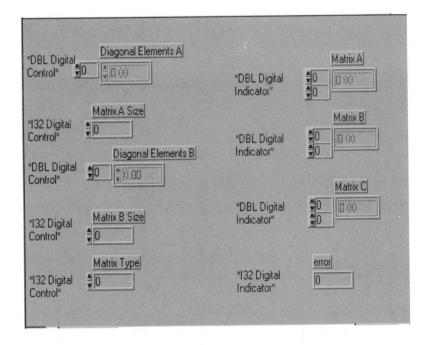

1. Build the front panel as shown above. You can resize the two-dimensional matrices **A, B,** and **C** to see all the elements in the matrix.

2. In this activity, you will experiment with three different types of matrices, namely upper triangular, lower triangular, and symmetric matrix. You will use the

Upper Lower Symmetric Matrix VI that you used in
the previous activity.

■ Block Diagram

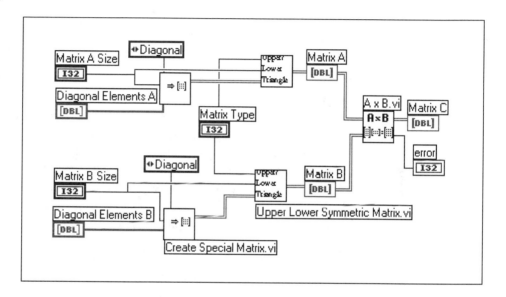

3. Build the block diagram as shown above.

Create Special Matrix VI (Functions » Analysis » Linear Algebra » Advanced Linear
Algebra subpalette). In this activity, this VI
creates special matrices.

AxB VI (Functions » Analysis » Linear Algebra subpalette) In this activity, this VI multiplies two matrices).

4. Create a diagonal matrix using the Create Special
Matrix VI. Select the Upper Lower Symmetric
Matrix VI from the library Activities.llb. This
VI will create the appropriate type of matrix depend-
ing on the Select Matrix Type control.

5. Select the AxB VI from the Analysis » Linear Algebra subpalette. You will use this VI to multiply

matrix **A** and matrix **B** and the result of this multiplication is stored in matrix **C**.

6. Return to the front panel. Choose **Matrix A Size** = 6 and **Matrix B Size** = 6. Set the **Diagonal Elements A** and **Diagonal Elements B** controls for both the matrices to some value (similar to what you did in the previous activity). The size of these arrays is equal to the matrix size.

7. Choose **Select Matrix Type** = 0 (Upper Triangular Matrix). Run the VI. Observe the structure of the upper triangular matrices **A** and **B**, whose upper nondiagonal elements are randomly chosen by the **Upper Lower Symmetric Matrix** VI. **Matrix C** is the product of these two matrices. Do you find anything interesting about the structure of this matrix?

8. Now choose **Select Matrix Type** = 1 (Lower Triangular Matrix). Run the VI. Observe the structure of the two lower triangular matrices **A** and **B**, whose lower nondiagonal elements are randomly chosen by the **Upper Lower Symmetric Matrix** VI. **Matrix C** is the product of these two matrices. Do you find anything interesting about the structure of this matrix?

9. Now choose **Select Matrix Type** = 2 (Symmetric Matrix). Run the VI. Observe the structure of the two symmetric matrices **A** and **B**.
Matrix **C** is the product of these two matrices. Do you find anything interesting about the structure of this matrix?

10. Change the **Matrix A Size** to 5. Keep the **Matrix B Size** at 6. Run the VI. Why did you get an error[5]?

11. Save the VI as **Matrix Multiplication.vi** in the library `Activities.llb` and close it.

■ End of Activity 8-3

---

5. See Appendix B for a list of error codes.

Activity 8-4

Objective: To study positive definite matrices

In many applications, it is advantageous to determine if your matrix is *positive definite*. This is because if your matrix is indeed positive definite, you can save a significant amount of computation time when using VIs to compute determinants, solve linear systems of equations, or compute the inverse of a matrix. All these VIs let you choose the matrix type, and properly identifying the matrix can significantly improve performance.

In this activity, you will learn about complex positive definite matrices. A complex matrix is positive definite if and only if it is Hermitian; that is, $\mathbf{A} = \mathbf{A}^{\mathbf{H}}$, and the quadratic form $\mathbf{x}^{\mathbf{H}}\mathbf{A}\mathbf{x} > 0$ for all nonzero vectors $\mathbf{x}$. Note that the quadratic form results in a scalar value.

■ Front Panel

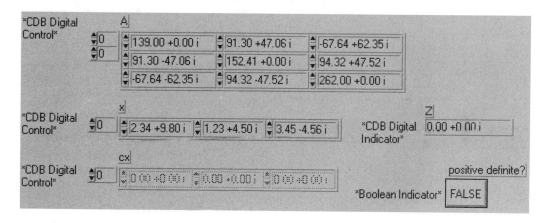

1. Build the matrix **A** and the rest of the front panel as shown in the front panel above.

2. Switch to the block diagram and use the **Test Complex Positive Definite** VI to check if this matrix is positive definite.

   Test Complex Positive Definite VI (Functions » Analysis » Linear Algebra » Complex Linear Algebra » Advanced Complex Linear

Algebra subpalette). Use this VI to check if the input matrix **A** is positive definite.

3. Enter the vector **x** shown below in the control **x**.

$$\mathbf{x} = \begin{bmatrix} 2.34 + 9.8i \\ 1.23 + 4.5i \\ 3.45 - 4.56i \end{bmatrix}$$

4. Enter the complex conjugate **cx** of **x** in the control **cx**. (This part has not been completed in the front panel.)

Complex conjugate of a+jb is a-jb

5. Switch to the block diagram. Compute the complex matrix vector multiplication **y** = **Ax**.

Use the Complex A x Vector VI

 Complex A x Vector VI (**Functions » Analysis » Linear Algebra » Complex Linear Algebra** subpalette). This VI multiplies a complex input matrix and a complex input vector.

6. Compute the complex dot product of the vectors **cx** and **y**. That is, $z = \mathbf{cx} \bullet \mathbf{y}$.

Use the Complex Dot Product VI.

 Complex Dot Product VI (**Functions » Analysis » Linear Algebra » Complex Linear Algebra** subpalette). Use this VI to compute the dot product of two complex vectors.

7. You have now computed the product $z = \mathbf{x}^H \mathbf{Ax}$. What is the value of z? Does this result verify the above definition of complex positive definite matrices?

8. Compare your block diagram with the diagram shown.

■ Block Diagram

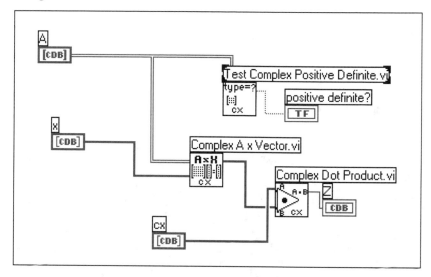

9. Save the VI as **Positive Definite Matrix.vi** in the library `Activities.llb` and close it.

■ End of Activity 8-4

Activity 8-5

Objective: To compute the eigenvalues and eigenvectors of a real matrix

As explained earlier, the eigenvalues-eigenvector problem is widely used in a number of different practical applications. For example, in the design of control systems, the eigenvalues help determine whether the system is stable or unstable. If all the eigenvalues have nonpositive real parts, the system is stable. However, if any of the eigenvalues have a positive real part, it means that the system is unstable. If the system is unstable, you can design a feedback system to obtain the desired eigenvalues and ensure stability of the overall system. In Chapter 10 (Control Systems), you will perform an activity that describes such an application.

In this activity, you will use the **EigenValues and Vectors** VI to compute all the eigenvalues and eigenvectors of a real matrix. You will also learn an alternative definition of eigenvalues and numerically verify this definition.

■ Front Panel

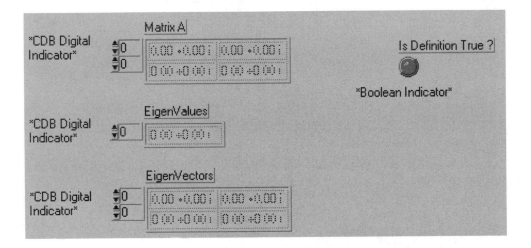

1. Build the front panel as shown above. You can resize the controls to view all the elements in the matrices **A** and **EigenVectors** and the array **EigenValues.** The size of the matrix **A** is 10 × 10.

   Eigenvalues is a one-dimensional complex array of size 10 containing all the computed eigenvalues of the input matrix.

   Eigenvectors is a 10 × 10 complex matrix containing all the computed eigenvectors of the input matrix. The *i*th column of **Eigenvectors** is the eigenvector corresponding to the *i*th component of the **Eigenvalues** vector.

2. Build the block diagram as shown below.

EigenValues and Vectors VI (**Functions » Analysis » Linear Algebra** subpalette). In this activity, you will use this VI to compute the eigenvalues and the eigenvectors of the input matrix.

■ Block Diagram

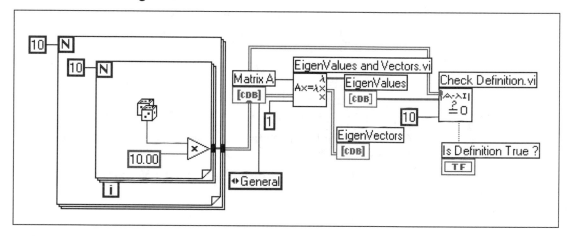

3. Matrix **A** is a real matrix consisting of randomly generated numbers. Use the **Random Number (0–1)** function from the **Functions » Numeric** subpalette to create this matrix. Use the **EigenValues and Vectors** VI to compute both the eigenvalues and the eigenvectors of this matrix. Remember to set the **output option** control to 1. Choose **matrix type** = *General*.

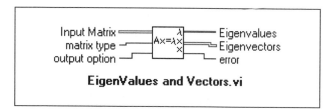

EigenValues and Vectors.vi

4. Earlier in this section, you looked at the classical definition of eigenvalues and eigenvectors. A different and widely used definition of eigenvalues is as fol-

lows. The eigenvalues of **A** are the values λ such that $\det(\mathbf{A} - \mathbf{I}\lambda) = 0$, where *det* stands for the determinant of the matrix. In this activity, you will numerically verify the validity of this definition. Select the **Check Definition** VI from the library `Activities.llb`. The Round LED on the front panel will glow green if the definition is true and turn to red if the definition fails.

5. Return to the front panel and run the VI several times. You will notice that the eigenvalues can be real or complex numbers although the matrix is purely real. The same holds true for the eigenvectors. Furthermore, you will notice that complex eigenvalues of a real matrix always occur in complex conjugate pairs (that is, if $\alpha + j\beta$ is an eigenvalue of a real matrix, so is $\alpha - j\beta$).

6. Did the definition stated above in step 4 always hold true?

7. Save the VI as **My EigenValues and Vectors.vi** in the library `Activities.llb` and close it.

■ **End of Activity 8-5**

Matrix Inverse and Solving Systems of Linear Equations

The *inverse*, denoted by $\mathbf{A}^{-1}$, of a square matrix **A** is a square matrix such that

$$\mathbf{A}^{-1}\mathbf{A} = \mathbf{A}\mathbf{A}^{-1} = \mathbf{I}$$

where **I** is the identity matrix. The inverse of a matrix exists if and only if the determinant of the matrix is not zero (that is, it is nonsingular). In general, you can find the inverse of only a square matrix. You can, however, compute the *pseudoinverse* of a rectangular matrix, as discussed in the next section.

■ Solutions of Systems of Linear Equations

In matrix-vector notation, a system of linear equations has the form $\mathbf{Ax} = \mathbf{b}$, where $\mathbf{A}$ is a $n \times n$ matrix and $\mathbf{b}$ is a given n-vector. The aim is to determine $\mathbf{x}$, the unknown solution n-vector. There are two important questions to be asked about the existence of such a solution. Does such a solution exist, and if it does is it unique? The answer to both of these questions lies in determining the singularity or nonsingularity of the matrix $\mathbf{A}$.

As discussed earlier, a matrix is said to be singular if it has any one of the following equivalent properties:

- The inverse of the matrix does not exist.
- The determinant of the matrix is zero.
- The rows (or columns) of $\mathbf{A}$ are linearly dependent.
- $\mathbf{Az} = 0$ for some vector $\mathbf{z} \neq 0$.

Otherwise, the matrix is nonsingular. If the matrix is nonsingular, its inverse $\mathbf{A}^{-1}$ exists, and the system $\mathbf{Ax} = \mathbf{b}$ has a unique solution: $\mathbf{x} = \mathbf{A}^{-1}\mathbf{b}$ regardless of the value for $\mathbf{b}$. On the other hand, if the matrix is singular, then the number of solutions is determined by the right-side vector $\mathbf{b}$. If $\mathbf{A}$ is singular and $\mathbf{Ax} = \mathbf{b}$, then $\mathbf{A}(\mathbf{x} + \gamma\mathbf{z}) = \mathbf{b}$ for any scalar γ, where $\mathbf{z}$ is a nonzero vector such that $\mathbf{Az} = 0$. Thus, if a singular system has a solution, then the solution cannot be unique.

It is not a good idea to explicitly compute the inverse of a matrix, because such a computation is prone to numerical inaccuracies. Therefore, it is not a good strategy to solve a linear system of equations by multiplying the inverse of the matrix $\mathbf{A}$ by the known right-side vector. The general strategy to solve such a system of equations is to transform the original system into one whose solution is the same as that of the original system but is easier to compute. One way to do so is to use the Gaussian elimination technique. The three basic steps involved in the Gaussian elimination technique are as follows. First, express the matrix $\mathbf{A}$ as a product $\mathbf{A} = \mathbf{LU}$ where $\mathbf{L}$ is a unit lower triangular matrix and $\mathbf{U}$ is an upper triangular matrix. Such a factorization is known as *LU factorization*. Given this, the linear system $\mathbf{Ax} = \mathbf{b}$ can be expressed as $\mathbf{LUx} = \mathbf{b}$. Such a system can then be solved by first solving the lower triangular system $\mathbf{Ly} = \mathbf{b}$, where $\mathbf{y} = \mathbf{Ux}$, for $\mathbf{y}$ by *forward-substitution*. This is the next step in the Gaussian elimination technique. For example, if

$$\mathbf{L} = \begin{bmatrix} a & 0 \\ b & c \end{bmatrix} \qquad \mathbf{y} = \begin{bmatrix} p \\ q \end{bmatrix} \qquad \mathbf{b} = \begin{bmatrix} r \\ s \end{bmatrix}$$

then $p = (r/a), q = (s - bp)/c$. The first element of $\mathbf{y}$ can easily be determined due to the lower triangular nature of the matrix $\mathbf{L}$. Then you can use this value to compute the remaining elements of the unknown vector sequentially. Hence, the name forward-substitution. The final step involves solving the upper triangular system $\mathbf{Ux} = \mathbf{y}$ by *back-substitution*. For example, if

$$\mathbf{U} = \begin{bmatrix} a & b \\ 0 & c \end{bmatrix} \qquad \mathbf{x} = \begin{bmatrix} m \\ n \end{bmatrix} \qquad \mathbf{y} = \begin{bmatrix} p \\ q \end{bmatrix}$$

then $n = q/c, m = (p - bn)/a$. In this case, this last element of $\mathbf{x}$ can easily be determined and then used to determine the other elements sequentially. Hence, the name back-substitution. So far, this chapter has discussed the case of square matrices. Because a nonsquare matrix is necessarily singular, the system of equations must have either no solution or a nonunique solution. In such a situation, you usually find a unique solution $\mathbf{x}$ that satisfies the linear system in an approximate sense.

The Analysis Library includes VIs for computing the inverse of a matrix, computing LU decomposition of a matrix, and solving a system of linear equations. It is important to identify the input matrix properly, as it helps avoid unnecessary computations, which in turn helps to minimize numerical inaccuracies. The four possible matrix types are general matrices, positive definite matrices[6], and lower and upper triangular matrices. If the input matrix is square, but does not have a full rank (a *rank-deficient matrix*), then the VI finds the *least square* solution $\mathbf{x}$. The least square solution is the one that minimizes the 2-norm of $\mathbf{Ax} - \mathbf{b}$. The same holds true also for nonsquare matrices.

---

6. A real matrix is positive definite if and only if it is symmetric and the quadratic form $x^T A x > 0$ for all nonzero vectors $\mathbf{x}$.

Activity 8-6

Objective: To compute the inverse of a matrix

You will build a VI that computes the inverse of a matrix **A**. Further, you will compute a matrix **B**, which is *similar* to matrix **A**. A matrix **B** is similar to a matrix **A** if there is a nonsingular matrix **T** such that $B = T^{-1}AT$ so that **A** and **B** have the same eigenvalues. Such a transformation, where $B = T^{-1}AT$, is known as a similarity transformation. In control theory, as you will see later in Chapter 10 (Control Systems), similarity transformations are used to convert the normal state-space representation to the modal form representation. Although such a transformation alters the states of a system, it does not alter the input-output model. You will verify the definition of similar matrices in this activity.

■ Front Panel

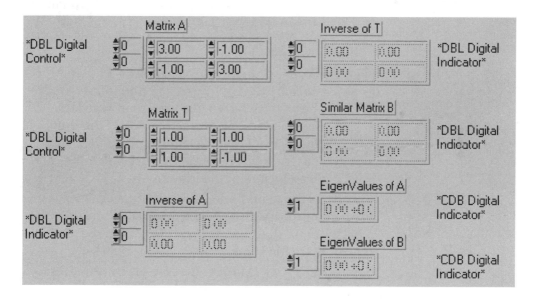

1. Build the front panel as shown above. **Matrix A** is a 2×2 real matrix. **Matrix T** is a 2×2 nonsingular matrix that will be used to construct the **Similar Matrix B**.

■ Block Diagram

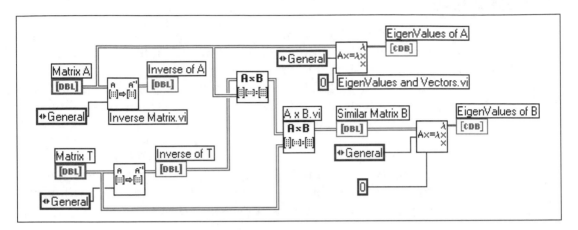

2. Construct the block diagram as shown above.

    Inverse Matrix VI (**Functions » Analysis » Linear Algebra** subpalette). In this activity, this VI computes the inverse of the input matrix.

    AxB VI (**Functions » Analysis » Linear Algebra** subpalette). In this activity, this VI multiplies two two-dimensional input matrices.

    EigenValues and Vectors VI (**Functions » Analysis » Linear Algebra** subpalette). In this activity, this VI computes the eigenvalues and eigenvectors of the input matrix.

3. Return to the front panel, enter the values as shown in the controls **Matrix A** and **Matrix T,** and run the VI. Check if the eigenvalues of **A** and the similar matrix **B** are the same.

4. Save the VI as **Matrix Inverse.vi** in the library `Activities.llb` and close it.

■ End of Activity 8-6

Activity 8-7

Objective: To solve a system of linear equations

Many practical applications require you to solve a system of linear equations. A very important area of application is related to military defense. This includes analysis of electromagnetic scattering and radiation from large targets, performance analysis of large radomes, and design of aerospace vehicles having low radar cross sections (the stealth technology). A second area of application is in the design and modeling of wireless communication systems such as hand-held cellular phones. This list of applications goes on and on, and therefore it is very important for you to properly understand how to use the VIs in the Analysis Library to solve a linear system of equations.

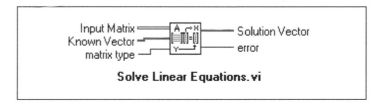

Solve Linear Equations VI

1. Use the **Solve Linear Equations** VI in the **Functions » Analysis » Linear Algebra** subpalette to solve the system of equations $\mathbf{Ax} = \mathbf{b}$ where the **Input Matrix A** and the **Known Vector b** are

$$\mathbf{A} = \begin{bmatrix} 2 & 4 & -2 \\ 4 & 9 & -3 \\ -2 & -1 & 7 \end{bmatrix}, \mathbf{b} = \begin{bmatrix} 2 \\ 8 \\ 10 \end{bmatrix}$$

   Choose **matrix type** equal to general.

2. Use **A x Vector.vi** to multiply the matrix **A** and the vector **x** (output of the above operation) and check if the result is equal to the vector **b** above.

3. Save the VI as **Linear System.vi** in the library `Activities.llb` and close it.

■ End of Activity 8-7

Matrix Factorization

The previous section discussed how a linear system of equations can be transformed into a system whose solution is simpler to compute. The basic idea was to factorize the input matrix into the multiplication of several, simpler matrices. You looked at one such technique, the *LU factorization* technique, in which you factorized the input matrix as a product of upper and lower triangular matrices. Other commonly used factorization methods are *Cholesky*, *QR*, and the *Singular Value Decomposition (SVD)*. You can use these factorization methods to solve many matrix problems, such as solving linear systems of equations, inverting a matrix, and finding the determinant of a matrix.

If the input matrix $\mathbf{A}$ is symmetric and positive definite, then an LU factorization can be computed such that $\mathbf{A} = \mathbf{U}^T\mathbf{U}$, where $\mathbf{U}$ is an upper triangular matrix and $\mathbf{L} = \mathbf{U}^T$. This is called *Cholesky factorization*. This method requires only about half the work and half the storage compared to LU factorization of a general matrix by Gaussian elimination. As you saw earlier in Activity 8-4, it is easy to determine if a matrix is positive definite by using the **Test Positive Definite** VI in the Analysis Library.

A matrix $\mathbf{Q}$ is *orthogonal* if its columns are *orthonormal*, that is, if $Q^TQ = \mathbf{I}$, the identity matrix. The *QR factorization* technique factors a matrix as the product of an orthogonal matrix $\mathbf{Q}$ and an upper triangular matrix $\mathbf{R}$. That is, $\mathbf{A} = \mathbf{QR}$. QR factorization is useful for both square and rectangular matrices. A number of algorithms are possible for $\mathbf{QR}$ factorization, such as the *Householder transformation*, the *Givens transformation*, and the *fast Givens transformation*.

The Singular Value Decomposition (SVD) method decomposes a matrix into the product of three matrices: $\mathbf{A} = \mathbf{USV}^T$. $\mathbf{U}$ and $\mathbf{V}$ are orthogonal matrices. S is a diagonal matrix whose diagonal values are called the *singular values* of A. The singular values of A are the nonnegative square roots of the eigenvalues of $A^T A$, and the columns of U and V, which are called left and right singular vectors, are orthonormal eigenvectors of AA^T and $A^T A$, respectively. SVD is useful for solving analysis problems such as computing the rank, norm, condition number, and pseudoinverse of matrices.

■ Pseudoinverse

The pseudoinverse[7] of a scalar σ is defined as $1/\sigma$ if $\sigma \neq 0$, and zero otherwise. You can now define the pseudoinverse of a diagonal matrix by taking the scalar pseudoinverse of each entry. Then the pseudoinverse of a general real $m \times n$ matrix A, denoted by $A^\dagger$, is given by

$$A^\dagger = VS^\dagger U^T$$

where S is a diagonal matrix. Note that the pseudoinverse exists regardless of whether the matrix is square or rectangular. If A is square and nonsingular, the pseudoinverse is the same as the usual matrix inverse. The Analysis Library includes a VI for computing the pseudoinverse of real and complex matrices.

---

7. In case of scalars, pseudoinverse is the same as the inverse.

Activity 8-8

Objective: To compute Cholesky decomposition and QR decomposition

Activity 8-4 discussed complex positive definite matrices. You verified that the matrix **A**

$$A = \begin{bmatrix} 139 & 91.30 + 47.06i & -67.64 + 62.35i \\ 91.30 - 47.06i & 152.41 & 94.32 + 47.52i \\ -67.64 - 62.35i & 94.32 - 47.52i & 262.00 \end{bmatrix}$$

is a positive definite matrix. In this activity, compute the Cholesky decomposition of this matrix. Also, compute the QR decomposition of a matrix **B**.

■ Front Panel

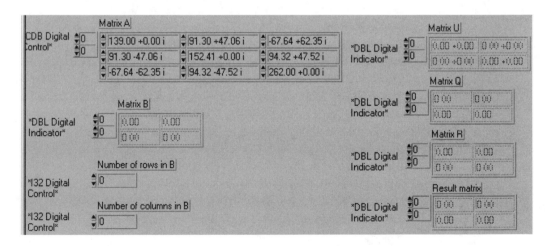

1. Build the front panel as shown above. You can resize the array of controls to see all the elements of the matrix.

2. The matrix **U** is an upper triangular matrix that is the result of the Cholesky Decomposition of the matrix **A**.

3. The matrix **B** is a rectangular matrix. That is, the number of rows m is different from the number of columns n. The result of the QR factorization is an $m \times m$ orthogonal matrix **Q** and an upper triangular matrix **R** of size $m \times n$.

4. In this activity, you will also verify the definition of orthogonal matrices, $Q^TQ = I$. The indicator **Result matrix** contains the product of the transpose of the orthogonal matrix and the orthogonal matrix itself.

■ Block Diagram

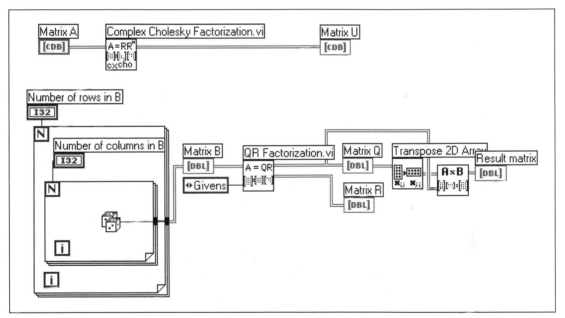

5. Build the block diagram as shown above.

   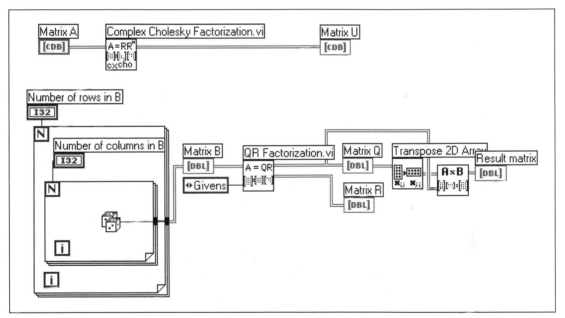 **Complex Cholesky Factorization** VI (**Functions » Analysis » Linear Algebra » Complex Linear Algebra » Advanced Complex Linear Algebra** subpalette). In this activity, this VI computes the Cholesky decomposition of the positive definite input complex matrix.

QR Factorization VI (**Functions » Analysis » Linear Algebra » Advanced Linear Algebra** subpalette). In this activity, this VI computes the QR factorization of the input real matrix.

6. Select the **Complex Cholesky Factorization** VI. The output terminal of this VI is connected to the indicator **Matrix U**.

7. Generate a random matrix **B**. Select the **QR Factorization** VI. Connect the matrix **B** as the input matrix for QR factorization. Connect the **Q** and **R** outputs of this VI to the indicators **Matrix Q** and **Matrix R**, respectively.

8. Compute the transpose of the matrix **Q** and then multiply this transpose with the original matrix **Q**. The result of this operation is connected to the indicator **Result Matrix**.

The ***Transpose 2D Array*** *function is in the* ***Functions » Array*** *subpalette.*

9. Choose a value for the number of rows and the number of columns. Run the VI. Notice the upper triangular structure of the output of the Cholesky decomposition and the **R** output of the QR factorization. You can run the VI a number of times to generate different matrices and check if this definition is true.

10. Save the VI as **QR Factor.vi** in the library `Activities.llb` and close it.

■ End of Activity 8-8

Activity 8-9

Objective: To compute singular value decomposition

In this activity, you will compute the singular value decomposition of the following matrix.

$$A = \begin{bmatrix} 1 & 2 & 3 \\ 4 & 5 & 6 \\ 7 & 8 & 9 \\ 10 & 11 & 12 \end{bmatrix}$$

Use the **SVD Factorization** VI from the **Analysis » Linear Algebra » Advanced Linear Algebra** subpalette. Compute the rank of the matrix **A** using the **Matrix Rank** VI (**Analysis » Linear Algebra » Advanced Linear Algebra** subpalette). Do you see an interesting relation between the rank of this matrix and the number of nonzero singular values? (The singular values are stored in the one-dimensional array **s**.) Now compute the 2-norm of this matrix using the **Matrix Norm** VI (**Analysis » Linear Algebra » Advanced Linear Algebra** subpalette). Do you see an interesting relation between this number and the largest singular value of the matrix?

The rank of a matrix is equal to the number of nonzero singular values, which in this example is equal to 2. Also, as discussed at the beginning of this chapter, the 2-norm of a matrix is equal to its largest singular value.

■ End of Activity 8-9

Wrap It Up!

■ A matrix can be considered as a two-dimensional array of m rows and n columns. Determinant, rank, and condition number are some important attributes of a matrix.

■ The condition number of a matrix affects the accuracy of the final solution.

■ The determinant of a diagonal matrix, an upper triangular matrix, or a lower triangular matrix, is the product of its diagonal elements.

■ Two matrices can be multiplied only if the number of columns in the first matrix is equal to the number of rows in the second matrix.

■ An eigenvector of a matrix is a nonzero vector that does not rotate when the matrix is applied to it. Similar matrices have the same eigenvalues.

■ The existence of a unique solution for a system of equations depends on whether the matrix is singular or nonsingular.

■ Review Questions

1. For the matrix given by $\mathbf{A} = \begin{bmatrix} 0 & 1 & 1 \\ 1 & 2 & 3 \\ 2 & 0 & 1 \end{bmatrix}$, calculate its rank, determinant, 1-norm, and inf-norm.

   Which LabVIEW/BridgeVIEW VI(s) could you use to check your answer?

2. The condition number of a matrix $\mathbf{B}$ is 15.3, and that of a matrix $\mathbf{C}$ is 30,532. Which of these matrices, $\mathbf{B}$ or $\mathbf{C}$, is closer to being singular?

3. Which of the following is true, and which is false?

   a) The eigenvalues of a real matrix are always real.
   b) The rank of an $m \times n$ matrix could at most be equal to the larger of m or n ($m \neq n$).

4. For the two vectors given by $\mathbf{x} = [1,2]$ and $\mathbf{y} = [3,4]$, calculate:

a) Their dot product

b) Their outer product

c) The angle between the two vectors

Which LabVIEW/BridgeVIEW VI(s) could you use to check your answer?

5. Which VI could you use to check if a matrix is positive definite? Why might you want to make such a check?

6. Why is matrix factorization important? Which are the matrix factorization VIs available in LabVIEW/BridgeVIEW?

■ Additional Activities

1. Use the **Create Special Matrix** VI in the Analysis Library to create a Toeplitz matrix and a Vandermonde matrix. Use the following data: **Input Vector 1** = [1.2 2.9 1.8 3.7 4.2]. Use the **Matrix Condition Number** VI, **Determinant** VI, and the **Matrix Norm** VI in the Analysis Library to compute the condition number, determinant, and norm of the Vandermonde matrix and the Toeplitz matrix.

2. Consider a plant design in which the matrix **A** of the state space representation $\dot{x} = Ax + Bu$ is given by

$$A = \begin{bmatrix} 1.8 & 5.4 & 5.0 \\ 2.9 & 2.7 & 3.2 \\ 1.6 & 0.8 & 1.1 \end{bmatrix}$$

(See Chapter 10 on control systems for more details.)

Use the **EigenValues and Vectors** VI to compute the eigenvalues and the eigenvectors of the matrix. Based on the location of the eigenvalues, comment on the stability of the system.

3. Consider the following set of equations:

$$x_1 + 3x_2 + 2x_3 + 5x_4 = 2$$
$$6x_1 + x_2 + 2x_3 + 3x_4 = 5$$
$$4x_1 + 6x_2 + 3x_3 + x_4 = 1$$
$$x_1 + 5x_2 + 2x_3 + 4x_4 = 1$$

This set of equations can be put in matrix form as $\mathbf{Ax} = \mathbf{b}$, where $\mathbf{A}$ is a 4 × 4 matrix, $\mathbf{b}$ is a 4 × 1 right-hand-side vector, and $\mathbf{x}$ is a 4 × 1 vector consisting of the unknowns x_1, x_2, x_3, x_4. Compute the determinant of the matrix $\mathbf{A}$. Is this matrix singular? Is there a unique solution for the system of equations? If yes, use the appropriate VI from the Analysis Library to compute the solution (method 1). As a second approach, use the appropriate VI to compute the inverse of matrix $\mathbf{A}$ and then multiply this inverse by the vector $\mathbf{b}$ to compute the vector $\mathbf{x}$ (method 2). Are both the solutions same? Which of the two methods is the proper technique for solving the linear system of equations and why?

4. Repeat Activity 8-9 with the following matrix:

$$\mathbf{A} = \begin{bmatrix} 1 & 2 & 1 & 5 \\ 6 & 1 & 2 & 3 \\ 2 & 1 & 4 & 1 \\ 1 & 1 & 3 & 8 \\ 1 & 2 & 1 & 1 \end{bmatrix}$$

Joint Time Frequency Analysis of Electromyographic Signals for Investigating Neuromuscular Coordination

Lawrence D. Abraham, Ed.D.
Associate Professor, Kinesiology & Health Education,
Biomedical Engineering, Neuroscience
The University of Texas at Austin,
Austin, TX 78712 USA

Ing-Shiou Hwang, M.S., P.T.
Doctoral Candidate in Biomedical Engineering
The University of Texas at Austin
Austin, TX 78712 USA

The Challenge Investigating changes in neural activation of individual muscles during dynamic human movement offers great potential benefit in basic research and clinical neurology and physical rehabilitation settings. However the most commonly available measure, surface electromyography, represents muscle activation as a complex combination of biosignal amplitude and frequency. The challenge is to determine instantaneous changes in signal frequency.

The Solution Using LabVIEW's JTFA Toolkit we are performing joint time frequency analysis to calculate the instantaneous median frequency of surface EMG signals, which represents an important factor in muscle activation.

Introduction

Human neuromuscular coordination is studied by scientists and clinicians who are seeking to understand, restore, and improve the way we move. A basic premise is that the specific temporal pattern of muscle activation throughout the body, regulated by the nervous system, determines the exact movement produced. Yet it is extremely difficult to accurately and quantitatively assess the mechanisms underlying movement. Whether the application is in rehabilitation medicine, sports, or ergonomics, there are no ideal methods for measuring the activation patterns of individual muscles or of the neural elements which regulate muscle function. The most common tool for approaching this problem has been electromyography (EMG), recording the weak bioelectrical signals generated by activated muscle fibers, to provide an indication of the intensity of muscle activation. However traditional methods of analyzing EMG signals have not been very successful in elucidating underlying mechanisms of coordination during dynamic movement. We describe here the application of a powerful new signal processing technique, using LabVIEW, which has the potential to greatly enhance our ability to use EMG to study patterns of neuromuscular coordination.

Traditional EMG Analysis

First, let's take a quick look at what EMG is. When skeletal muscle contracts, each of the many fibers activated generates a small voltage potential. Greater contraction effort involves recruitment of more fibers as well as a more rapid "firing" rate in active fibers, and the individual potentials sum at any recording site to a signal of varying amplitude and frequency. There are several recording techniques for acquiring EMG signals, including intramuscular needles, intramuscular fine wires, and surface recording electrodes. Usually these electrodes (signal detection surfaces) are either bipolar or multipolar, yielding a differential, biphasic signal which reflects the instantaneous difference in electrical potential between two or more sites. Since the bioelectric potentials generated in the active muscle fibers spread easily throughout most body tissues, proper arrangement of the recording lead surfaces can help to localize the area from which signals are detected. Intramuscular electrodes tend to have the recording sites closer together, yielding a higher frequency signal and detecting signals from a smaller volume of muscle. However surface electrodes are far more commonly used, particularly during analysis of voluntary movement, because of ease of application, less risk of medical complications, and less interference with actual muscular contraction and movement. Surface EMG typically contains lower frequency signals because of spatial filtering of the volume-conducted potentials between the muscle fibers of origin and the electrode surface affixed to the overlying skin. Surface EMG records also tend to include signals from a larger volume of muscle tissue, since the electrodes are larger and farther apart than those used intramuscularly. So surface EMG signals are biphasic signals originating in underlying muscle and representing the intensity of activation by modulations in amplitude and frequency (for a more detailed discussion of surface EMG, see DeLuca, 1997).

To study human neuromuscular coordination we must simultaneously record from several muscles and examine the patterns of activation. The central nervous system has many interconnections among its active elements which provide for unique temporal and spatial patterns of muscle activation for each unique movement. Since it is extremely difficult and/or risky to record the activity of these individual neural elements during voluntary movement, we normally must infer from EMG records how the underlying neural elements are orchestrating the muscular action. Specific patterns of activity which are used repeatedly are commonly referred to as synergies and are thought to reflect patterns of connectivity among underlying neural elements. One way to investigate human coordination is by looking for common aspects of activation among multichannel EMG recordings. Two aspects of EMG signals are fairly easily studied: relative timing and amplitude. We study relative timing by simply using threshold detectors to determine when the signal rises above or returns to some baseline level of background activity. This information reveals which muscles are activated simultaneously, sequentially, or perhaps alternately. We study amplitude (as a measure of intensity of contraction) using the root mean square (RMS) value of the signal (with a window of length around 25ms) with respect to the baseline value. Intensity of activity is also studied by examining the frequency of the signal, but traditional techniques for EMG frequency analysis have not been particularly satisfactory.

Problems With Traditional Analysis of EMG Frequency

During the course of a movement, activation of each muscle increases and decreases as the demand for muscle force varies. Early attempts to study EMG signal frequency using traditional Fourier transform-based power spectral analysis reported average frequency characteristics during an entire contraction. The results clearly showed that as the level of muscle activation increased, measures such as the median frequency also increased, reflecting (we assume) both increased recruitment of motor units and also increased individual unit firing rate. Using such data several investigators have shown a linear relationship between frequency and isometric (constant muscle length) force production (e.g. Woods & Bigland-Ritchie, 1983). Such data have been useful in assessing relationships between activation level and force output during a steady, invariant contraction, but do not provide any information about changes in recruitment during typical dynamic movement. To try to address this latter question, some investigators have tried short time Fourier transform (STFT) techniques (e.g. Hannaford et al., 1986). Unfortunately this approach often yields poor resolution of the spectrum and biased instantaneous frequency estimation because of a fixed moving window regardless of the signal properties. In addition, it fails to preserve finite support in the time domain, an important requirement for understanding motor synergy (Cohen, 1995). More recently several short-time-based parametric approaches have been tried, such as the auto-regression (AR) model (Kiryu, 1992). However AR analysis is best suited to reflect drastic changes in signal frequency (Devedeux et al., 1994) and requires numerous repeated measures and a lengthy data segment to ensure converged AR parameters (Box et al, 1994).

Our solution to these problems is to employ a relatively new dynamic spectral analysis technique known as joint time frequency analysis (JTFA) which is being used to analyze non-stationary signals, including other biosignals such as the electroencephalogram (EEG) and heart sounds (Williams et al., 1995; Wood et al., 1995; Sun et al., 1996). In addition to allowing us to examine modest changes in EMG signal spectrum over time, this method also allows us to calculate the time history of the median frequency to represent the changing frequency instantaneously.

Research Methods

We describe here an initial experiment to examine two or more EMG signals for evidence of synergistic control, reflected in similar simultaneous patterns of EMG signal frequency changes over time. The subjects are strapped in a comfortable seat with one leg fixed to an isokinetic dynamometer (Biodex Medical Systems, Inc.) which allows movement against resistance at only one joint (in our case, the ankle). The dynamometer is set to allow a specific joint angular velocity and records the torque generated by the subject trying to move against the velocity-regulating resistance. The EMG signals are collected using preamplified surface electrodes (Motion Control, Iomed Corp., gain = 360, bandwidth = 8Hz – 26Khz, CMRR = 100dB, DC input impedance = 100MΩ). The analog electrode output signals are led through an analog bandpass filter (6Hz – 400Hz) to limit artifactual features in the frequency content, and then sampled at 1Khz using a National Instruments AT-MIO-16E board in a PowerMac 7100/66 computer. The data are stored in raw data files, using a data collection virtual instrument (VI) developed in LabVIEW 4, and then analyzed off-line using a second LabVIEW VI. The data are collected during voluntary maximal isokinetic contraction of one muscle group at the ankle, through the full range of ankle motion, while EMG signals are collected from muscles controlling both the ankle and the knee. We then attempt to determine whether the involuntary activity which usually appears in the knee muscles is modulated in the same way as the activity in the primary muscle used to perform the task.

The JTFA method requires selection of an appropriate kernel function to reflect the fundamental waveform shape. We have chosen a cone-shaped kernel based on typical values for peak width and amplitude of a single motor unit action potential "signature", the fundamental element in EMG signals (Zhao, Atlas & Marks, 1990). We also have chosen an interference suppression parameter for the EMG time-frequency distribution of 1, to minimize distortion of the typically highly variable EMG signal. Furthermore, and importantly, a cone-shaped kernel allows preservation of finite support in the time domain for expression of the EMG time-frequency distribution. From the JTFA results, we then calculate the instantaneous median frequency of the signal, which we can compare statistically with the median frequency of other muscles to determine whether they exhibit similar time-varying patterns. The instantaneous median frequency indicates the recruitment strategy of motor units in an activated muscle (Solomonow et al., 1987). The resolution in the frequency domain of the time-varying spectrum analysis is equal to the Nyquist frequency divided by 2^n (n=8). The resolution in the time domain is 2^n ms (in our case we choose n=4), constrained by the uncertainty principle. The mathematical expression of JTFA with a cone-shaped kernel is expressed as follows:

The general form of JTFA is

$$C(t, w) = \frac{1}{4\pi^2} \iiint s^*\left(u - \frac{1}{2}u\right)s\left(u + \frac{1}{2}\right)\varphi(\theta, \tau)e^{-j\theta t - j\tau\omega + j\theta u}\,du\,d\tau\,d\theta$$

(8.1)

where $\varphi(\theta, \tau)$ is a two dimensional function called the kernel function.

The selection of a cone-shaped distribution refers to Devedeux and Duchene's work (1994) which compares different EMG time frequency distributions.

The cone-shaped function (φ_c) is defined as

$$\varphi_c(t, \tau) = \xi(\tau) \qquad\qquad |\tau| > 2\,|t| \tag{8.2}$$

$$= 0 \qquad\qquad\qquad \text{otherwise}$$

In the kernel function domain, the form is expressed mathematically (Zhao et. al, 1990) as

$$\varphi(\theta, \tau) = \zeta(\tau)\int_{-\tau/2}^{\tau/2} e^{-j\theta t}\,dt = 2\zeta(\tau)\frac{\sin(\theta\tau/2)}{\theta} \tag{8.3}$$

Let

$$\zeta(\tau) = \frac{1}{\tau}e^{-\alpha\tau^2} \tag{8.4}$$

Then,

$$\varphi(\theta, \tau) = \frac{\sin(\theta\tau/2)}{\theta\tau/2}e^{-\alpha\tau^2} \tag{8.5}$$

Design of the Virtual Instrument

The data collection VI is built around several basic sub-VIs from the standard library. The flow diagram (Figure 8–6) shows the path of the surface EMG signals from electrodes to storage on the hard disk. The primary elements of the VI are signal digitization, display, and storage, using the analog input, waveform

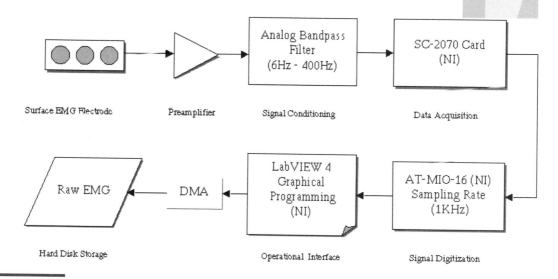

Figure 8–6

Schematic diagram of the data acquisition process from surface EMG electrodes to storage of digitized data

graph, and write to spreadsheet sub-VIs provided by LabVIEW 4. User-adjustable features include number of active channels, sample rate and duration, display gain, a trial-by-trial acceptance option (to screen out unacceptable trials), and ASCII text labels to be stored with each set of raw data indicating subject, trial conditions, and trial number.

The data analysis VI contains three basic elements (Figure 8–7). The raw data are retrieved from storage and any DC bias is removed (detrending). Then the traditional RMS (25ms window) of the signal from each muscle and trial is computed, normalized, and made available for subsequent statistical comparison with the JTFA data. Finally the JTFA is performed, using the National Instruments JTFA Toolkit, and then the median frequency trace is calculated for each muscle and trial. The main features of the data analysis VI wiring diagram are shown in Figure 8–8.

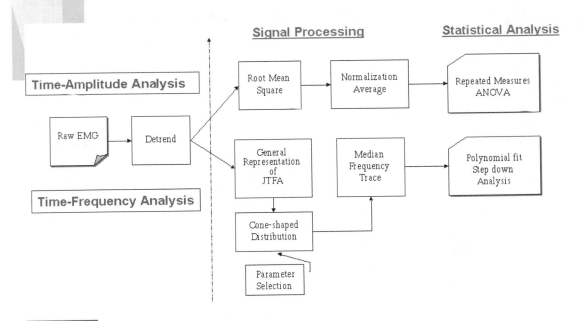

Figure 8–7

Schematic diagram of the data analysis process. The raw data are retrieved from storage, detrended, and then processed using traditional time-amplitude methods (top row) or using JTFA (bottom row).

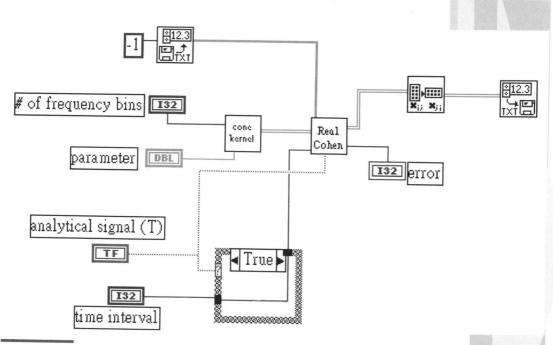

Figure 8–8
Simplified wiring diagram of the data analysis VI. The raw data are retrieved from storage (top left) and passed to the Real Cohen element of the JTFA Toolkit, to be processed according to several user-selected parameters. The "cone kernel" element, also from the JTFA Toolkit, allows specification of a waveform specific to surface EMG parameters. The "number of frequency bins" determines the resolution in the frequency domain, the "time interval" allows selection of temporal resolution of the analysis, and the "analytical signal" option permits additional analytic resolution. The dynamic EMG power spectrum output is stored for subsequent statistical analysis.

Sample Data

Figure 8–9 shows typical (a) raw data and (b) EMG RMS collected from both knee and ankle muscles simultaneously during a slow (30 deg/sec) dorsiflexion effort (pulling the toes up toward the knee) through the full range of ankle motion. Figure 8–10 shows the relationship between raw EMG data, traditional power density frequency analysis, and joint time frequency analysis (JTFA). Figure 8–11 shows the unsmoothed and smoothed median fre-

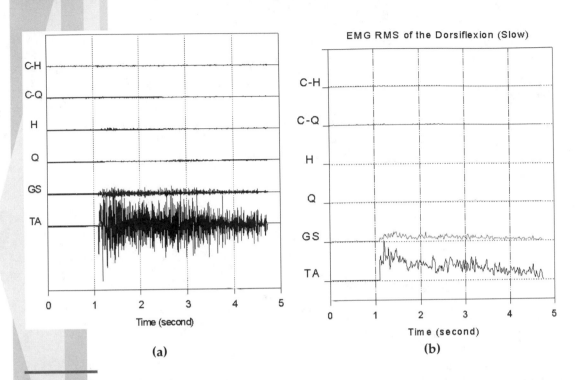

Figure 8–9

(a) Raw surface EMG signals from six muscles in the ankle-knee complex during isokinetic dorsiflexion at 30 degrees/second (TA: tibialis anterior, a shin muscle which acts to pull the toes up; GS: gastrocnemius, a calf muscle which acts to pull the toes down; Q: quadriceps, a thigh muscle which acts to extend the knee; H: hamstring, a thigh muscle which acts to flex the knee; C-Q: contralateral quadriceps, located in the opposite leg; C-H: contralateral hamstring; located in the opposite leg). (b) The EMG RMS of the data shown in (a), typically used to represent extent of muscle activity over time..

quency traces derived from a time-varying spectrum. The smoothed median frequency trace is determined by a best-fit statistical procedure. In this case it demonstrates that the EMG signal increased in frequency during the first two-thirds of the motion, and declined slightly toward the end of the movement. This result can be readily compared to that of another muscle or to the torque history to determine the extent of correlation.

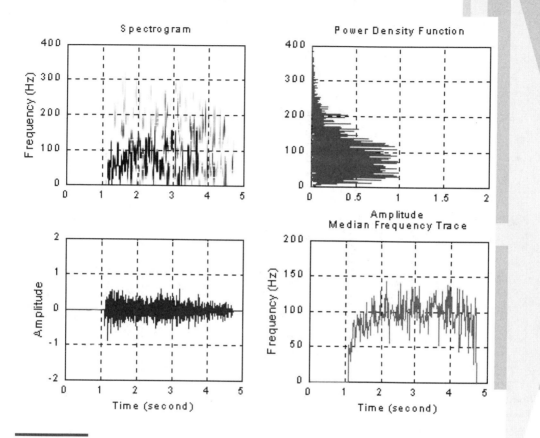

Figure 8–10

Graphical representation of the relationships between raw surface EMG, a traditional power density function, and the dynamic EMG spectrum. The data are derived from surface EMG recording from the muscle tibiulis anterior (TA) during isokinetic ankle dorsiflexion at 30 degrees/ second. (Lower left: raw surface EMG; upper right: power spectrum; upper left: dynamic EMG spectrum; lower right: median frequency trace, the unbiased estimate of the median power frequency of the dynamic EMG spectrum over time).

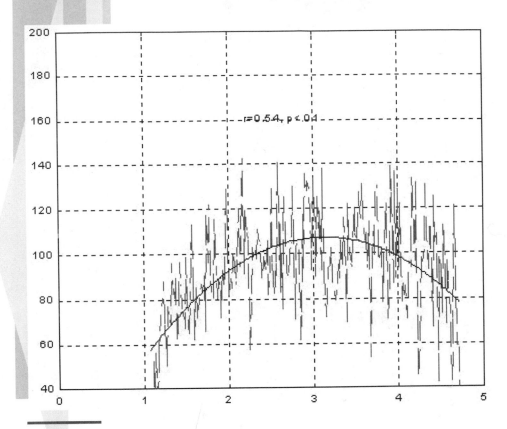

Figure 8–11
Smoothed and unsmoothed median frequency traces of tibialis anterior (TA) during isokinetic dorsiflexion movement at the ankle through the full range of motion. The unsmoothed median frequency trace was best fitted by a smooth second-order polynomial function. The simple correlation coefficient and significance level are shown on the plot.

Summary

The analysis of EMG signals has long offered the promise of a window into the control and coordination of human movement. Yet for the most part researchers have been unable to do much more than to estimate static levels of muscular effort, because only amplitude of the signals could be studied as a time-varying parameter. Using JTFA, we hope to elucidate time-varying changes in EMG signal frequency, revealing much more information about the time history of neural regulation of muscle activity. Since changes in EMG signal frequency are determined largely by alterations in motor unit recruitment and firing rate, this tool should greatly facilitate our investigation of coordination, synergy, and other problems of neuromotor control.

References

Box, G.P., Jenkins, G.J., and Reinsel, G.C. (1994). Time series analysis: forecasting and control. 3d ed. Prentice-Hall, Inc. Englewood Cliffs.

Cohen, L. (1995). Time-frequency analysis. 1st ed. Prentice-Hall, Inc. Englewood Cliffs.

DeLuca, C.J. (1997). The use of surface electromyography in biomechanics. *Journal of Applied Biomechanics*, 13, 135–163.

Devedeux, D., and Duchene, J. (1994). Comparison of various time/frequency distributions (classical and signal dependent) applied to synthetic uterine EMG signals. *Proceedings of the IEEE-SP International Symposium on Time-Frequency and Time-Scale Analysis*, 572–575.

Hannaford, B., and Lehman, S. (1986). Short time Fourier analysis of the electromyogram: fast movements and constant contraction. *IEEE Transactions on Biomedical Engineering*, 33(12), 1173–1181.

Kiryu, T., Saitoh, Y., and Ishioka, K. (1992). Investigation on parametric analysis of dynamic EMG signals by a muscle-structured simulation model. *IEEE Transactions on Biomedical Engineering*, 39(2), 280–288.

Solomonow, M., Baratta, R., Zhou, B.H., et al. (1987). The EMG-force model of electrically stimulated muscles: dependence on control strategy and predominant fiber composition. *IEEE Transactions on Biomedical Engineering*, 34(9), 692–703.

Sun, M., Qian, S., Yan, X., Baumann, S.B., et al. (1996). Localizing functional activity in the brain through time-frequency analysis and synthesis of the EEG. *Proceedings of the IEEE, v 84*, p 1302–1311.

Williams, W.J., Zaveri, H.P., and Sackellares, C. (1995). Time-frequency analysis of electrophysiology signals in epilepsy. *IEEE Engineering in Medicine and Biology, v 14,* p 133—143.

Wood, J.C., and Barry, D.T. (1995). Time-frequency analysis of the first heart sound. *IEEE Engineering in Medicine and Biology, v 14,* p 144–151.

Woods, J.J., and Bigland-Ritchie, B. (1983). Linear and non-linear surface EMG/force relationships in human muscles. *American Journal of Physical Medicine,* 62, 287-299.

Zhao, Y., Atlas, L.E., and Marks, R.J. (1990). The use of cone-shaped kernels for generalized time-frequency representations of nonstationary signals. *IEEE Transaction on Acoustic, Speech, Signal Processing, 38(7),* 1084-1091.

Dr. Lawrence D. Abraham is an associate professor in the Kinesiology and Health Education Department, and the Biomedical Engineering and Neuroscience graduate programs at The University of Texas at Austin and also is a research associate at Healthcare Rehabilitation Center in Austin, TX. His specialization is in biomechanics and motor control. His research focuses on human motor coordination in settings such as rehabilitation, sport, and microgravity.

Ing-Shiou Hwang is a doctoral candidate in Biomedical Engineering at The University of Texas at Austin. He previously earned a bachelor's degree in physical therapy and a master's degree in bioengineering in Taiwan. His doctoral dissertation is titled "Quantitative electromyographic analysis of synergy in neuromuscular control."

■ **Contact Information**

Lawrence Abraham, Ed.D.
Kinesiology & Health Education
The University of Texas at Austin
Austin, TX 78712
(512) 232-2684
l.abraham@mail.utexas.edu

The Science ToolBox ®
An Attempt to Deliver a Turnkey Science Instrumentation Workstation

Scott Myers
DeskTop Laboratories
New York

The Challenge To provide a turnkey desktop science workstation from the authoring software and data acquisition boards.

The Solution Use LabVIEW to provide good analog sensors, a good signal conditioning circuit, and a handy container along with completed software libraries.

Introduction

All scientific thinking involves use of the experimental method, which requires an ever tightening circle of hypothesis, testing the hypothesis, and analysis of the testing. The investigator often becomes bogged down in the testing phase.

Computers entered the schools first through the principal's office, primarily for attendance gathering and statistics. Then computers made it to the library as a replacement for the card files. Eventually a few computers made it to a teacher's desk. Finally, some are gradually becoming available for the students.

All quality science education includes some hands-on lab. These lab experiences are where most students report their greatest interest and retention of the subject matter. But science labs (elementary through graduate school) have not changed significantly over the last twenty years.

The promise (and inspiration) of LabVIEW® is the offer of a flexible science environment. Software instruments can be rearranged much more quickly than independent stand alone hardware instruments.

LabVIEW is facile, and although the data acquisition boards one uses with LabVIEW have progressed considerably, software and a DAQ card are not at all enough. A science teacher cannot begin to deliver 'the science experience' with software and a DAQ card. The educator, engineer of researcher would still need sensors, signal conditioners and well developed software instruments for them.

The following is our contribution to a complete, turnkey (plug&play) science workstation.

The Science ToolBox

The Science ToolBox supplies a DAQ card (your choice of LabPC+, LabNB, DAQpad 1200, DAQCard-1200, etc.), over a dozen high quality analog sensors, and a signal conditioner which is also the container for all supplies. Everything is provided in one box to permit the educator or researcher to set up a small very capable science workstation.

For economy, this device is built around the less expensive Lab cards. The signal conditioning (Figure 8–12) on the science toolbox enhances the Lab cards by providing a) power for the sensors b) independent software controlled instrumentation amplifiers c) convenient color coded connectors. There's a place on the mother board for adding a pair of high isolation amps which permits safe human physiology (EKG, etc.).

Part of the added value is in the choice of analog sensors from various manufacturers. Many sensors are monolithic, i.e. formed on a single chip. Some of these can be quite complex like the 0-5 G accelerometer (available to basic science because the need for cheap accurate accelerometers for automobile airbags). The pressure sensor (0 - 15 psi, absolute) is also monolithic. Some sensors are more basic, like thermocouples, or the silversilver-chloride pH probe. These sensors have delicate signals which need more careful handling than the more refined monolithic devices. Often two of the same kind of sensor are provided. Why not? Some of the sensors themselves are quite inexpensive, and with two sensors differential readings can be made, comparisons can be made.

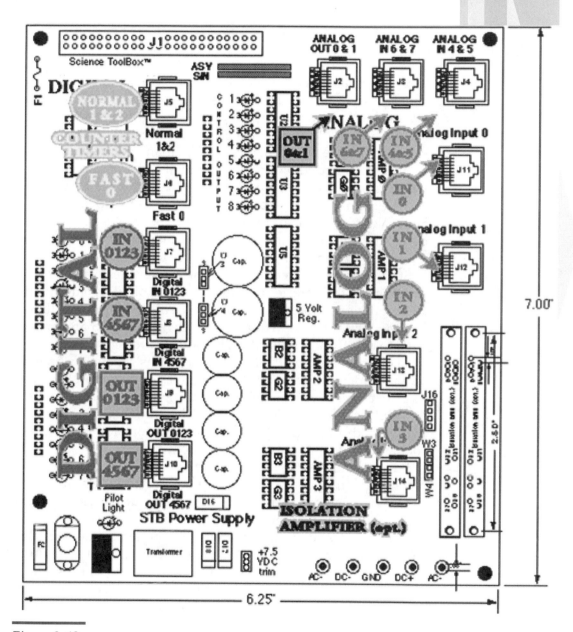

Figure 8–12
Signal conditioning board. Note easy to understand layout—digital on the left and analog on the right.

The Virtual Scientist

The term Science ToolBox (STB) refers to a box of hardware, sensors and software which enables students to explore a wide variety of science and math phenomenon. STB is also called The Virtual Scientist® (TVS). The term Virtual Scientist (Figure 8–13) refers to the use of simulations rather than actual data acquisition. All supplied STB virtual instruments can operate in a simulation mode. They can operate the VI without hardware. This assists them become familiar with the software instrument. Some VIs are not meant to acquire data, but are used to analyze data. Students or researchers can take data and the proper VI to another (non-hands-on) workstation for analysis.

Students and staff can modify the VI libraries for their purposes or use the libraries as they come. The ability to modify the VI libraries adds value to this solution. All VIs shipped with STB include the source code (diagrams). Owners of STB are encouraged to modify and reuse the VIs for their projects.

STB can be used by the teacher in demonstrations, by an individual student or by small study groups. Students can take data in class and analyze the data later using their own copy of the LabVIEW student edition or one at the school library. Contemplation and the preparation of lab reports outside of the classroom is therefore possible with this approach.

Projects can last 10 minutes, or extend over months. Data recording can usually be interrupted and restarted at a later date, which allows STB hardware to be put to use in many projects at the same time.

Most science learning materials are pre-made, canned. But, STB provides a low cost high quality programming environment. Students and teachers aren't even required to program in order to use STB. The supplied virtual instruments are complete and ready to use. But with LabVIEW the instruments can be modified and improved upon.

Most calibration is via software control. There are no buttons or dials on the STB hardware, onboard digital potentiometers are controlled by digital lines via LabVIEW. A clear cover over the electronics allows users to see the electronics and to take note of the LED indicators as they check the state of a digital

The Virtual Scientist®

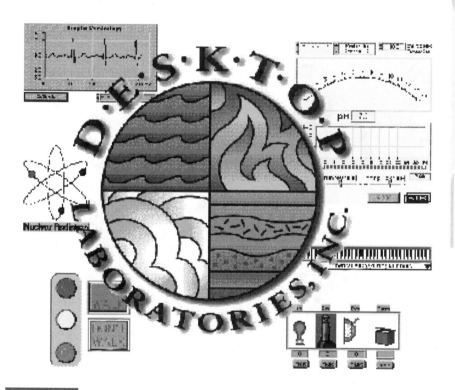

Figure 8–13
Cover art for software libraries

line, or set an amplifiers gain or offset. Multicolored labels provide visual clues to the position of a sensor's hookups.

Some science teaching 'probe ware' uses the computer's serial port. STB is much more robust and uses a special card (supplied) which is installed inside the computer. Since everything is centralized, curriculum can originate within the computer. Often, professional researchers focus intensely on a single area. This allows progress in our knowledge of that area. STB can be used in a dedicated focused research project. But for the young student, an open platform rewards curiosity and contributes to a good general science education.

A simulation mode is part of each software instrument. Simulations provide ideal examples, but simulations are also practical and provide a way to become familiar with an instrument before using it in hands-on projects. A good simulation can provide a significant amount of insight into a physical phenomena. The Geiger Counter.vi is a good example. A radioactive event is discrete, a count. When the Geiger tube detects a radioactive particle a pulse is sent to the software instrument (Figure 8–14). The

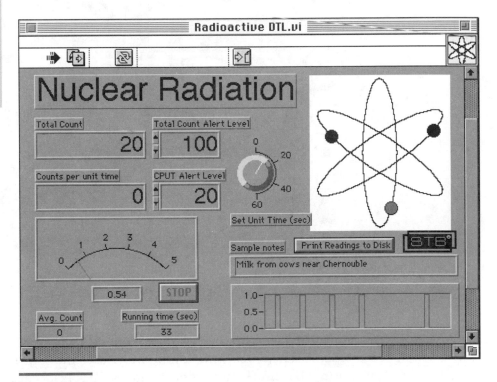

Figure 8–14
Nuclear radiation front panel: A good example of a Gieger counter simulation in LabVIEW

counts per unit time provide a sense of the magnitude of these events, the accumulated effects. In the case of radioactivity, it's certainly better to use simulations rather than handling genuine samples. Figures 8–15, 8–16, and 8–17 show examples of several other features of the STB.

The STB package replaces and surpasses many pieces of traditional lab equipment at a lower cost and higher effectiveness than the individual pieces. This combination of items is a way to provide everything needed to get up and running with software, sensors, signal conditioning.

The basic ingredients of software, and signal conditioning will always be useful. As new sensors become available, software can be written or modified and the sensors can be added.

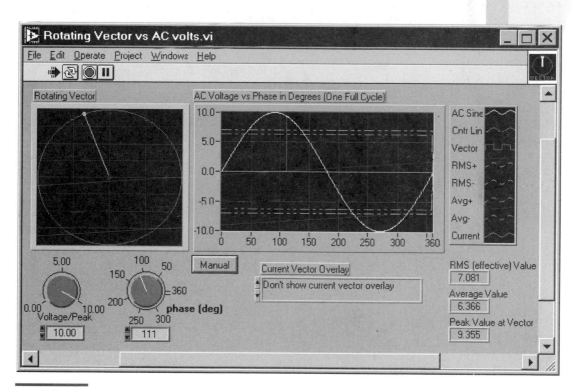

Figure 8–15
Rotating vector and its corresponding AC voltage

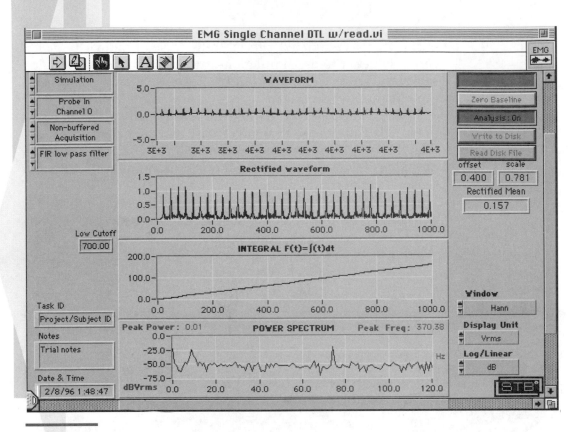

Figure 8–16
Front panel of a VI to process electromyograms (EMG's)

In this way a DAQ card and good software will continue to provide value and find many future uses.

Figure 8–17
Front panel of the Virtual
Instrument Digital Multimeter

■ Contact Information

Scott Myers
DeskTop Laboratories
12 John Street DP12
New York, NY 10038
212 619-3021 fax 0821
email: scottmyers@desktoplabs.com
web site: http://www.desktoplabs.com

Overview

In this chapter, you will learn some of the fundamental concepts in probability and statistics and will become familiar with the most common statistics such as the mean, variance, and histogram. This chapter describes different LabVIEW/BridgeVIEW VIs that compute these quantities and demonstrates how they can be used in different applications. Some such applications are in the fields of economics, weather forecasting, biomedical signal processing, and manufacturing.

GOALS

- Grasp the concepts of *probability* and *statistics* and see their relevance in different applications.
- Learn about the most commonly used concepts in statistics and how to use the statistics VIs.
- Understand the most commonly used concepts in probability and how to use the probability VIs.

KEY TERMS

- probability
- statistics
- mean
- median
- standard deviation
- histogram
- variance
- probability density function
- normal distribution
- inverse normal distribution

Probability and Statistics

9

Probability and Statistics

We live in an information age in which facts and figures form an important part of life. Statements such as "There is a 60 percent chance of thunderstorms," "Joe was ranked among the top five in the class," "Michael Jordan has an average of 30 points per game this season," and so on are common. These statements give a lot of information, but we seldom think about how this information was obtained. Was there a lot of data involved in obtaining this information? If there was, how did someone condense them to single numbers such as *60 percent chance* and *average of 30 points* or terms such as *top five*? The answer to all these questions brings up the very interesting field of statistics.

First, consider how information (data) is generated. Consider the statistics of part of the 1997 basketball season. Michael Jordan of the Chicago Bulls played 51 games, scoring a total of 1568 points. This includes the 45 points he posted, including the game-winning buzzer three-pointer, in a 103–100 victory over the Charlotte Hornets; his 36 points in an 88–84 victory over the Portland Trail Blazers; a season high of 51 points in an 88–87 victory over the New York Knicks; 45 points, seven rebounds, five assists, and three steals in a 102–97 victory over the Cleveland Cavaliers; and his 40 points, six rebounds, and six assists in a 107–104 victory over the Milwaukee Bucks. The point is not about Jordan being a great player, but that a single player can generate lots of data in a single season. The question is, how do you condense all the data so that all the essential information is brought out and is yet easy to remember? This is where the term *statistics* comes into the picture.

To condense all the data, single numbers must make it more intelligible and help draw useful inferences. For example, consider the number of points that Jordan scored in different games. It is difficult to remember how many points he scored in each game. But if you divide the total number of points that Jordan scored (1568) by the number of games he has played (51), you have a single number of 30.7 and can call it points per game *average*.

Suppose you want to rate Jordan's free-throw shooting skills. It might be difficult to do so by looking at his performance in each game. However, you can divide the number of free throws he has scored in all the games by the total number of free throws he was awarded. This shows he has a free throw *percentage* of 84.4 percent. You can obtain this number for all the NBA players and then rank them. Thus, you can condense the information for all the players into single numbers representing free-throw percentage, points per game, and three-point average. Based on this information, you can rank players in different categories. You can further weigh these different numbers and come up with a single number for each player. These single numbers can then help in judging the Most Valuable Player (MVP) for the season. Thus, in a broad sense, the term statistics implies different ways to summarize data to derive useful and important information from them.

The next question is, What is probability? You have looked at ways to summarize lots of data into single numbers. These numbers then help draw conclusions for the present. For example, looking at Jordan's statistics for the 1996 season helped the NBA officials elect him the MVP for that season. It also helped people to infer that he is one of the best players in the game. But can you say anything about the future? Can you measure the degree of accuracy in the inference and use it for making future decisions? The answer lies in the theory of probability. Whereas, in lay terms, one would say that it is *probable* that Jordan will continue to be the best in the years to come, you can use different concepts in the field of probability, as discussed later in this chapter, to make more quantitative statements.

In a completely different scenario, there may be certain experiments whose outcomes cannot be predetermined, but certain outcomes may be more probable. This once again leads to the notion of probability. For example, if you flip an unbiased coin in the air, what is the chance that it will land heads up? The chance or probability is 50 percent. That means, if you repeatedly flip the coin, half the time it will land heads up. Does this mean that 10 tosses will result in exactly 5 heads? Will 100 tosses result in exactly 50 heads? Probably not. But in the long run, the probability will work out to be 0.5.

To summarize, whereas statistics allows you to summarize data and draw conclusions for the present, probability allows you to measure the degree of accuracy in those conclusions and use them for the future.

Statistics

In this section, you will look at different concepts and terms commonly used in statistics and see how to use the analysis VIs in different applications.

■ Mean

Consider a data set $\mathbf{x}$ consisting of n samples x_0, x_1, x_2, x_3, ..., x_{n-1}. The mean value (aka average) is denoted by $\bar{x}$ and is defined by the formula

$$\bar{x} = \frac{1}{n}(x_0 + x_1 + x_2 + x_3 + ... + x_{n-1})$$

In other words, it is the sum of all the sample values divided by the number of samples. As you saw in the Michael Jordan example, the data set consisted of 51 samples. Each sample was equal to the number of points that Jordan scored in each game. The total of all these points was 1568, divided by the number of samples (51) to get a mean or average value of 30.7.

The **Mean** VI in the Analysis Library can be used to calculate the mean of a data set. The input-output connections for this VI are shown in Figure 9–1.

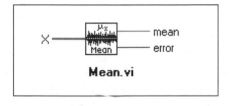

Figure 9–1
Mean VI with Input–Output terminals

■ Median

Let $\mathbf{s} = [s_0, s_1, s_2, \ldots, s_{n-1}]$ represent the *sorted* sequence of the data set $\mathbf{x}$. The sequence can be sorted either in ascending order or in descending order. The median of the sequence is denoted by x_{median} and is obtained by the formula

$$x_{median} = \begin{cases} s_i & n \text{ is odd} \\ 0.5(s_{k-1} + s_k) & n \text{ is even} \end{cases}$$

where $i = (n-1)/2$ and $k = n/2$.

In words, the median of a data sequence is the *midpoint* value in the sorted version of that sequence. For example, consider the sequence [5, 4, 3, 2, 1] consisting of five (odd number) samples. This sequence is already sorted in descending order. In this case, the median is the midpoint value, 3. Consider a different sequence [1, 2, 3, 4] consisting of four (even number) samples. This sequence is already sorted in ascending order. In this case,

there are two midpoint values, 2 and 3. The median is equal to
$0.5 \times (2 + 3) = 2.5$. If a student X scored 4.5 points on a test and
another student Y scored 1 point on the same test, the median is
a very useful quantity for making qualitative statements such as
"X lies in the top half of the class" or "Y lies in the bottom half of
the class."

The **Median** VI in the Analysis Library can be used to calculate the median of a data set. The input-output connections for
this VI are shown in Figure 9–2.

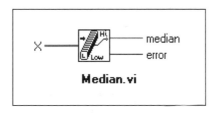

Figure 9–2
Median VI with Input–Output terminals

Sample Variance

The sample variance of the data set **x** consisting of n samples is
denoted by s^2 and is defined by the formula

$$s^2 = \frac{1}{n-1}[(x_1 - \bar{x})^2 + (x_2 - \bar{x})^2 + \dots + (x_n - \bar{x})^2]$$

where $\bar{x}$ denotes the mean of the data set. Hence, the sample
variance is equal to the sum of the squares of the deviations of
the sample values from the mean divided by $n - 1$.

*The above formula does not apply for n = 1. However, sample
variance is undefined if there is only one sample in the data set.*

The **Sample Variance** VI in the Analysis Library can be used to
calculate the sample variance of a data set. The input-output
connections for this VI are shown in Figure 9–3.

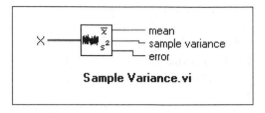

Figure 9–3
Sample Variance VI with Input–Output terminals

In other words, the sample variance measures the spread or dispersion of the sample values. If the data set consists of the scores of a player from different games, the sample variance can be used as a measure of the consistency of the player. It is always positive, except when all the sample values are equal to each other and in turn equal to the mean.

There is one more type of variance called population variance. The formula to compute population variance is similar to the one to compute sample variance, except that the $(n - 1)$ in the denominator is replaced by n.

The **Variance** VI in the Analysis Library can be used to calculate the population variance of a data set. The input-output connections for this VI are shown in Figure 9–4.

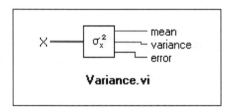

Figure 9–4
Variance VI with Input–Output terminals

The **Sample Variance** VI computes sample variance, whereas the **Variance** VI computes the population variance. Whereas statisticians and mathematicians prefer to use the latter, engineers prefer to use the former. It really does not matter for large values of n, say $n \geq 30$.

Use the proper type of VI suited for your application.

■ Standard Deviation

The positive square root of the sample variance s^2 is denoted by s and is called the standard deviation of the sample.

The input-output connections for the **Standard Deviation** VI are shown in Figure 9–5.

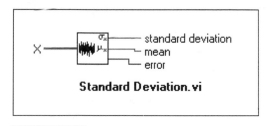

Figure 9–5
Standard Deviation VI with Input–Output terminals

■ Mode

The mode of a sample is a sample value that occurs most frequently in the sample. For example, if the input sequence **x** is

$$\mathbf{x} = [0, 1, 3, 3, 4, 4, 4, 5, 5, 7]$$

then the mode of **x** is 4, because that is the value that most often occurs in **x**.

The input-output connections for the **Mode** VI are shown in Figure 9–6.

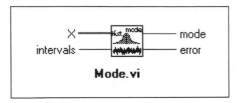

Figure 9–6
Mode VI with Input–Output terminals

■ Moment about Mean

If **x** represents the input sequence with n number of elements in it, and $\bar{x}$ is the mean of this sequence, then the mth-order moment can be calculated using the formula

$$\sigma_x^m = \frac{1}{n} \sum_{i=0}^{n-1} (x_i - \bar{x})^m$$

In other words, the moment about mean is a measure of the deviation of the elements in the sequence from the mean. Note that for $m = 2$, the moment about mean is equal to the population variance.

The input-output connections for the **Moment About Mean** VI are shown in Figure 9–7.

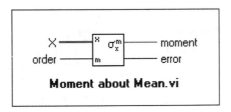

Figure 9–7
Moment about Mean VI with Input–Output terminals

■ Histogram

So far, this chapter has discussed different ways to extract important features of a data set. The data is usually stored in a table format, which many people find difficult to grasp. It is generally useful to display the data in some form. The visual display of data helps us gain insights into the data. A histogram is one such graphical method for displaying data and summarizing key information. Consider a data sequence $x = [0, 1, 3, 3, 4, 4, 4, 5, 5, 8]$. Divide the total range of values into 8 intervals. These intervals are 0–1, 1–2, 2–3, ..., 7–8. The histogram for the sequence x then plots the number of data samples that lie in that interval, not including the upper boundary (except for the last interval).

Figure 9–8 shows that one data sample lies in the range 0–1 and 1–2, respectively. However, there is no sample in the interval 2–3. Similarly, two samples lie in the interval 3–4, and three samples lie in the range 4–5. Examine the data sequence x above and be sure you understand this concept.

The **Histogram** VI in the Analysis Library is used to calculate the histogram of a data sequence. The input-output connections to this VI are shown in Figure 9-9.

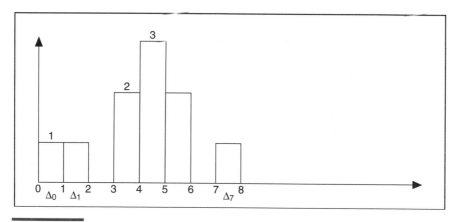

Figure 9–8
Histogram plot for the data sequence **x**

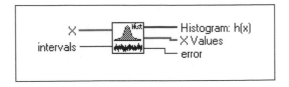

Figure 9–9
Histogram VI with Input–Output terminals

The Analysis Library also has a **General Histogram** VI that is more advanced than the **Histogram** VI. Please refer to the Lab-VIEW Online Help for detailed information.

■ Mean Squared Error (MSE)

If **x** and **y** represent two input sequences, the mean squared error is the average of the sum of the square of the difference between the corresponding elements of the two input sequences. The following formula is used to find the MSE:

$$\text{MSE} = \frac{1}{n} \sum_{i=0}^{n-1} (x_i - y_i)^2$$

where *n* is the number of data points.

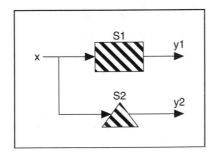

Figure 9–10
A typical application for the MSE VI

As shown in Figure 9–10, consider a digital signal **x** fed to a system, S_1. The output of this system is $\mathbf{y_1}$. Now you acquire a new system, S_2, which is theoretically known to generate the same result as S_1 but has two times faster response time. Before replacing the old system, you want to be absolutely sure that the output response of both the systems is the same. If the sequences $\mathbf{y_1}$ and $\mathbf{y_2}$ are very large, it is difficult to compare each element in the sequences. In such a scenario, you can use the **MSE** VI to calculate the mean square error (MSE) of the two sequences $\mathbf{y_1}$ and $\mathbf{y_2}$. If the MSE is smaller than an acceptable tolerance, the system S_1 can be reliably replaced by the new system S_2.

The input-output connections for the **MSE** VI are shown in Figure 9–11.

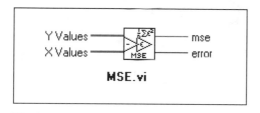

Figure 9–11
MSE VI with Input–Output Terminals

■ Root Mean Square (RMS)

The root mean square Ψ_x of a sequence **x** is the positive square root of the mean of the square of the input sequence. In other words, you can square the input sequence, take the mean of this new squared sequence, and then take the square root of this quantity. The formula used to compute the RMS value is

$$\Psi_x = \sqrt{\frac{1}{n}\sum_{i=0}^{n-1} x_i^{\,2}}$$

where n is the number of elements in **x**.

RMS is a widely used quantity in signal analysis. For a sine voltage waveform, if V_p is the peak amplitude of the signal, then the root mean square voltage V_{rms} is given by $V_p/\sqrt{2}$. Figure 9–12 shows a voltage waveform of peak amplitude = 2 V and the RMS value of $\sqrt{2} \approx 1.41$ V computed using the **RMS** VI from the Analysis Library.

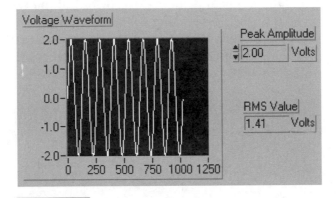

Figure 9–12
Graph of a Voltage Waveform, Peak Amplitude and RMS value

The input-output connections for the **RMS** VI are shown in Figure 9–13.

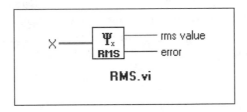

Figure 9–13
RMS VI with Input–Output terminals

Activity 9-1

Objective: To use different Statistics VIs in an example using Michael Jordan's basketball scores

In this activity, you will learn how to use different statistics VIs in an interesting example. The data set consists of the number of points scored by Michael Jordan of the Chicago Bulls in each of the 51 games he played during part of the 1997 NBA season. You will use some of the concepts discussed in the previous section to decipher this information and create single numbers that are easy to remember and yet reveal all the information that the entire data set provides.

■ Front Panel

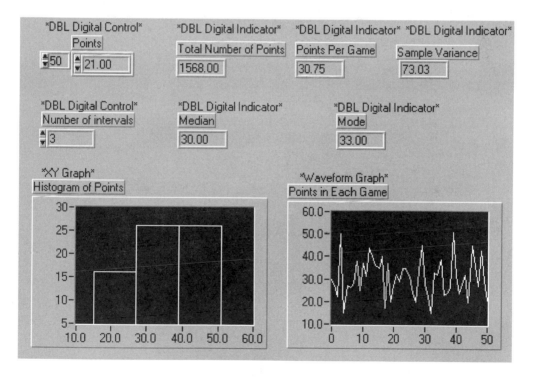

1. Open the **Statistics** VI from the library `Activities.llb`.
2. The front panel is as shown above. The **Points** control is the data set consisting of the number of points scored by Jordan in a season of 51 games.

The **Total Number of Points** digital indicator is the sum of all the elements in the given data set.

The **Points Per Game** digital indicator is the average of Jordan's scores in the season.

The **Sample Variance**, **Median**, and **Mode** digital indicators are the sample variance, median, and mode of the data set.

Histogram of Points is an XY graph that shows the distribution of points in different intervals. The number of intervals is set using the **Number of intervals** digital control. **Points in Each Game** is a waveform graph that plots the number of points scored in each game.

■ Block Diagram

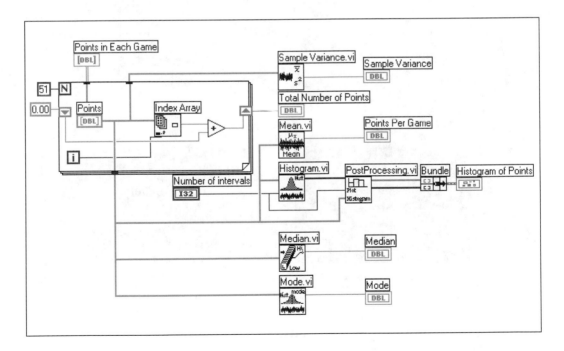

3. Switch to the block diagram. Edit and build it as shown above.

 Sample Variance VI (**Functions » Analysis » Probability and Statistics** subpalette). In this activity, this VI computes the sample variance of the data set **Points**.

 Mean VI (**Functions » Analysis » Probability and Statistics** subpalette). In this activity, this VI computes the mean value (average) of the data set **Points**.

 Histogram VI (**Functions » Analysis » Probability and Statistics** subpalette). In this activity, this VI computes the histogram of the data set **Points**.

 Median VI (**Functions » Analysis » Probability and Statistics** subpalette). In this activity, this VI computes the median of the data set **Points**.

 Mode VI (**Functions » Analysis » Probability and Statistics** subpalette). In this activity, this VI computes the mode of the data set **Points**.

 Bundle function (**Functions » Cluster** subpalette). In this activity, this function assembles the outputs of the **Post Processing** VI (explained later) to plot on the XY graph **Histogram of Points**.

4. Follow the instructions in steps 5 through 10 and build the block diagram as shown.

5. You will compute the **Points Per Game** (Jordan's average) by using the **Mean** VI in the Analysis Library. Connect the **Points** control to the input ter-

minal **X** and the output **mean** to the **Points Per Game** indicator.

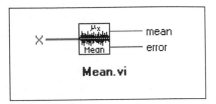

6. You will compute the sample variance using the **Sample Variance** VI in the Analysis Library. Once again, connect the **Points** control to the input terminal **X** and the output **sample variance** to the **Sample Variance** indicator.

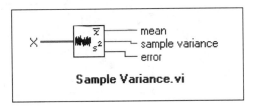

 This VI also computes the mean of the data set.

7. You will compute the median using the **Median** VI in the Analysis Library. Connect the **Points** control to the input terminal **X** and the output **median** to the **Median** indicator.

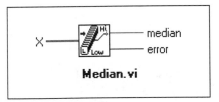

8. You will compute the mode using the **Mode** VI in the Analysis Library. Connect the **Points** control to the input terminal **X**. Choose the number of intervals equal to 3. Connect the output **mode** to the **Mode** indicator.

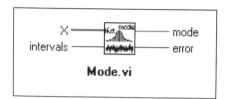

Mode.vi

9. You first will compute the data for the histogram using the **Histogram** VI. Connect the control **Points** to the input terminal **X** and set the number of intervals equal to 3. This VI generates histogram values and X values, which are the midpoints of the different intervals as discussed above. You can plot the histogram by using the X values for the X-axis and the histogram values for the Y-axis. If you are interested, try doing this and observe the histogram.

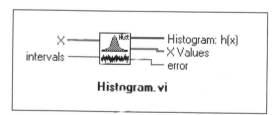

Histogram.vi

10. Generally, you may want to view the histogram in a different way. Select the **PostProcessing** VI from the library Activities.11b. This subVI can be found below the top level VIs. Connect the **Points** control to the **Points** input terminal, the **Histogram: h(x)** output of the **Histogram** VI to the **Histogram** input terminal, and connect the **Number of intervals** control to the **Number of intervals** input. This VI generates the data for plotting the histogram in a better way. The X-axis values are stored in the **Boundaries** out-

put terminal, and the Y-axis values are stored in the **PlotValues** output terminal. You can use the output of this VI for plotting the histogram on the **Histogram of Points** XY graph.

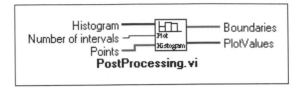

11. Return to the front panel. Right-click on the Legend for the **Histogram of Points** graph, select **Common Plots** to be of type vertical lines, and select the **Interpolation type** on the lower right. Set the number of intervals equal to 3 and run the VI. Study the different output values. See how the histogram (on the left) provides more information than just plotting the points in each game (on the right).
12. Save the VI and close it.

■ End of Activity 9-1

Probability

In any random experiment, there is always a chance that a particular event will or will not occur. A number between 0 and 1 is assigned to measure this chance, or probability, that a particular event occurs. If you are absolutely sure that the event will occur, its probability is 100 percent or 1.0, but if you are sure that the event will not occur, its probability is 0.

Consider a simple example. If you roll a single unbiased die, there are six possible events that can occur—either a 1, 2, 3, 4, 5, or 6 can result. What is the probability that a 2 will result? This probability is one in six, or 0.16666. You can define probability in simple terms as the probability that an event A will occur is the ratio of the number of outcomes favorable to A to the total number of equally likely outcomes.

■ Random Variables

Many experiments generate outcomes that you can interpret in terms of real numbers. Some examples are the number of cars passing a stop sign during a day, number of voters favoring candidate A, and number of accidents at a particular intersection. The values of the numerical outcomes of this experiment can change from experiment to experiment and are called random variables. Random variables can be discrete (if they can take on only a finite number of possible values) or continuous. As an example of the latter, weights of patients coming into a clinic may be anywhere from, say, 80 to 300 pounds. Such random variables can take on any value in an interval of real numbers. Given such a situation, suppose you want to find the probability of encountering a patient weighing exactly 172.39 pounds. You will see how to calculate this probability next using an example.

Consider an experiment to measure the life lengths x of 50 batteries of a certain type. These batteries are selected from a larger population of such batteries. The histogram for the observed data is shown in Figure 9–14.

This figure shows that most of the life lengths are between zero and 100 hours, and the histogram values drop off smoothly as you look at larger life lengths.

You can approximate the histogram shown above by an exponentially decaying curve. You could take this function as a mathematical model for the behavior of the data sample. If you want to know the probability that a randomly selected battery will last longer than 400 hours, this value can be approximated by

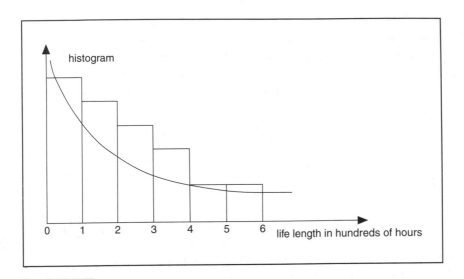

Figure 9–14
Histogram plot for data obtained from the experiment to measure life length
of 50 batteries of a certain type

the area under the curve to the right of the value 4. Such a function that models the histogram of the random variable is called the *probability density function*.

To summarize all the information above in terms of a definition, a random variable X is said to be *continuous* if it can take on an infinite number of possible values associated with intervals of real numbers, and there is a function $f(x)$, called the *probability density function*, defined as

$$P(a \leq X \leq b) = \int_a^b f(x)dx \qquad (9.1)$$

Probabilities are always nonnegative, and so we have $f(x) \geq 0$ for all x. The area under the probability density function graph is equal to one.

Notice from Equation 9.1 that for a specific value of the continuous random variable, that is for $X = a$,

$$P(X = a) = \int_{a}^{a} f(x)dx = 0$$

It should not be surprising that you assign a probability of zero to any specific value, because there are an infinite number of possible values that the random variable can take. Therefore, the chance that it will take on a specific value $X = a$ is extremely small.

The previous example used the exponential function model for the probability density function. There are a number of different choices for this function. One of these is the Normal distribution, which we will discuss next.

■ Normal Distribution

The Normal distribution (as you have already seen in the chapter on Curve Fitting) is one of the most widely used continuous probability distributions. This distribution function has a symmetric bell shape, as shown in Figure 9–15. The curve is centered at the mean value $\bar{x} = 0$, and its spread is measured by the variance $s^2 = 1$. These two parameters completely determine the

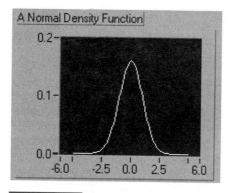

Figure 9–15
Graphical representation of a Normal density function

shape and location of the Normal density function, whose functional form is given by

$$f(x) = \frac{1}{s\sqrt{2\pi}}e^{-(x-\bar{x})^2/(2s^2)}$$

Suppose a random variable Z has Normal distribution with mean equal to zero and variance equal to one. This random variable is said to have *standard normal distribution.*

The **Normal Distribution** VI in the Analysis Library computes the one-sided probability, p, of the normally distributed random variable x,

$$p = \text{Prob}(X \le x)$$

Where X is a standard Normal distribution with the mean value equal to zero and variance equal to one, p is the probability, and x is the value.

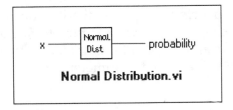

Normal Distribution.vi

Normal Distribution VI with Input–Ouptut terminals

Suppose you conduct an experiment in which you measure the heights of adult males. You conduct this experiment on 1000 randomly chosen men and obtain a data set S. The histogram distribution has many measurements clumped closely about a mean height, with relatively few very short and very tall males in the population. Therefore, the histogram can be closely approximated by a normal distribution. Now suppose that, among a different set of 1000 randomly chosen males, you want to find the probability that the height of a male is less than or

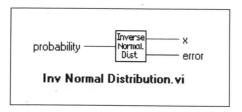

Inverse Normal Distribution VI with Input–Output terminals

equal to 170 cm. You can use the **Normal Distribution** VI to find this probability. Set the input $x = 170$. Thus, the choice of the probability density function is fundamental to obtaining a correct probability value.

The **Inverse Normal Distribution** VI performs exactly the opposite function as the **Normal Distribution** VI. Given a probability p, it finds the values x that have the chance of lying in a normally distributed sample. For example, you might want to find the heights that have a 60 percent chance of lying in a randomly chosen data set.

As mentioned earlier, there are different choices for the probability density function. The well-known and widely used ones are the chi-square distribution, the F distribution, and the T-distribution. The Analysis Library includes VIs that compute the one-sided probability for these different types of distributions. In addition, it also has VIs that perform the corresponding inverse operation.

Activity 9-2

Objective: To understand key probability concepts

In this activity, you will first generate a data sample with standard normal distribution and then use the **Normal Distribution** VI to check the probability of a random variable x.

■ Front Panel

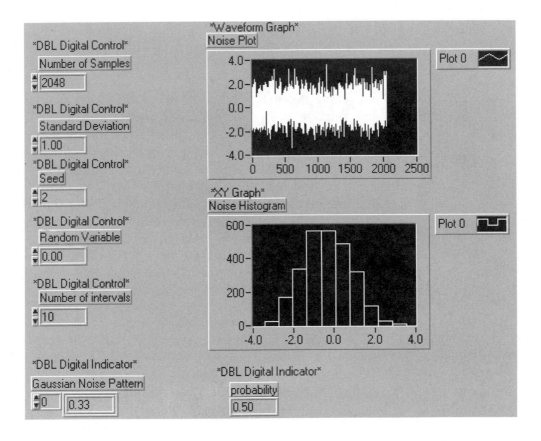

1. Build the front panel as shown above. **NoisePlot** is a waveform graph, whereas **NoiseHistogram** is an XY graph.

■ Block Diagram

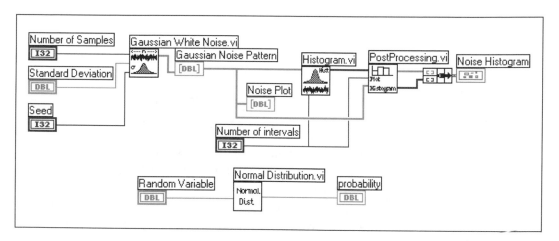

2. Build the block diagram as shown above. The **Gaussian White Noise** VI generates a Gaussian-distributed pattern with mean value equal to 0 and standard deviation set by the user using the input **Standard Deviation. Samples** is the number of samples of the Gaussian noise pattern. **Seed** is the seed value used to generate the random noise.

 White Noise VI (**Functions » Analysis » Signal Generation** subpalette). In this activity, this VI generates a Gaussian white noise pattern.

 Histogram VI (**Functions » Analysis » Probability and Statistics** subpalette). In this activity, this VI computes the histogram of the Gaussian noise pattern.

 Normal Distribution VI (**Functions » Analysis » Probability and Statistics » Probability** subpalette). In this activity, this VI computes the one-sided probability of the normally distributed random variable **Random Variable**.

Connect the **Gaussian Noise Pattern** to the waveform graph **Noise Plot**.

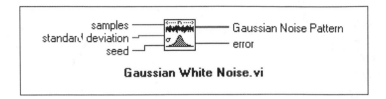

3. You will compute the histogram of the Gaussian noise pattern using the **Histogram** VI used in the previous activity.

4. As discussed earlier, do some postprocessing to plot the histogram in a different way. Select the **PostProcessing** VI from the library `Activities.llb`.

5. Bundle the output of this VI and connect it to the **Noise Histogram**.

6. Select the **Normal Distribution** VI. Connect the **Random Variable** control to the input terminal and connect the output to the **probability** indicator.

7. Return to the front panel. Set the **Number of Samples** control to 2048, **Standard Deviation** to 1, **Seed** to 2 and **Number of intervals** to 10.

8. Right-click on the Legend for the **Histogram of Points** graph, select **Common Plots** to be of type vertical lines, and select the **Interpolation Type** on the lower right.

9. Run the VI. You will see the Gaussian white noise on the **Noise Plot** graph. It is difficult to tell much from this plot. However, the histogram plot for the same noise pattern provides a lot of information. It shows that most of the samples are centered around the mean value of zero. From this histogram, you can approximate this noise pattern by a **Normal Distribution** function (Gaussian distribution). Because the mean value is zero and you set the standard deviation equal to one, the probability density function is actually a standard Normal distribution.

It is very important that you carefully choose the proper type of distribution function to approximate your data. In this activity, you actually plotted the histogram to make this decision. Many times, you can make an intelligent decision based solely on prior knowledge of the behavior and characteristics of the data sample.

10. Return to the front panel and enter a value for **Random Variable**. This VI will compute the one-sided probability of this normally distributed random variable. Remember, you have assumed that the variable is normally distributed by looking at the histogram.

11. Save the VI as **Probability.vi** in the library `Activities.llb` and close it.

■ End of Activity 9-2

Wrap It Up!

- Different concepts in statistics and probability help decipher information and data to make intelligent decisions.

- Mean, median, sample variance, and mode are some of the statistics techniques to help in making inferences about samples from a data set.

- Histograms are widely used as a simple but informative method of data display.

- Using the theory of probability, you can make inferences about samples from a data set and then measure the degree of accuracy in those inferences.

■ Review Questions

1. What is the difference between probability and statistics? Which VI(s) would you use in each case?

2. What is the difference between

   a) Mode and median?
   b) Sample variance and population variance?

   Which VI(s) would you use in each case?

3. Name some real-world practical applications of using:

   a) Histograms
   b) The Gaussian probability density function

4. What is the difference between the **Normal Distribution** VI and the **Inv Normal Distribution** VI?

■ Additional Activities

1. The following data set contains the scores (out of 100) of 30 students on a test:

   X = [40 75 32 98 43 67 73 82 91 77 67 32 93 23 35 88 45 86 12 87 44 76 59 67 30 88 60 45 98 76].

   Use the proper VIs from the Analysis Library to compute the mean, median and sample variance of the data set. Plot the histogram of the data set using the **Histogram VI** and the **Plot Histogram** VI.

2. Run Activity 9-2 with the **Standard Deviation** input equal to 5.0. Closely study the noise histogram plot. Then enter a value in the **Random Variable** control and look at the probability assuming a normal distribution as before. Do the results look correct to you? Explain your answer.

3. Man-made observatories fitted with robots are often sent to other planets to perform different experiments. One example of such an experiment would be to measure the temperature on that planet during different times. The robot, which is fitted with a sensor, would measure the temperature and send the data back to the control station on earth via a satellite as shown in the following figure. This type of data is

usually random in nature. Simply plotting the data on a graph does not reveal any useful information. Scientist usually perform statistical analysis on the data. They then use this statistical data to extract important information about the planet such as its atmospheric content, presence of water, and presence of life.

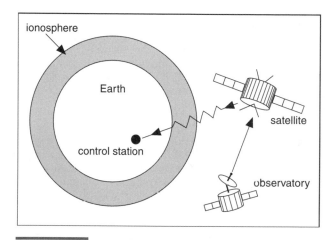

An observatory transmitting temperature data to the control station on the earth via a satellite.

In this activity, you will see how you can use the VIs that you learned in this chapter to compute different statistics associated with random data. Open the **Noise Statistics** VI from the library `Activities.llb`. Run the VI with default values. Observe different statistics for the Gaussian noise such as mean, variance and mode. Observe the histogram plot for the noise waveform. Vary the number of intervals and study the histogram plot.

Generate uniform noise and observe the statistics for different numbers of samples. Also vary the number of intervals and observe the histogram plot.

Close the VI. Do not save any changes.

Signal Processing of Biomedical Data Using PC

F. Clemente
C. DeLazzari
G. Ferrari
A. Nicoletti
Institute of Biomedical Technologies
National Research Council of Italy

The Challenge Because of the growing up of instrumentation, often medical staff have to pay attention during diagnosis and therapy to a multiplicity of monitoring and alarm signals. Data from different apparatus may be presented on a single screen using new elaboration and presentation procedures for innovative trials. This situation is useful in most industrial sectors but it is more critical in the medical environment because of the consequence on human life due to some kind of monitoring failure.

The Solution Rapid progress in the development of computer systems and new software packages such as LABVIEW offer to the unspecialized staff the possibility of setting up flexible PC based acquisition and monitoring workstations. However, knowledge of the signal source, characterization of the signal, and definition of processing and presentation procedures are, for instance, necessary to implement the systems that can be useful in clinical practice. This article gives an overview of biomedical signal processing and shows an application in intensive care during experimental surgery. The considerations presented here are applicable in other sectors too.

Introduction to Biomedical Data

Nowadays signal processing is useful in clinical practice. Looking around in hospitals and clinical departments it is common to find data acquisition systems.

As acquired data can be used to determine the status of different human "systems", analysis and computation are performed over a wide part of them by biochemical tests on blood and other fluids, or analyzing electrical signals to monitor, for instance, cardiovascular and respiratory parameters.

Patients can be "measured" in different ways and diagnosis are performed using data coming from instrumentation (such as thermometers) to electrical amplifiers, or from more complex Doppler ultrasound flowmeters. Sometimes medical doctors can extract information just by looking at the shape of the raw signals, but sometimes diagnosis requires a kind of processing made through analog systems or by means of digital signal processing.

Biomedical signal processing means the manipulation of biological data. It requires to apply to physiology the know-how derived from mathematics and different branches of engineering. However visualisation and simple manipulation of biomedical data coming from different devices, even if on a large amount of data, can be easily performed using dedicated software for virtual instrumentation. In this way it is possible to create open systems, that can be improved depending on the needs of the users, and also that can be used to test new simple multimodal approach analysis.

This article points out some of the problems arising from the acquisition and study of biological signals using PC's and virtual instrumentation. An application for an intensive care unit in experimental surgery is reported.

■ The Need of Feasibility in Biomedical Signal Processing

As in all data acquisition and processing from physical phenomena, the set up of a biomedical signal processing workstation requires proceeding through the primary steps as shown in Figure 9–16 and later on described.

Data Collection

Data come in our case from medical instrumentation. Of course, these devices must be supported with analog output. Digital output based on standard protocol (RS232 or IEEE P1073) is sometimes present.

Particular attention must be paid to the safety aspect. Connecting other electrical systems to medical devices can generate undesirable leakage current on patients. So the connections must be executed according to electrical safety practice (i.e. technical standard IEC 601).

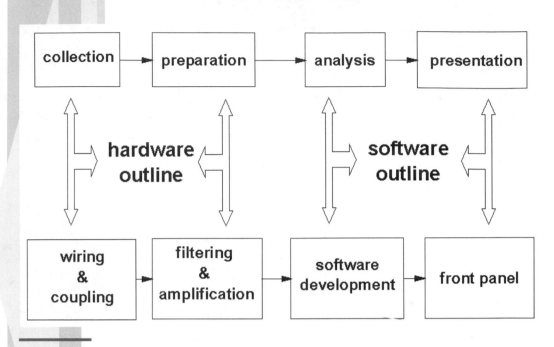

Figure 9–16
Steps in set up data acquisition and processing system design

Data Preparation

Raw data are usually available as voltage signals so that they need a sort of editing. It starts from pre-analysis operation carried out through traditional oscilloscopes or digital virtual instrumentation set at a very high sampling frequency. Raw analysis can be used to pre-process a signal in order to avoid aliasing of noise components (through analog and/or digital filtering) and quantization errors due to the improper fitting between the level of signals and the voltage range of A/D conversion. To solve the first problem take note that there are now available on board PC expansion modules performing upto 8 poles analog filtering. For the latter problem, the solution can be provided by amplifying the signal levels acting both on the transducers and its electronic unit and antialiasing filter devices. Of course, the sampling frequency must be set according to the

Nyquist Theorem by considering the maximum frequency component present in the signal.

Data Qualification

Theoretically speaking, biomedical data are of random nature. Therefore, before acquisition, an elaboration test or some bibliographic studies on their stationarity and/or on the linearity of the systems under examination must be carried out. As linearity, stationarity, periodicity (or quasi periodicity) are tested for a large number of them, powerful tools of signal processing can be utilised. In this section, only techniques for stationary data are presented. If the signals are not stationary, special analysis techniques are required.

Practical consideration in a first approach suggests considering the physical phoenomenon generating the signals. Evaluation might range from visual inspection of time series to the application of statistical algorithms. For instance, periodic or quasi-periodic signals can be analysed in the time series by including at least one cycle. However, particular attention has to be paid to interaction between different physiological systems. For example, in cardiovascular applications, the pressure waveform is a typical quasi-periodic signal. As intrathoracic pressure is influenced by pulmonary respiratory activity, attention must be taken in overlapping different frequency components derived from different physiological activity.

Definition of long-time stationarity or periodicity of the signal and tests on it lead to the definition of sample record length on which to perform further analysis. Therefore linear system theory and spectral analysis have been successfully applied, for instance, to circulatory and respiratory systems as well as to stimulus related activity in neurological systems.

Data Analysis

The choice of the signal processing approach is a final strategy considering the system, the useful parameters necessary for characterising it, and the type of investigation that is carried out.

Considering a set of records, the study can be performed on individual records, between records, and between functions derived from two or more sets of data. It is useful to trigger signal acquisition to biological activity (i.e. cardiovascular signal can be triggered by electrocardiographic—ECG—signal) and or external events (as for instance neurological activities are related to external events such as stimulus or movements picked up with special devices as accelerometers). Coherent average may be helpful in order to avoid uncorrelated noise. Time series analysis is very useful to extract a set of records to be studied in either the time or in the frequency domain.

The main strategy is to define the time domain or the frequency domain analysis. Time domain analysis includes mean, variance, mean square and RMS analysis which are very early calculated on time series. Furthermore, calculation of the short time averaged mean and mean square value estimates provides a basis to evaluate the stationarity of the data. Trends and other statistical computation can be useful in extracting powerful information from series as for instance on biochemical data. The frequency domain approach technically does not yield any new information about the signal. However, especially in digital signal processing, it offers the possibility to study the signal and the system in a more convenient form. The FFT is the basis for autocorrelation and mutual correlation computation as well as for the coherence function. Very useful analyses based on the study of the frequency response are transfer function and network analysis.

Data Presentation

Last but not least, is data presentation. Biomedical signal processing represents a border between the technical activity of engineers and the diagnosis capability of medical doctors. The background of clinicians to observe and evaluate data derived from patients, as well as common standards used in clinical activity, must be taken into account in the presentation of data.

For instance, in intensive care it is useful to overlap different pressure signals whereas a split presentation can be more instinctive from an engineers point of view. The immediate visualization of mean values added to a flow chart of some pressure channels is a must in clinical monitoring. Some acoustic alarms are also needed. So in defining a system, particular attention must be paid to already existing medical standards, which sometimes are far away from technical tradition.

■ Technical Characteristic of a Workstation for Cardiovascular Applications

Different kinds of transducers, through appropriate instrumentation, can be connected to a computer which can display data, perform on line processing, store, and, last but not least, drive other instrumentation or alarms for clinicians.

Signal processing can give objective information and quantitative assessment of the patient during therapy. Therefore, PC based systems can be useful in medical departments applying simple concepts of signal processing as in the following examples.

- Data storage on a personal computer offers the possibility to realise a mixed database containing traditional data with the addition of a simple ECG recording. Recorded signal can be displayed and added via software to a written medical diagnosis using a standard word processor.

- In an intensive care unit, the medical staff has to pay attention to a multiplicity of monitoring apparatus and signals. For instance, a blood pressure waveform can be made to match respiratory variation signals in order to better define the status of the patients. To show and record monitored data in an integrated unit is useful in routine activity as well as in research applications requiring comparison between different biological functions.

- ECG peak analysis is necessary to determine cardiac frequency and mean values of flow and pressure; it is useful to divide cardiac correlated signals in coherent series for studies such as phase shift analysis or frequency domain analysis too.

- Different kind of data (as weight, high, mean pressure and flow) are used to compute parameters and index of common use in medical practice (vascular resistance, cardiac index, etc).

- Frequency analysis and the consequent study of frequency parameters, such as the mean frequency, are largely used in cardiovascular applications.

In all cases, development of computer equipment and software techniques offers to users and researchers a large possibility in investigating biological data. Of course general-purpose applications are now useful for this scope.

The simultaneous increase in PC and add on board capability and the availability of powerful software devoted to data acquisition and virtual instrumentation offer the possibility of setting up PC workstations. These kinds of systems are implemented with simple and cheap expansion modules; from 8/16 channels A/D conversion boards working up to 1 Msamples/sec to specialised DSP boards executing more complicated computing in a very short time. Such systems are able to acquire and store data and to compute simple elaboration on biological signals even in real time. Using these systems, users not familiar with signal processing software can directly develop acquisition and simple elaboration of data. More expert users can easily use virtual instrumentation as a basis for more complicated elaboration. Therefore, more complicated and on line computing requires the development of dedicated software using traditional programming languages (C^{++}, Visual Basic, etc.).

Last but not least it is possible to create a standard in the set up and use of data acquisition systems.

Simple stations for biomedical signal processing require:

- Number of channels up to 16, and triggered acquisition.

- Acquisition frequency up to 500 Hz is enough for cardiovascular applications as signals are in most cases included in a frequency range from DC to 100 Hz or less; of course signals included in a wide frequency range as Doppler ultrasound require a higher value of the sampling rate during A/D conversion.

■ To monitor physical phenomena for a period of up to several hours, one must take into account the amount of data to be stored and the sampling rate; sometimes, continuous (circular) data acquisition and keys to start/end storing during monitoring is a good choice; stored data must be of a format useful for further off line computation using popular programs and worksheets.

■ A PC based work station for experimental surgery

This part is devoted to the presentation of a data acquisition and monitoring system developed for in vivo experimental research activity.

The Experimental Activity

A system was built to perform medium and short term monitoring of circulation using pneumatic left ventricular assist devices (LVAD—NIPPON ZEON 906 control unit connected to its experimental pump) on anaesthetised sheep.

Figure 9–17 shows the general layout of the experimental setup. All animals received care in accordance with the Guide for the Care and Use of Laboratory Animals prepared by the National Academy of Sciences and published by the National Institute of Health (NIH Publication No. 85–23, revised 1985). The Italian Health Ministry approved the experimental protocol.

Transducer

Flow was measured by a thermodilution method (monitor HP56S) and an ultrasonic flowmeter (Transonic model T106). Gould pressure transducers (amplifier mod.13–4615–50) were used to measure all pressures.

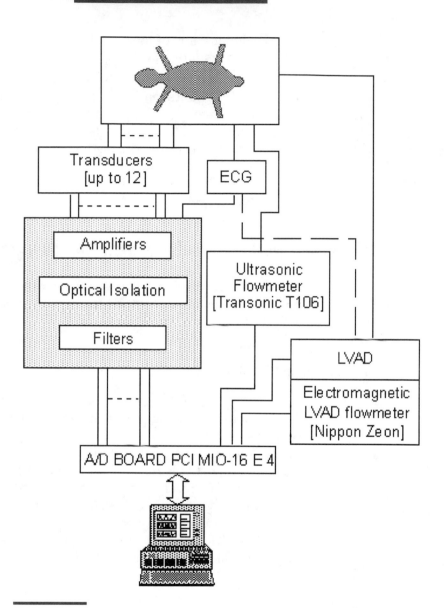

Figure 9–17
Set up of the system

Coupling

In biomedical signal processing considerable attention must be paid to minimize the risk of electric shock to the patient. Optical isolators were used in order to avoid dangerous leakage currents on the sheep.

Acquisition unit

The hardware of the system in its actual configuration consists of a PC Pentium 166 MHz, equipped with a National Instrument PCI-MIO-16E-4 data acquisition board.

Software description

Driving and visualisation software has been realised using LAB-VIEW 4.1. The developed software acquires up to 12 pressure channels (usually 6), 2 flow channels and one ECG derivation and offers the possibility to store data and compute simple parameters required in intensive care monitoring.

Main panel includes (Figure 9-18, from left to right): waveform display, output of computed data, control panel.

Waveform Display

Channels are displayed on three separate waveform charts used for (Figure 9–18, from top to bottom):

Systemic circulation (higher pressure) + ECG

Pulmonary circulation (lower pressure)

Flow

Computed Data

On line computation was performed on the data acquired from each cardiac cycle to give:

Heart rate (HR) in beats/sec

For systemic arterial pressure (Pas) maximum, mean and minimum values useful to have information about heart pumping activity

For other pressure channels (Pxx) and flow (Qxx) channels mean values over the entire cardiac cycle

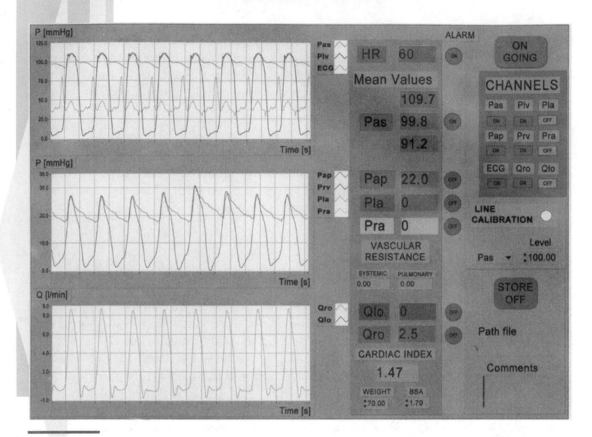

Figure 9–18
Front panel

Vascular resistance for pulmonary and systemic circulation computed from mean values of measured flow and pressures.

Cardiac index as the ratio between flow and body surface area (BSA)

Control panel

Controls are used for:

Switch on/off alarms on each acquired channels

Run/stop

Switch on/off channels

Run service routines (for calibration of the transducer amplifier)

Store data in ASCII files for off line computation.

References

The following references are useful to the reader who wants to go deeper into the different topics treated in this article.

Bendat, J. S., and Pierrsol, A. G. (1986). Random data. *Analysis and measurement procedures,* John Wiley & Sons, New York.

Bronzino, J. D. (1995). *The biomedical engineering handbook,* IEEE Press Piscatway, NY.

Challis, R. E., and Kitney, R. I. Biomedical signal processing, *Medical & Biological Engineering & Computing,* 28, pp. 509–524, 1990, 29, pp. 1–17, 1991, 29, pp. 225–241, 1991.

Grossman, W. (1986). *Cardiac catheterization and angiography,* Lea & Febiger, Philadelphia.

IEC (International Organisation for Standardisation). Publications IEC 601–1 and IEC 601-1-1.

Webster, J. G. (ed.). (1988). *Encyclopedia of medical devices and instrumentation,* John Wiley & Sons, New York.

■ **Contact Information**

F. Clemente
Institute of Biomedical Technologies
National Research Council of Italy
Via G. B. Morgagni 30/e–00161
Roma, Italy
www.cardio.irmkant.rm.cnr.it
E-mail: clemente@color.irmkant.rm.cnr.it

Overview

In this chapter, we will study linear, time-invariant (LTI) control systems. We will discuss different types of representations for these systems and conversion from one type to another, frequency-domain analysis of such systems, and the design of linear state feedback systems. We will describe different G VIs that are part of the Control and Simulation Software (GSim) and show how they can be used in the design and analysis of control systems.

GOALS

- Learn about different types of representations for linear, time-invariant (LTI) systems and conversion from one form to another.
- Understand frequency-domain analysis tools such as Bode plot, Nyquist plot and the root-locus plot.
- Discuss about the stability of such systems and the design of linear state feedback.

KEY TERMS

- linear
- time-invariant
- series connection
- parallel connection
- zeros
- poles
- state-space model

- transfer function model
- conversion
- Bode Plot
- root locus
- Nyquist Plot
- linear state feedback
- controllability

Control Systems

10

Linear, Time-Invariant Control Systems

In the simplest form, a system consists of an input, an output, and a transfer function which relates the input to the output. For example, consider a system as shown in Figure 10–1.

$$X \longrightarrow \boxed{\text{SQUARE}} \longrightarrow Y = X^2$$

Figure 10–1
A non-linear system with input-output
relationship $Y = X^2$

The input is X and the output is Y and the input and the output are related by $Y = X^2$. This is a very simple example. In practice, physical systems are much more complicated. Modeling of such physical systems is a very tricky job and can be carried out in

many ways. A brute force method is to apply various input signals to the system and to measure its output response. We can
then average the response over many input signals and come up
with an empirical system representation. This approach may be
restrictive in many instances, especially if the physical systems
are too complicated and expensive to be experimented. In such
cases we have to resort to the analytical methods for modeling
physical systems. As shown in Figure 10–2 below, the analytical
method roughly consists of the following parts:

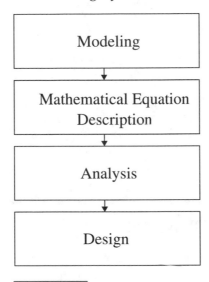

Figure 10–2
Components of the analytical method for modeling physical systems

Modeling involves determining features which describe the
operation of a system. These features are based on a thorough
understanding of the physical system and its operational range.
One can come up with different models for a physical system
depending on where the system is used, operating ranges, and so
on. For example, a transistor amplifier in the Common Emitter (CE)
configuration can be modeled as the hybrid parameter (h-model) at
low frequencies and as the hybrid-Π model at high frequencies.

Once we have selected a model for a real system of interest,
we need to describe it in a mathematical sense. For this purpose,
a mathematical equation-based description of the system is
required. A set of equations describing a physical model is

referred to as the mathematical model of the system. For example, if the model of the system consists of electrical components such as a resistor and an inductor, then we can apply Kirchhoff's law to describe the input-output behavior of the system. On the other hand, if the physical model consists of mechanical components such as mass and springs, then we can apply Newton's laws of motion to such a model. These equations which describe the system can be linear equations, nonlinear equations, integral equations, difference equations, differential equations, or others. A system may have different mathematical equation descriptions, just as a physical system may have different models.

Once a mathematical model of a system is obtained, the next step involves analysis. Analysis is concerned with the determination of the response of the system to certain inputs and initial conditions. This type of analysis is often referred to as quantitative analysis. On the other hand, we are also interested in the general properties of the system such as stability, controllability, and observability. This type of analysis is often referred to as qualitative analysis. Analysis is a very important part because design techniques often evolve from this study.

The final stage of the analytical approach involves design of the system. If the response of the system does not meet the requirements, then the system has to be improved or optimized. In some cases, this can be done by adjusting certain parameters of the system. In other cases, an additional system must be introduced. This additional system changes the behavior of the original system so that it meets the necessary specifications. Note that the design is carried out on the model of a physical system. If the model is properly chosen, the performance of the physical system can be correspondingly improved by introducing the required adjustments or compensators. In this chapter, we will concentrate mostly on the analysis of systems. In order to simplify the analysis process, we introduce certain approximations. We will discuss these approximations next.

■ Time-Invariant Systems

If the characteristics of a system do not change with time, then the system is said to be time invariant or stationary. Otherwise, the system is time varying. Figure 10–3 shows the response of both these types of systems. Observe that for the time-invariant system, the output is not dependent on when the input signal is applied.

Any system which undergoes slow variations in characteristics with time is, in a strict sense, a time-varying system. This applies to most of the physical systems. However, if the characteristics of a physical system change very slowly in relation to the variations in the input, then the system can be approximated as a time-invariant system. In this chapter, we will concentrate mostly on the analysis of purely time-invariant systems (very few in nature) or systems which can be approximated as time-invariant systems.

■ Linear Systems

We will now discuss linear approximations of physical systems. Most of the physical systems are linear within some range of one or more of their parameters. For example, consider the characteristics of a spring as shown in Figure 10–4.

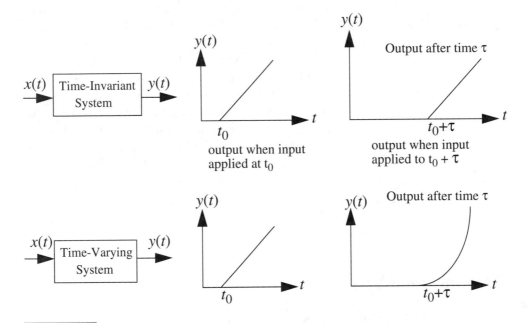

Figure 10–3
Response of a time-varying and a time-invariant system

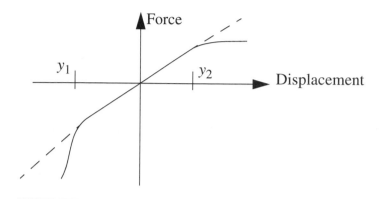

Figure 10–4
Characteristics of a spring

The spring has linear characteristics within the displacement range y_1 to y_2. Outside this range, the spring has nonlinear characteristics. So, in case of most physical systems, we can restrict the operation range of the system and approximate it as a linear system. A linear system must satisfy two important properties:

Property of superposition: If the input x_1 to the system produces a response y_1 and the input x_2 produces a response y_2, then an input $x_1 + x_2$ must result in a response $y_1 + y_2$.

Property of homogeneity: If the input x to the system produces a response y, then an input βx must result in a response βy.

A linear system satisfies both these properties. A system characterized by the relation $y = x^2$ is not linear, whereas a system characterized by the relation $y = x$ is a linear system.

There are two different models for such systems.

■ Transfer Function Model

If you are interested only in the input-output response of the system, then the transfer function model is the best for you. Consider the continuous system in Figure 10–5 with input $x(t)$ and output $y(t)$.

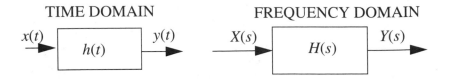

TIME DOMAIN   FREQUENCY DOMAIN

Figure 10–5
Time domain and frequency domain representation of a continuous system

This model relates the input and the output by the transfer function. In the Laplace transform representation (see Appendix A), the transfer function of the system can be written as

$$H(s) = \frac{Y(s)}{X(s)} = \frac{\beta_n s^n + \beta_{n-1} s^{n-1} + \ldots + \beta_1 s + \beta_0}{\alpha_n s^n + \alpha_{n-1} s^{n-1} + \ldots + \alpha_1 s + \alpha_0} \tag{10.1}$$

This method is particularly attractive in the frequency domain because it provides a practical approach for the analysis of continuous, linear, time-invariant systems.

We will now discuss two other types of representations in the frequency domain.

■ Zero-Pole Representation

When we set the denominator polynomial in Equation 10.1 equal to zero, we get the characteristic equation of the system. The roots of this equation are called the *poles* of the system. These values are also referred to as singularities of the system because the transfer function becomes singular at the poles.

The roots of the numerator polynomial in Equation 10.1 are called the *zeros* of the system. They are called zeros because at these values the transfer function of the system becomes equal to zero. Poles and zeros are critical frequencies.

Consider a system with the transfer function

$$H(s) = \frac{s+3}{s^2 + 3s + 2} \tag{10.2}$$

The characteristic equation of the system is $s^2 + 3s + 2 = 0$, the poles of the system are at $s = -1$ and $s = -2$ and the zero of the system is at $s = -3$.

In Figure 10–6, we have plotted the poles and the zeros of the system on the pole-zero plot. This plot is in the complex s-plane, where $s = \sigma + j\omega$. The poles are represented by an "x" and the zeros are represented by a "o".

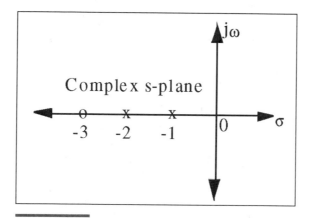

Figure 10–6
Pole zero plot for the transfer function in Equation 10.2

The transfer function representation can be converted into a zero-pole representation by expressing the numerator and denominator in Equation 10.1 in terms of its zeros and poles, respectively

$$H(s) = \frac{K \prod_{k=0}^{nz-1} (s - z_k)}{\prod_{k=0}^{np-1} (s - p_k)} \tag{10.3}$$

Equation 10.3 represents the zero-pole representation of the system. K represents the gain constant of the system, and nz and np are the number of zeros and poles, respectively. The locations of the zeros and poles are given by z_k and p_k.

■ Residual-Pole Representation

If we expand Equation 10.1 in a partial fraction expansion, we get

$$H(z) = \sum_{k=0}^{np-1} \frac{K_k}{(s-p_k)}$$

where K_1, K_2, ..., K_n are the coefficients of the expansion. These coefficients are called the residues. For example, for

$$H(s) = \frac{(s+3)}{(s+1)(s+2)}$$

$$K_1 = \frac{(s+3)}{(s+2)}\bigg|_{s=-1} = 2$$

$$K_2 = \frac{(s+3)}{(s+1)}\bigg|_{s=-2} = -1$$

the residual pole representation of the system can then be written as

$$H(s) = \frac{2}{(s+1)} + \frac{-1}{(s+2)}$$

■ State Variable Model

We will now discuss a different method of system analysis using time-domain methods. We will first define the state of a system as a set of numbers such that the knowledge of these numbers and input functions, along with the equations describing its internal behavior, will provide the future state and output of the system. The state of such a system, also known as a dynamic system, is described by a set of state variables $[x_1(t), x_2(t), ..., x_n(t)]$. These state variables determine the future behavior of a system when the present state of the system and the excitation signals are known.

Consider a dynamic system with state variables $x_1, x_2, ..., x_n$. It has two inputs u_1, u_2 and two outputs y_1, y_2 as shown in Fig-

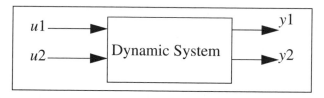

Figure 10–7
A dynamic system with two inputs and two outputs

ure 10–7. The state of this system can be described by the set of first-order differential equations written in terms of the state variables. These first-order differential equations can be written in general form as

$$\dot{x}_1 = a_{11}x_1 + a_{12}x_2 + a_{13}x_3 + b_{11}u_1 + b_{12}u_2$$

$$\dot{x}_2 = a_{21}x_1 + a_{22}x_2 + a_{23}x_3 + b_{21}u_1 + b_{22}u_2$$

$$\dot{x}_3 = a_{31}x_1 + a_{32}x_2 + a_{33}x_3 + b_{31}u_1 + b_{32}u_2$$

where $\dot{x}(t) = dx/dt$. This set of simultaneous equations can be written in matrix form as follows

$$\begin{bmatrix} \dot{x}_1 \\ \dot{x}_2 \\ \dot{x}_3 \end{bmatrix} = \begin{bmatrix} a_{11} & a_{12} & a_{13} \\ a_{21} & a_{22} & a_{23} \\ a_{31} & a_{32} & a_{33} \end{bmatrix} \begin{bmatrix} x_1 \\ x_2 \\ x_3 \end{bmatrix} + \begin{bmatrix} b_{11} & b_{12} \\ b_{21} & b_{22} \\ b_{31} & b_{32} \end{bmatrix} \begin{bmatrix} u_1 \\ u_2 \end{bmatrix}$$

The column vector consisting of the state variables is called the state vector and is written as

$$\mathbf{x} = \begin{bmatrix} x_1 \\ x_2 \\ x_3 \end{bmatrix}$$

The vector of input signals is referred to as **u.** Then the system can be represented by the compact notation of the state differential equation as

$$\dot{x} = Ax + Bu \tag{10.4}$$

This differential equation is also commonly called the *state equation*. This equation relates the rate of change of the state of the system to the current state of the system and the input signals. In general, the outputs of the linear system can be related to the state variables and the input signal by the state equation

$$y = Cx + Du \tag{10.5}$$

The set of Equations 10.4 and 10.5 that describes the unique relations between the input, output, and state is often referred to as the *dynamical equation*.

One example of a dynamic system can be an RLC circuit with voltage and current sources. In this case, the inputs can be the voltage and current, the states can be chosen as the voltage across the capacitor and the current flowing through the inductor, and the output can be chosen as the voltage across the resistor.

In this section, you have learned about different types of representations for linear, time-invariant systems. The type of representation chosen depends on the application and how the system is defined. As shown in the activities that follow, you can use the VIs in the Control and Simulation Software to convert from one type of representation to another. Follow the instructions in the preface to install this software which can be found on the CD-ROM at the back of this book.

Activity 10-1

Objective: To draw the pole-zero plot for a system given its transfer function representation

In the analysis of control systems, it is often required to determine the location of the poles and zeros of the system on the pole-zero plot. The location of the poles of the system gives us valuable information about the stability of the system. (We will discuss stability in more detail later on in this chapter.)

In this activity, you will learn how to use the VIs in the Control and Simulation Software to compute the poles and zeros of the system. We will assume that we have the transfer function representation of the system in the Laplace domain.

■ Front Panel

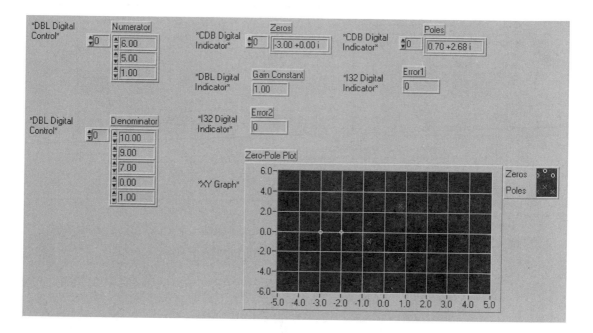

1. Build the front panel as shown in the figure above.

2. The array control **Numerator** will hold the coefficients of the numerator of the transfer function of the system. The first element in the array will contain the lowest-order (constant term) coefficient, the second element will contain the first-order coefficient, and so on.

3. The array control **Denominator** will hold the coefficients of the denominator of the transfer function of the system. The first element of the array will contain the lowest-order (constant term) coefficient, the second element will contain the first-order coefficient, and so on.

4. The array indicator **Zeros** and **Poles** will contain the zeros and poles of the system after the VI has been run.

5. The XY Graph **Zero-Pole Plot** will plot the poles and zeros of the system in the complex s-plane.

■ Block Diagram

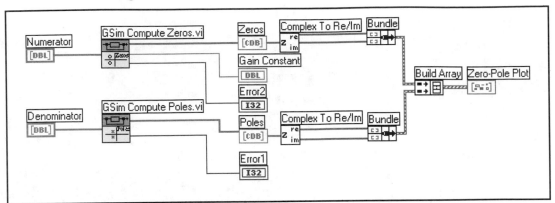

6. Switch to the block diagram and complete it as explained in the next three steps.

7. Choose the **GSim Compute Poles** VI (**Functions >>GSim** subpalette). Wire the array control **Denominator** to the **Alphas** input. Wire the output **Poles** to the array indicator **Poles** and the output **Error** to the indicator **Error1**.

GSim Compute Poles.vi

8. Choose the **GSim Compute Zeros** VI (**Functions >> GSim** subpalette). Wire the array control **Numerator** to the **Betas** input. Wire the output **Zeros** to the array indicator **Zeros**, the output **Gain Constant** to the indicator **Gain Constant,** and the output **Error** to the indicator **Error2**.

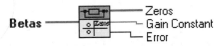

GSim Compute Zeros.vi

9. Build the rest of the block diagram as shown in the figure above. This is needed for separating the real and imaginary parts of the complex poles and zeros and then plotting them on the XY Graph **Zero-Pole Plot.**

10. Increase the size of the Legend of the XY Graph, so that it shows two plots, Plot 0 and Plot 1. Pop up on Plot 0 in the Legend, and select **Common Plots** so that only the points are displayed. Also, in the Legend of Plot 0, select **Point Style** so that the points are displayed as circles. Similarly, pop up on Plot 1 in the Legend, and select **Common Plots** so that only the points are displayed. Also, in the Legend of Plot 1, select **Point Style** so that the points are displayed as x's.

11. You will compute the poles and zeros of the system whose transfer function representation is

$$H(s) = \frac{s^2 + 5s + 6}{s^4 + 7s^2 + 9s + 10}$$

Switch to the front panel and enter the numerator and denominator coefficients as shown on the front panel. As explained earlier, the first element of the array **Numerator** is 6 (the constant term in the numerator), the second element of the array is 5, and so on. Similarly, the first element of the array **Denominator** is 10 (the constant term in the denominator), the second element of the array is 9, the third element is 7, and the fifth element is 1. Note that the fourth element of the array is zero (since the coefficient of s^3 is zero).

12. Run the VI. The VI computes the poles and zeros of the system and then plots them on the **Zero-Pole Plot.** The zeros are represented by a circle and the poles are represented by an "x." You will see that the system has two zeros at –2 and –3 and four poles. You will also observe that complex valued poles appear in complex conjugate pairs.

13. Save this VI as **Zeros Poles.vi** in the library `Activi-ties.11b` and close it.

■ End of Activity 10-1

Activity 10-2

Objective: To study conversion from one type of system representation to another

In many applications, it is often needed to convert from one type of system representation to another. In this activity you will start with the following system representation in the state space form:

$$\begin{bmatrix} \dot{x}_1 \\ \dot{x}_2 \\ \dot{x}_3 \end{bmatrix} = \begin{bmatrix} 1 & 1.5 & 1 \\ 3.2 & 0 & 1 \\ 0.5 & 1.5 & 2.0 \end{bmatrix} \begin{bmatrix} x_1 \\ x_2 \\ x_3 \end{bmatrix} + \begin{bmatrix} 1 & 1 \\ 3 & 1 \\ 1 & 5 \end{bmatrix} \begin{bmatrix} u_1 \\ u_2 \end{bmatrix}$$

$$y = \begin{bmatrix} 1 & 3 & 1 \end{bmatrix} \begin{bmatrix} x_1 \\ x_2 \\ x_3 \end{bmatrix} + \begin{bmatrix} 1 & 2 \end{bmatrix} \begin{bmatrix} u_1 \\ u_2 \end{bmatrix}$$

In order to plot the poles and zeros of the system on the zero-pole plot, you will convert the system representation from the state space form to the zero-pole form. Next, in order to compute system parameters such as settling time, you will require the system representation in the transfer function form. For this, you will convert the system representation from the zero-pole form to the transfer function form. In this activity, you will learn how to use VIs in the Control and Simulation Software to do such conversions.

■ Front Panel

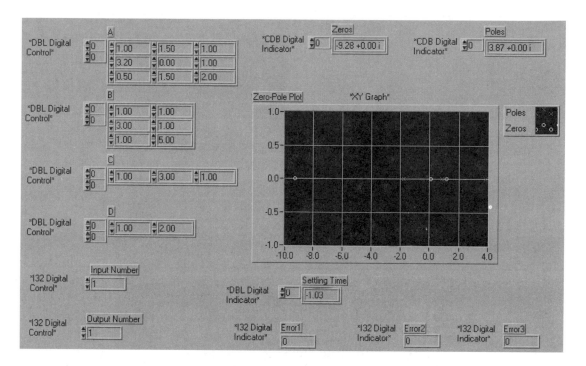

1. Build the front panel as shown in the figure above.

2. The array controls **A, B, C,** and **D** contain matrices **A, B, C,** and **D** of the state space representation as explained in the previous section.

3. In case of multiple input systems, the control **Input Number** contains the index of the input for which the conversion is required. In case of multiple output systems, the control **Output Number** contains the index of the output number for which the conversion is required.

4. After the VI has been run, the array indicators **Zeros** and **Poles** contain the zeros and poles of the system for the input and output number specified. The XY Graph **Zero-Pole Plot** will plot the poles and zeros of the system in the complex s-plane.

■ Block Diagram

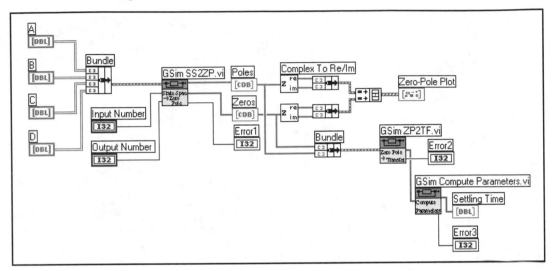

5. Switch to the block diagram and complete the block diagram as shown above and explained in the following four steps.

6. Choose the **GSim SS2ZP** VI (**Functions >> GSim** subpalette). You will use this VI to convert the system representation from the state space form to the transfer function form. Bundle the controls **A, B, C,** and **D** into a cluster. Wire this cluster to the **System** input of the VI. Wire the controls **Input Number** and **Output Number** to the inputs **Input Number** and **Output Number** of the VI, respectively. Wire the output terminals **Poles** and **Zeros** to the array indicators **Poles** and **Zeros,** respectively.

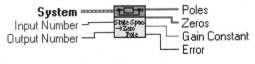

GSim SS2ZP.vi

7. Choose the **GSim ZP2TF** VI (**Functions >> GSim** subpalette). You will use this VI to convert the system representation from the zero pole form to the transfer function form. Bundle the **Zeros** and **Poles** indicators to a cluster and wire the cluster to the **System** input of this VI.

GSim ZP2TF.vi

8. Choose the **GSim Compute Parameters** VI (**Functions >> GSim** subpalette). You will use this VI to compute the settling time, given the denominator coefficients of the transfer function of the system. Wire the **Alphas** output of the **GSim ZP2TF** VI to the **Alphas** input of the **GSim Compute Parameters** VI. Wire the output **Settling Time** to the array indicator **Settling Time**.

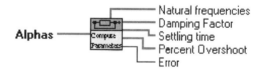

GSim Compute Parameters.vi

9. Build the rest of the block diagram as shown in the figure on the previous page. This is needed for separating the real and imaginary parts of the complex poles and zeros and then plotting them on the XY Graph **Zero-Pole Plot**.

10. Switch to the front panel and enter the elements of matrices **A, B, C,** and **D** of the state space representation as shown in the equations at the beginning of this activity. There are two inputs and one output. Let us first look at the response at the output terminal due to the first input. Set both the **Input Number** and **Output Number** equal to 1. Run the VI and look at

the poles and zeros on the **Zero-Pole Plot.** The array indicator **Settling Time** contains the settling time for the transfer function between the first input and the output.

Follow the instructions as in step 10 of Activity 10–1 in order to see the poles and zeros as x's and circles respectively on the XY Graph.

11. Set the control **Input Number** equal to 2. Run the VI. Study the **Zero-Pole Plot** and the settling time for the transfer function between the second input and the output.
12. Save this VI as **Conversion.vi** in the library `Activities.llb` and close it.

■ End of Activity 10-2

Frequency Domain Analysis

In this section, we will briefly describe frequency-domain analysis tools such as Bode plot, Nyquist plot and root-locus plot. We will then show how VIs in the Control and Simulation Software can be used to analyze systems using these tools.

In the frequency domain, the transfer function $G(s)$ of a system can be described by the relation

$$G(j\omega) = G(s)|_{s = j\omega} = R(\omega) + jX(\omega)$$

where

$$R(\omega) = \text{Re}[G(j\omega)]$$

and

$$X(\omega) = \text{Im}[G(j\omega)].$$

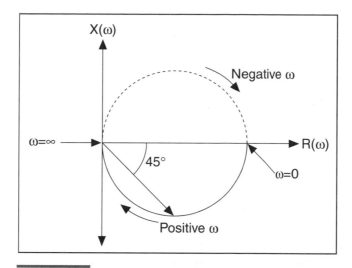

Figure 10–8
Polar plot representation for an RC filter

As shown in Figure 10–8, in a polar plot representation of an RC filter, (a filter consisting of a resistor (R) and capacitor (C) in series), the imaginary part of the transfer function is plotted versus the real part. This is the simplest form of the frequency-domain analysis. However, the polar plot representation has its own limitations. The calculation of the frequency response is tedious and does not indicate the effect of individual poles or zeros.

It is sometimes more convenient to plot the magnitude and phase as the function of frequency. In terms of magnitude and phase, the transfer function can be represented as

$$G(j\omega) = |G(j\omega)|e^{j\phi(\omega)} \tag{10.6}$$

where

$$|G(j\omega)| = \sqrt{R(\omega)^2 + X(\omega)^2} \quad \text{and} \quad \phi(\omega) = \tan^{-1}\frac{X(\omega)}{R(\omega)}$$

The analysis procedure is further simplified if the magnitude and phase are plotted on a logarithmic plot. The introduction of such logarithmic plots, often called Bode plots, simplifies the calculation for plotting the frequency response. Such plots were introduced by H.W. Bode who used them extensively in his studies of feedback amplifiers. Equation 10.6 can be written in natural logarithm form as

$$\ln G(j\omega) = \ln|G(j\omega)| + j\phi(\omega)$$

where $\ln|G(j\omega)|$ is the magnitude in nepers. The logarithm of the magnitude is normally expressed in terms of the logarithm to the base 10, so that we use Logarithmic Gain = 20 $\log_{10}|G(j\omega)|$ where the units are in decibels(dB). The magnitude is plotted as the function of frequency on one graph and the phase is plotted as the function of frequency on a different graph. The phase is usually plotted in degrees. In the next activity, we will compute the data and graphically portray the Bode plot for two different systems.

Activity 10-3

Objective: To plot and study the Bode diagram for a given system

In this activity, we will use the Bode Plot to analyze the classical control system shown in Figure 10–9.

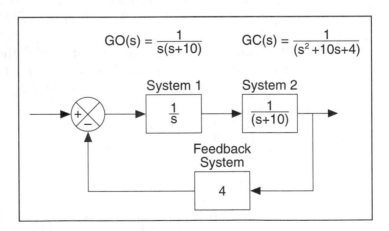

Figure 10–9
Plant design

The control system consists of two systems in a series connection and a feedback system in a closed loop connection. System 1 in the main loop is an integrator and System 2 in the main loop is a motor. The feedback system is a constant gain *K*, which is equal to 4 in this example. In this activity, you will learn how to use the VIs in the Control and Simulation Software to analyze the system.

■ Front Panel

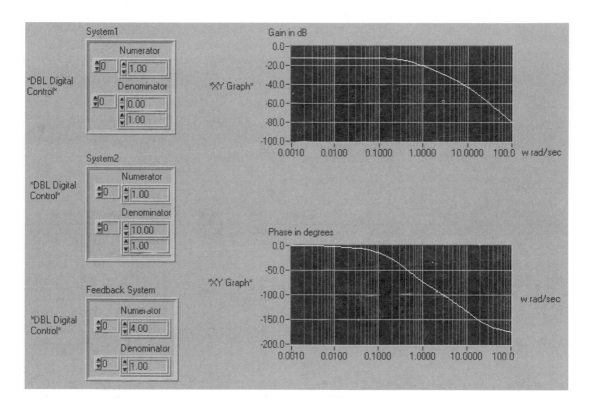

1. Build the front panel as shown in the above figure. **System1** is a cluster containing the following two elements. The array control **Numerator** will hold the coefficients of the numerator of the transfer function of System 1 in the plant design. The first element in

the array will contain the lowest order (constant term) coefficient, the second element will contain the first order coefficient, and so on. The array control **Denominator** will hold the coefficients of the denominator of the transfer function of System1. The first element of the array will contain the lowest order (constant term) coefficient, the second element will contain the first order coefficient, and so on.

2. **System2** is a cluster containing the numerator and denominator coefficients of the transfer function of System2 in the plant design. **Feedback System** is a cluster containing the numerator and denominator coefficients of the transfer function of the feedback system in the plant design.

3. **Gain in dB** and **Phase in degrees** are XY Graphs. When the VI is run, gain in decibels is plotted as a function of frequency on the **Gain in dB** plot and phase in degrees is plotted as a function of frequency on the **Phase in degrees** plot.

■ Block Diagram

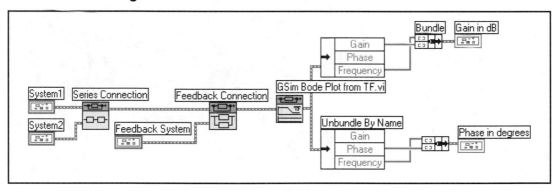

4. Switch to the block diagram and complete it as explained in the following steps:

5. Choose the **GSim Series** VI (**Functions >> GSim** subpalette). You will use this VI to compute the overall transfer function of the two systems, System 1 and

System 2, connected in series. Wire the cluster

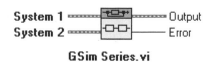

GSim Series.vi

System1 to the input **System 1** and the cluster **System2** to the input **System 2.**

6. Choose the **GSim Feedback** VI (**Functions >> GSim** subpalette). You will use this VI to compute the closed loop transfer function. Wire the **Output** cluster from the VI described in step 5 to the input **System 1** and the cluster **Feedback System** to the input **System 2.** On execution, the cluster **Output** contains the transfer function of the closed loop transfer function.

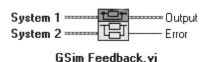

GSim Feedback.vi

7. Choose the **GSim Bode Plot from TF** VI (**Functions >> GSim** subpalette). You will use this VI to compute data for the Bode plot using the transfer function representation of the closed loop system computed in step 6. Wire the **Output** cluster from the VI described in step 6 to the input **System.** Use the **Unbundle By Name** VI to unbundle the **Bode Plot Data** cluster which contains three components: Gain, Phase, and Frequency. Bundle the Gain and Frequency together and plot them on the **Gain in dB** plot. Bundle the

Phase and Frequency together and plot them on the **Phase in degrees** plot.

GSim Bode Plot from TF.vi

8. Switch back to the front panel. Enter the numerator and denominator coefficients of the transfer functions of System 1, System 2, and Feedback System in the respective clusters. Remember that the constant term is the first element of the array and the first order term is the second element of the array.

9. Run the VI and study the Bode plots for the plant design.

To see a logarithmic x-scale, pop up on the XY graph, select X Scale >> Formatting… and choose the Mapping Mode as Log.

10. Save the VI as the **BodePlot.vi** in the library `Activities.llb` and close it.

■ End of Activity 10-3

Nyquist Plot

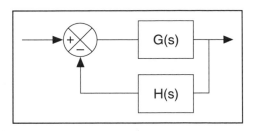

Figure 10–10
A typical closed loop feedback system

The Nyquist stability criteria was first proposed by H. Nyquist in 1932. It is a very valuable tool that determines the degree of stability, or instability, of a typical feedback control system as shown in Figure 10–10 above. Control engineers often use the information obtained from the Nyquist plot to improve both the steady state and the transient response of a feedback control system. In order to apply the Nyquist stability criterion, engineers often need a polar plot of the open-loop transfer function, $G(s)H(s)$, which is usually referred to as the *Nyquist diagram*. The polar plot is in the complex plane of $G(s)H(s)$ as s follows the contour shown in Figure 10–11.

Notice that the locus of s avoids poles of $G(s)H(s)$ that lie anywhere on the imaginary axis by small semicircular paths passing to the right. These semicircular paths are assumed to have radii of infinitesimal magnitude. Any roots of the characteristic equation having positive real parts will lie within the contour shown in the figure.

The Nyquist criterion determines the number of roots of the characteristic equation that have positive real parts from a polar plot of the open-loop transfer function. System stability can be determined by locating the roots of the closed loop characteristic equation, $1 + G(s)H(s)$, in the complex s-plane.

The Control and Simulation Software contains VIs that can be used to compute the data for the Nyquist plot given the numerator and denominator coefficients of the transfer function of the system.

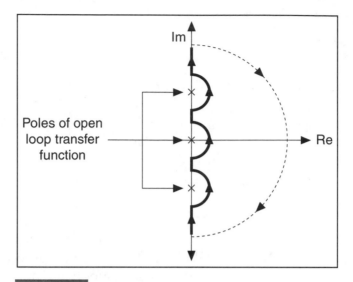

Figure 10–11
Nyquist contour

Root-Locus

The root-locus method is a graphical technique for determining the roots of the closed-loop characteristic equation of a system as a function of the gain, K, associated with $H(s)$ in the feedback loop. As the gain is varied, it is of interest to examine how the poles of the closed-loop system change. The location of these poles give us a lot of information about the behavior of the closed-loop system. The root-locus method has several advantages. Knowledge of where the closed-loop roots are located permits accurate determination of a control system's relative stability and transient performance. If accurate solutions are not required, the root-locus method can be used to obtain approximate solutions with less work.

We will demonstrate the graphical technique for plotting the root-locus with the aid of an example. Figure 10–12 shows the

root-locus of a system for which the open-loop transfer function
is given by

$$G(s)H(s) = \frac{K}{s(s+4)(s+5)}$$

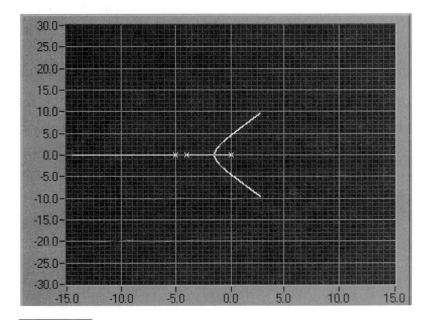

Figure 10–12
Root locus plot of the system

The number of branches of the locus equals the order of the
characteristic equation. Each branch of the root-locus describes
the motion of a particular pole of the closed-loop system as the
gain is varied. As seen in Figure 10–12, the number of branches
is equal to three. The open-loop poles define the start of the root-
locus and the open-loop zeros define the termination of the root-
locus. In this example, the branches start at the location of the
poles which are at 0, −4 and −5 on the real axis. All the zeros are
at infinity, so the branches terminate at infinity. Complex portions of the root-locus always occur as complex-conjugate pairs.

Therefore the root-locus is symmetric with respect to the real axis.

Sections of the real axis are part of the root locus if the number of poles and zeros to the right of an exploratory point along the real axis is odd. Thus, in Figure 10–12, the sections of the real axis between 0 and –4, and between –5 and –∞, are part of the root locus. If the locus has a branch on the real axis between two poles, the branch then *breaks* away from the real axis at the *break-point* and enters the complex region of the s-plane in order to approach the zeros which are finite or are located at infinity. In this example, the branch between the poles at zero and –4 breaks away at –1.470. For the case where the locus has branches on the real axis between the two zeros, branches come from poles in the complex region and merge onto the real axis.

The Control and Simulation Software provides VIs for computing data for the root-locus plot. This VI also generates a text file which contains very valuable information about the root-locus plot. The text includes information about the number of branches of the root-locus, section of the root-locus on the real axis, and the breakpoints. The toolkit also contains a utility VI for plotting the root-locus. In the next activity, we will show how the VI can be used to obtain data for the root-locus and plot it.

Activity 10–4

Objective: To analyze a negative feedback system using the root-locus plot

In this activity, we will analyze a negative feedback system using the root-locus method. The open-loop transfer function of the system is given by

$$G(s)H(s) = \frac{1.6K(s + 10)}{s^4 + 9s^3 + 24s^2 + 16s + 0}$$

■ Front Panel

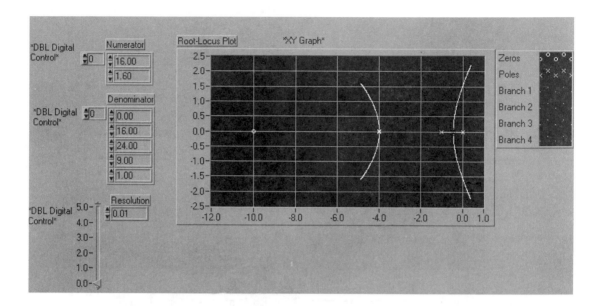

1. Build the front panel as shown in the above figure.
2. The array control **Numerator** will hold the coefficients of the numerator of the transfer function of the system. The first element in the array will contain the lowest-order (constant term) coefficient, the second element will contain the first-order coefficient, and so on.
3. The array control **Denominator** will hold the coefficients of the denominator of the transfer function of the system. The first element of the array will contain the lowest-order (constant term) coefficient, the second element will contain the first-order coefficient, and so on.
4. The vertical slide control **Resolution** decides the resolution of the root-locus plot. You can observe very minute details of the root-locus plot using small values of resolution.

5. After you run the VI, the root locus will be plotted on the XY Graph **Root-Locus Plot.**

Use legends for different plots as shown on the front panel.

■ Block Diagram

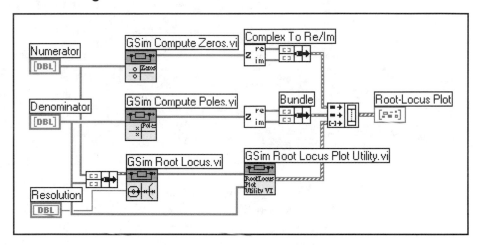

6. Switch to the block diagram and complete it as explained in the following steps.

7. Compute the poles and zeros of the system. Follow the same steps as you did in Activity 10-1.

8. Choose **GSim Root Locus** VI (**Functions >> GSim** subpalette). Use the **Bundle** VI to bundle the **Numerator** and **Denominator** array controls. Wire the cluster to the **System** input of the **GSim Root Locus** VI. Wire the vertical slide control **Resolution** to the **Resolution** input. Use the default values for the remaining two inputs.

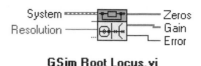

GSim Root Locus.vi

9. Choose the **GSim Root Locus Plot Utility** VI (**Functions >> GSim** subpalette). You will use this VI to plot the root locus. Wire the **Zeros** output of the **GSim Root Locus** VI to the **Zeros** input of the **GSim Root Locus Plot Utility** VI. Wire the control **Denominator** to the **Alphas** input of the VI. Make the rest of the connections as shown in the block diagram.

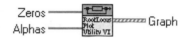

GSim Root Locus Plot Utility.vi

10. Switch to the front panel. Enter the numerator and denominator coefficients of the transfer function in the controls **Numerator** and **Denominator,** respectively. The constant term is the first element of the arrays.

11. Plot the root-locus with three different values of resolution 0.05, 0.5, and 3.0. Closely observe different features of the root-locus as described in the previous section.

When you run the VI, you will be prompted to save information about the root-locus plot in a file. You may either hit the "Save" or "Cancel" button.

12. Save the VI as **Root Locus.vi** in the library `Activities.llb` and close it.

■ End of Activity 10-4

Linear State Feedback

In this section, we will briefly discuss about stability of linear, time-invariant systems. We will then study the design of linear state feedback. This design process consists of finding feedback gains such that the poles of the closed loop system can be placed at desired locations.

■ The Concept of Stability

Stability is an important attribute of all practical systems. Stability in a system implies that small changes in the input, initial conditions, or system parameters do not cause large changes in the system output. Almost all working systems must be designed to operate in the stable region. Design of linear state feedback is the exercise done to improve the system performance within the boundaries of parameter variations imposed by stability considerations.

Stability for LTI systems can be defined in two ways:

1. Bounded-Input-Bounded-Output (BIBO) stability: A system is said to be BIBO stable if a bounded input to the system produces a bounded output. What this means is that if the input has finite amplitude, the output also has a finite amplitude.

2. Asymptotic stability: A system whose response to an arbitrary set of initial conditions with zero input (force-free system) tends toward zero is said to be asymptotically stable.

These two notions of stability are essentially equivalent for linear, time-invariant systems. This is not necessarily true in nonlinear systems.

For a LTI system, whose state space representation is given by

$$\dot{x} = Ax + Bu \tag{10.7}$$

$$y = Cx + Du \tag{10.8}$$

the equilibrium state is asymptotically stable if and only if all the eigenvalues of the matrix **A** have strictly negative parts. In other words, if the poles of the system lie strictly in the negative half of the s-plane, then the system is asymptotically stable.

■ Design of Linear State Feedback

Before we dwell into the topic of linear state feedback, let us quickly discuss the notion of controllability.

Let t_0 : inital time

t_f : finite final time

x_0 : initial state at time t_0

x_f : final state at time t_f

$u_{to, tf}$: Input from time t_0 to time t_f.

A system is said to be controllable at time t_0, if there exists a finite time t_f and input $u_{to, tf}$, such that for any x_0, $x_{t_f} = x_f$, for any x_f.

The definition of controllability for LTI systems is in terms of the matrices $\mathbf{A}$ and $\mathbf{B}$. Define the controllability matrix $\mathbf{C}$ of the LTI system whose state space representation is given by Equations 10.7 and 10.8 as

$$\mathbf{C} = [\mathbf{B} \quad \mathbf{AB} \quad \mathbf{A^2B} \quad \dots \quad \mathbf{A^{n-1}B}].$$

Such a system is controllable if the controllability matrix has full rank n.

We discussed earlier that a LTI system is not stable if some of the poles of the system lie in the right half of the s-plane. If the original system is controllable, then we can design a state feedback such that the poles of the closed loop system can be arbitrarily assigned.

Consider the single-input, single-output, linear, time-invariant system

$$\dot{\mathbf{x}} = \mathbf{Ax} + \mathbf{B}u$$

$$y = \mathbf{Cx} + \mathbf{D}u$$

where **x** is the state vector, u is the scalar input, and y is the scalar output. In state feedback, every state variable is multiplied by a gain and is fed back into the input terminal. Let the gain between the ith state variable and the input be k_i. Define $\mathbf{k} = [k_1 \quad k_2 \quad \ldots \quad k_n]$. The state feedback system is shown in Figure 10–13:

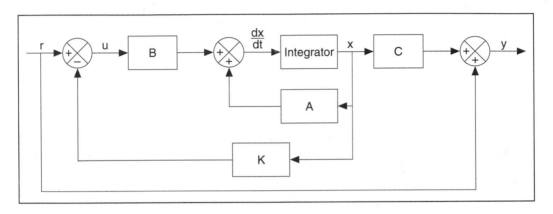

Figure 10–13
State feedback for a LTI system

The state space representation of the state feedback system shown in Figure 10–13 is obtained by replacing u by $r - kx$, where r is the reference input and is given by

$$\dot{\mathbf{x}} = (\mathbf{A} - \mathbf{Bk})\mathbf{x} + \mathbf{B}r$$

$$y = (\mathbf{C} - \mathbf{Dk})\mathbf{x} + \mathbf{D}r$$

These equations have the same dimension and the same state space as Equations 10.7 and 10.8. The controllability of a linear, time-invariant dynamical equation is invariant under any linear state feedback. It can be shown that the eigenvalues of (**A-Bk**), that is, the poles of the closed loop system, can be placed arbitrarily if and only if, the open loop system is controllable.

The Control and Simulation Software contains VIs that can be used to design linear state feedback from the state space representation or the transfer function representation. In the next activity, you will see how you can use these VIs.

Activity 10-5

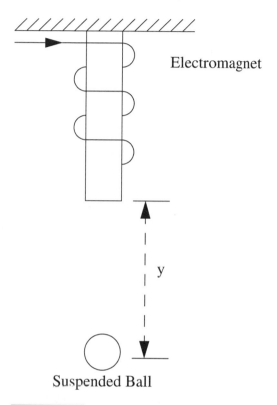

Electromagnet

y

Suspended Ball

Figure 10–14
An electromagnet and magnetically suspended ball

Objective: To design linear state feedback

Consider a magnetically suspended ball as shown in Figure 10–14. The aim is to control the distance of the ball from the electromagnet. The motion of the ball is controlled by Newton's laws. We will define state variables x_1 and x_2 as displacement and velocity, respectively. The actual state space representation is nonlinear in nature. We can approximate the system by a linear time-invariant state space representation given by the following equations:

$$\begin{bmatrix} \dot{x}_1 \\ \dot{x}_2 \end{bmatrix} = \begin{bmatrix} 0 & 1 \\ \dfrac{2cu}{mx_1^3} & 0 \end{bmatrix} \begin{bmatrix} x_1 \\ x_2 \end{bmatrix} + \begin{bmatrix} 0 \\ \dfrac{-2cu}{mx_1^2} \end{bmatrix} u$$

$$y = \begin{bmatrix} 1 & 0 \end{bmatrix} \begin{bmatrix} x_1 \\ x_2 \end{bmatrix} + \begin{bmatrix} 0 \end{bmatrix} u$$

where c is a constant, u is the input, m is the mass of the ball, and y is the output.

For this exercise, assume that the variables are such that the matrices **A, B, C,** and **D** are given by

$$\mathbf{A} = \begin{bmatrix} 0 & 1 \\ 1 & 0 \end{bmatrix} \tag{10.9}$$

$$\mathbf{B} = \begin{bmatrix} 0 \\ 1 \end{bmatrix} \tag{10.10}$$

$$\mathbf{C} = \begin{bmatrix} 1 & 0 \end{bmatrix} \tag{10.11}$$

$$\mathbf{D} = [0] \tag{10.12}$$

In this activity, you will first determine if the system is stable. If not, you will determine if the system is controllable. If the system is controllable, then as discussed earlier in this section, you can design a state feedback vector such that the poles can be arbitrarily assigned, for example at $-1 \pm j$, in the complex s-plane.

■ Front Panel

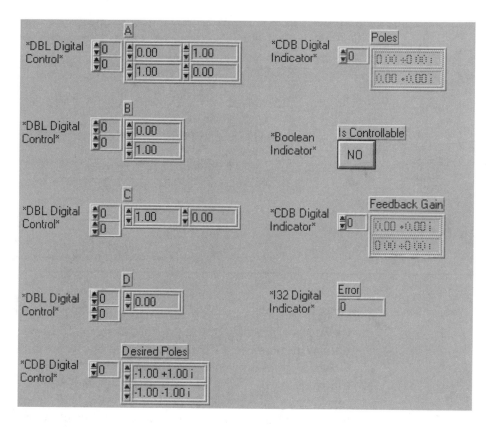

1. Build the front panel as shown in the figure above.
2. Array controls **A, B, C,** and **D** contain the matrices **A, B, C,** and **D** of the state space representation of the system as shown in Equations 10.9–10.12. Enter the matrices in the array controls as shown in the front panel.
3. The array control **Desired Poles** contains the desired pole locations.

■ Block Diagram

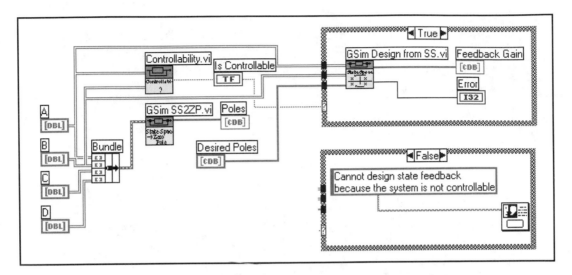

4. Switch to the block diagram. You will first compute the zeros and poles of the original system. To do this, you will use the **GSim SS2ZP** VI (**Functions >> GSim** subpalette). Use the **Bundle** VI to bundle the matrices **A, B, C** and **D**. Wire the cluster to the **System** input of the VI. Since the system is a single-input, single-output system, use the default values for the other two inputs.

5. To determine if the system is controllable, you will use the **Controllablity** VI from the library `Activities.llb`. Wire the controls **A** and **B** to the inputs A and B, respectively. Wire the output to the **Is Controllable** Boolean indicator.

6. If the system is controllable, then the output **Is Controllable** is TRUE; else it is set to FALSE. Use the

Case Structure to perform one of the two operations. If the output is FALSE, then display an error message "Cannot design state feedback because the system is not controllable" by using the **One Button Dialog** function (**Functions >> Time and Dialog** subpalette) and a string constant (**Functions >> String** subpalette).

7. If the output is TRUE, you will design the linear state feedback. To do this, you will use the **GSim Design from SS** VI (**Functions >> GSim** subpalette). Wire the controls **A, B,** and **Desired Poles** to the respective inputs of this VI. Wire the output **Feedback_Gain** to the array indicator **Feedback Gain.**

GSim Design from SS.vi

8. Make the rest of the connections as shown in the block diagram.

9. Run the VI. Look at the poles of the original system in the array indicator **Poles**. One of the poles lies in the right half of the s-plane, so the system is not stable. The Boolean indicator **Is Controllable** is set to YES (True) which implies that the system is controllable. Therefore, it is possible to design a state feedback such that the poles can be arbitrarily assigned. To assign the pole locations to the desired pole locations, we will need the linear state feedback gain as shown in the array indicator **Feedback Gain.**

10. Save the VI as **Suspended Ball.vi** in the library `Activities.llb` and close it.

■ End of Activity 10-5

Wrap It Up!

- Analytical methods for studying a control system consist of modeling, mathematical equation description, analysis, and system design.

- Almost all systems are nonlinear and time varying in nature, but many of them can be approximated as Linear, Time-Invariant (LTI) systems.

- LTI systems can be represented using the transfer function model or the state space model. Other variations of the transfer function model are the zero-pole model and the residual-pole model. You can convert from one form of representation to another.

- The transfer function model describes only the input-output relationship whereas the state space model gives a lot of information about the internal behavior (states) of the system.

- Bode plot, Nyquist plot and root-locus plot are three widely used frequency domain analysis tools.

- LTI systems are asymptotically stable if all the poles of the system (eigenvalues of the matrix **A** of the state space representation) strictly lie in the negative half of the s-plane.

- If a LTI system is controllable, then its poles can be arbitrarily assigned.

- If the original system is not stable (has poles with positive real parts), then we can move these poles to the negative half of the s-plane and make the system stable. This can be done by the aid of a linear state feedback.

■ Review Questions

1. Is the following statement true or false? The transfer function representation gives us information about the internal behavior of the system.

2. The input u_1 to a system produces an output y_1 and the input u_2 produces an output y_2. If the input $3u_1 + 8u_2$ produces the output $7y_1 + 9y_2$, is this system linear?

3. Compute the poles and zeros of the system with the transfer function representation given by

$$H(s) = \frac{s-9}{s^2 + 7s + 10}$$

Which VIs will you use to compute the poles and zeros?

4. The easiest way to determine system stability is by locating the roots of the characteristic equation in the complex s-plane. To do this graphically, will you use the Bode plot or the Nyquist plot?

5. List two advantages of the root-locus plot as a graphical tool for frequency domain analysis.

6. Consider a closed loop system where

$$G(s)H(s) = \frac{K}{(s+4)(s+9)}$$

Answer the following questions related to the root-locus plot for such a system, without actually plotting the root-locus.

   a) How many branches does the root-locus plot contain?

   b) Where do the branches start from and where do they end?

   c) Will the root-locus have complex portions?

7. Compute the poles of the system whose transfer function representation is given by

$$G(s) = \frac{s + 9}{s^2 + s - 2}$$

Where do the poles lie on the complex s-plane? Based on the location of the poles, what can you say about the stability of the system?

8. Consider the state space representation of a system where

$$\mathbf{A} = \begin{bmatrix} 1 & 6 \\ 4 & 2 \end{bmatrix} \qquad \mathbf{B} = \begin{bmatrix} 4 \\ 1 \end{bmatrix}$$

Construct the controllability matrix $\mathbf{C}$ for the system. Is this system controllable? Can you arbitrarily move the poles of the system to a desired location on the complex s-plane?

■ Additional Activities

1. The system whose transfer function is given in step 10 of Activity 10-1 is connected in **series** with another system whose transfer function is given by

$$K(s) = \frac{(s + 9)}{s^3 + 9s^2 + 8s + 1}$$

The transfer function of the overall series connection is given by $G(s)K(s)$. Use the **GSim Series** VI (**Functions >> GSim** subpalette) to compute the transfer function of the overall system. Use the **GSim Compute Parameters** VI (**Functions >> GSim** subpalette) to compute the parameters' natural frequencies, damping ratio and maximum percent overshoot. Compute the poles and zeros of the system and comment on the stability of the overall system.

2. In Activity 10-2, you first converted the state space representation to the zero-pole representation and

then converted from the zero-pole representation to the transfer function representation. There is a VI in the **Functions >> GSim** subpalette which lets you do the conversion to the transfer function model directly from the state space representation. Identify this VI and verify if the results are the same as those obtained in Activity 10-2.

The modal form representation is useful in many applications such as the design of linear quadratic regulators. Use the **GSim Modal** VI (**Functions >> GSim** subpalette) to convert the state space representation in Activity 10-2 to the modal form representation.

3. Consider a system whose open-loop transfer function representation is given by

$$G(s)H(s) = \frac{0.9501s^3 + 0.2311s^2 + 0.6068s + 0.486}{0.8913s^4 + 0.7621s^3 + 0.4565s^2 + 0.0185s + 0.8214}$$

Use the **GSim Nyquist Plot** VI (**Functions >> GSim** subpalette) to draw the Nyquist plot and analyze the Nyquist plot.

4. The lateral-direction perturbation equations for a Boeing 747 on landing approach in horizontal flight at sea level with velocity of 221 ft/seconds and a weight of 564,000 lbs are

$$\begin{bmatrix} \dot{\beta} \\ \dot{r} \\ \dot{p} \\ \dot{\phi} \end{bmatrix} = \begin{bmatrix} -0.0890 & -0.989 & 0.1478 & 0.1441 \\ 0.168 & -0.217 & 0.166 & 0 \\ -1.33 & 0.327 & -0.975 & 0 \\ 0 & 0.149 & 1 & 0 \end{bmatrix} \begin{bmatrix} \beta \\ r \\ p \\ \phi \end{bmatrix} + \begin{bmatrix} 0.0148 \\ -0.151 \\ 0.0636 \\ 0 \end{bmatrix} u$$

$$y = \begin{bmatrix} 0 & 1 & 0 & 0 \end{bmatrix} \begin{bmatrix} \beta \\ r \\ p \\ \phi \end{bmatrix}$$

where β is the velocity coordinate, p is the roll rate, and r is the yaw rate. Use the VIs in the Control and Simulation Software to determine the poles of the system. Comment on the system stability. Is this system controllable? Design a linear state feedback to place the closed-loop poles at

$$-1.12, -0.165, (0.162 \pm j0.681) \ .$$

5. In this activity, you will interactively design a linear state feedback for the suspended ball problem discussed in Activity 10.5 using a graphical user interface. Open the **Suspended Ball Example** VI from the library `Activities.llb`. Run the VI. You will see the poles and zeros of the original system on the POLE-ZERO plot. You will also see an indication that the system is unstable. Now move the poles on the pole-zero plot using the cursor. Carefully make the following observations:

   a) In case of complex poles, both the poles are complex conjugates of each other.

   b) As long as the poles continue to be in the right half of the s-plane (the POLE-ZERO plot), the VI continues to indicate that the system is unstable. In this case, you cannot control the position of the ball.

   c) When you move both the poles to the left half of the s-plane, the system becomes stable. You can then design a linear state feedback and control the distance of the ball from the electromagnet.

   d) Close the VI. Do not save any changes.

Development of Full Order Observers and State Feedback Control for the Inverted Pendulum using the GSim Software for LabVIEW

Luis V. Meléndez González
Dr. Gerson Beauchamp Báez, Advisor
Process Control and Instrumentation Laboratory
Electrical and Computer Engineering Department
University of Puerto Rico—Mayaguez Campus

The Challenge To develop a full order observer and state feedback control for the inverted pendulum system.

The Solution As part of the study of several control schemes for the inverted pendulum system, a set of three VI libraries were developed on LabVIEW with the use of the GSim Control and Simulation Software. Two simulation libraries will present simulations of the system under state feedback control, one assuming all state variables are available for measurement and the other using a full order observer provided that the only measurable variables of the system are the cart position and the pendulum angle. The third VI library contains two VI's, one that implements the state feedback control by using backward differentiation while the other implements the state feedback control using the full order observer.

Introduction

The inverted pendulum system has been widely used as an example for undergraduate students in the understanding of several control strategies. As useful as the development of controllers for this highly unstable system, are the development of software which can simulate and implement these types of systems with ease. On this article, we will present the development of simulations of state feedback controllers and full order observers with the use of the GSim software for LabVIEW. We will also present the implementations of the developed control schemes on that same software. A description of the system's model will be presented in the second section. The third section will be devoted to the methodology behind these control schemes. The fourth and fifth section will then explain the way these control schemes were simulated and implemented on Lab-VIEW with GSim. Finally, we will present our results on the use of GSim for the simulation and implementation of these control schemes.

System Description

The inverted pendulum system consists of a rod on a pivot which rotates on top of a cart. The cart moves horizontally on a rail while the pendulum's angle of rotation is perpendicular to the direction of motion of the cart. This system is a simplified model of the attitude control of a space booster on takeoff. The objective of the attitude control problem is to keep the space booster in a vertical position[1]. This model is simplified in the inverted pendulum system since the pendulum can only rotate on one axis. The system used for this implementation is the Quanser Consulting IP-01 inverted pendulum setup. In this setup the cart contains a motor which supplies the input force and a potentiometer which indicates its position with respect to the rail. Another potentiometer is mounted on the axis of rotation of the pendulum to measure the angle with respect to the vertical. Figure 10–15 provides an illustration of the setup.

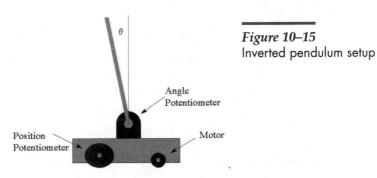

Figure 10–15
Inverted pendulum setup

In order to obtain a state-space representation of the system, we define the following state variables:

x_1 = cart position x_3 = pendulum angle
x_2 = cart linear velocity x_4 = pendulum angular velocity

For this setup, the state-space representation of the system is given by [2]

$$\begin{bmatrix} \dot{x}_1 \\ \dot{x}_2 \\ \dot{x}_3 \\ \dot{x}_4 \end{bmatrix} = \begin{bmatrix} 0 & 1 & 0 & 0 \\ 0 & \dfrac{-K_m^2 K_g^2}{MR_a r^2} & \dfrac{-mg}{M} & 0 \\ 0 & 0 & 0 & 1 \\ 0 & \dfrac{K_m^2 K_g^2}{MLR_a r^2} & \dfrac{(M+m)g}{ML} & 0 \end{bmatrix} \begin{bmatrix} x_1 \\ x_2 \\ x_3 \\ x_4 \end{bmatrix} + \begin{bmatrix} 0 \\ \dfrac{K_m K_g}{MR_a r} \\ 0 \\ \dfrac{-K_m K_g}{MLR_a r} \end{bmatrix} u$$

where:
K_m = motor torque constant K_g = gear ratio of gearbox
      (N • m/Amp)   (dimensionless)
M = mass of cart(kg) m = mass of pendulum(kg)
L = center of gravity of pendulum(m) g = acceleration due to
                                            gravity(m/s^2)
R_a = motor's armature resistance (Ω) r = radius of output gear(m)

Control Schemes for the System

The control scheme to be studied for this system will be state-feedback control. However, since this scheme implies that the state variables of the system are available for measurement, we will treat the system as if all state variables are measured and then as if some of the state variables were present. At this point, we will discuss the concept of state observation.

State-Feedback Control

Consider a linear time invariant system of the form

$$\dot{x}(t) = Ax(t) + Bu(t)$$

where $x(t)$ is a $n \times 1$ state vector, A is a constant $n \times n$ matrix, B is a constant $n \times m$ matrix and $u(t)$ is a $m \times 1$ input vector.[3] The objective of state feedback control is to obtain a gain matrix **K** that shifts the closed loop eigenvalues of the system into a defined location that meets the performance requirements. A condition that is sufficient to achieve this task is for the system to be controllable. That is, the rank of the controllability matrix given by

$$\text{rank}[B \quad AB \ldots A^{n-1}B]$$

must be equal to n, the order of the state vector $x(t)$. If the controllability condition is met, then there exists a gain matrix **K** such that the input $u(t)$ becomes

$$u(t) = -Kx(t)$$

and the closed loop system under state feedback takes the form

$$\dot{x}(t) = (A - BK)x(t)$$

where the eigenvalues of the system will depend directly upon the chosen values of the gain matrix **K**. Figure 10–16 shows a block diagram for the implementation of state feedback control. Although there are several pole placement techniques available to design a gain matrix **K** such that eigenvalue assignment is accomplished, state feedback is seldom used on real-world applications. The reason for this is that not all state variables of the system are available for measurement. In this case, in order to implement the state feedback control, it is necessary to estimate the state variables that are not measurable.

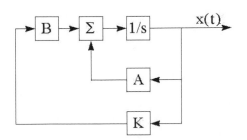

Figure 10–16
Block diagram representation for state feedback control

Full Order State Observers

A state observer is a subsystem which performs an estimation of the state variables of a system given the output variables and the control signal[4]. An observer that estimates all state variables of the system is known as a full order observer while one that estimates only the unmeasurable states is considered a minimum order observer. In our case, we will address the full order observer.

Let us consider the system shown in the previous section but now the measured signals are given as an output $y(t)$. The system representation is now of the form

$$\dot{x}(t) = Ax(t) + Bu$$

$$y(t) = Cx(t)$$

where matrix C is an $m \times n$ matrix. In the case of the full order observer, the condition necessary for state observation is for the system to be observable, that is, the rank of the observability matrix given by

$$\text{rank}[C^T \quad (CA)^T...(CA^{n-1})^T]$$

is equal to n, the order of the state vector $x(t)$. The full order observer has a dynamic model given by [1]

$$\dot{x}_e(t) = Ax_e(t) + Bu(t) + L(Cx(t) - Cx_e(t))$$

Defining the observer error as

$$\dot{x}(t) - x_e(t) = Ax(t) - Ax_e(t) - L(Cx(t) - Cx_e(t))$$

$$\dot{x}(t) - x_e(t) = (A - LC)(x(t) - x_e(t))$$

we can see that the observer error dynamics take a form similar to that of the closed loop form of the state feedback control scheme. This means that if we choose matrix **L** in such a way that the matrix (A − LC) is stable, then the error will converge to zero and the observer will be able to estimate the corresponding state variables. The diagram for the full order observer is given by Figure 10–17.

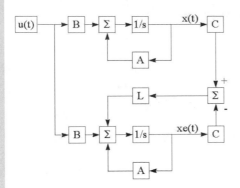

Figure 10–17
Block diagram representation of a full
order observer

If the observability condition is met and a suitable state observer is implemented, then the state feedback control scheme can be implemented by using the observer states $x_e(t)$ as feedback to the gain matrix **K.** The block diagram for the state feedback controller using a full order observer is given in Figure 10–18.

Controller Simulations on LabVIEW

For performance evaluation, two VI libraries were developed for the system. These VI libraries are *Ipend_CT.llb* and *Ipend_DT.llb*, where CT and DT refer to continuous time and discrete time,

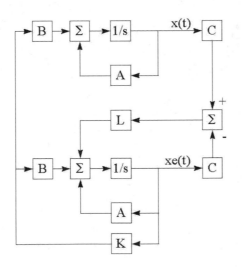

Figure 10–18
Block diagram for state feedback
using a full order observer

respectively. On each of these libraries, we have a VI which simulates the state feedback control and another which simulates the state feedback control with the use of a full order observer. We will explain the construction of the VI's in continuous time and then show the modifications made to implement the discrete time versions.

■ State Feedback Control

The *Full State Feedback(Continuous Time).vi* simulates the state feedback control of the inverted pendulum in continuous time. The front panel of the VI receives from the user the model and continuous time integration parameters and provides the response of the state variables on those intervals. This allows for the VI to be used not only to model the response under state feedback for the inverted pendulum, but also for any other system with four state variables and one input. The model parameters that the VI receives are the state matrices **A** and **B**, the initial state vector $x(0)$, the gain matrix **K** and a reference input u which can be used to model disturbances or reference inputs. The continuous time integration parameters for the VI are the time interval dt, the end time of the simulation and the continuous integrator to be used (RK Fixed, Euler or Adams). Finally, the stop button allows termination of the VI at any time in the simulation and the timing button indicates if the simulation could be run in real time.

The block diagram of the VI is shown in Figure 10–19.

On the first iteration, the initial conditions on the *GSim integration(array).vi* are set to $x(0)$ as well as the shift register that carries the values of $x(t)$. Using matrix algebra manipulation, we form the vector corresponding to the differential equation of the closed loop system described previously as

$$\dot{x}(t) = (A - BK)x(t) + Bu(t)$$

which is then integrated by the *GSim integration(array).vi*. The integration parameters are specified to this VI by the *GSim Initialize.vi* in the time interval dt with integration method specified by the user. Finally, the new state vector $x(t)$ is plotted as well as fed to the shift register for the next iteration.

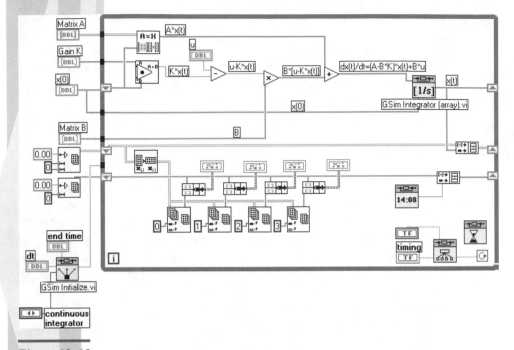

Figure 10–19
Block diagram for state-feedback in continuous time

The discrete time version of this VI is found in the *Full State Feedback(Discrete Time).vi*. The model parameters on the front panel do not change, since the order of the system does not change upon discretization, just the value of the state matrices. However, the integration parameters do change in the form that the time interval dt is replaced by the control Sample Time and the integration methods are replaced. The block diagram for the discrete time version of the VI, shown in Figure 10–20, changes only in the conversion from $x(k + 1)$ to $x(k)$.

On the continuous time case, we had to integrate the vector $\dot{x}(t)$ to obtain $x(t)$. In this case, the resulting expression is of the form

$$x(k + 1) = (A - BK)x(k) + Bu(k)$$

and in order to obtain the next value of $x(k)$, we used the GSim Unit Delay. vi once for each state variable. Since the general structure of the process is the same, the only section that gets altered is the conversion from $x(k + 1)$ to $x(k)$.

State Feedback Control using a Full Order Observer

The *Full State Observer(Continuous Time).vi* simulates the state feedback control using a full order observer. The observer will estimate the state variables of the system provided that the only measurable variables are the cart position and the pendulum angle. On the front panel of the VI, the model representation of

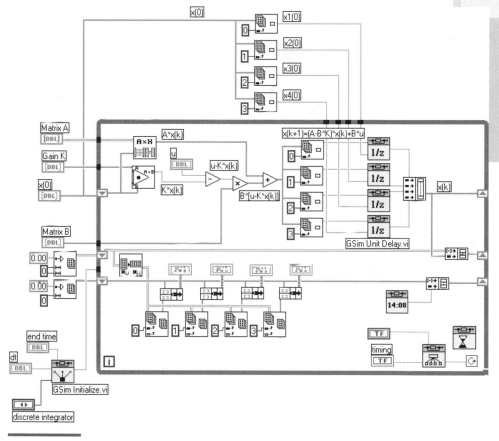

Figure 10–20
Block diagram for state feedback in discrete time

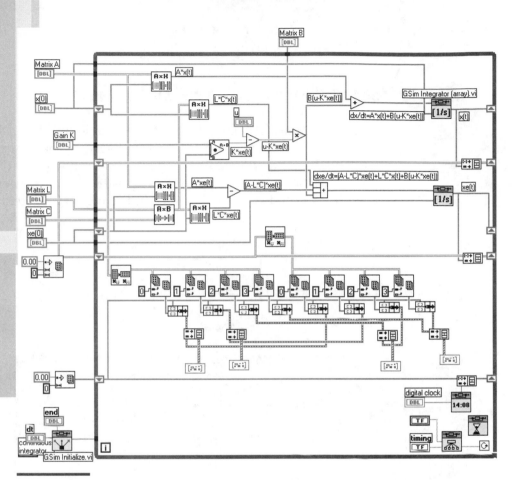

Figure 10–21
Block diagram for the state feedback with full order observer in continuous time

the system contains the same parameters of the state feedback version and adds three parameters, the output matrix **C**, the observer gain matrix **L** and the observer initial state vector $x_e(t)$. Also, notice that the graphs of the state variables show both the state variables $x(t)$ and their estimated counterparts $x_e(t)$.

The block diagram of the VI is shown in Figure 10–21.

On the first iteration, the initial conditions on the two *GSim integration(array).vi* are set to $x(0)$ and $x_e(0)$ respectively, as well as their corresponding shift registers that carry the values of $x(t)$

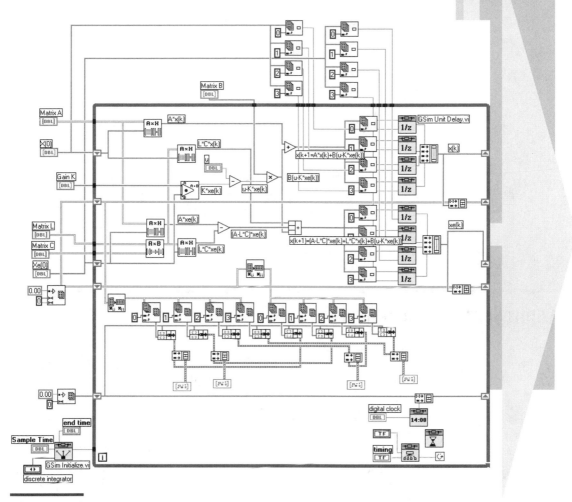

Figure 10–22
Block diagram of full order observer in discrete time

and $x_e(t)$. Using matrix algebra, we form the two differential equations for the state and the observer of the form

$$\dot{x}(t) = Ax(t) - BKx_e(t) + Bu(t) \rightarrow System$$

$$\dot{x}_e(t) = (A - LC)x_e(t) + LCx(t) - BKx_e(t) + Bu(t) \rightarrow Observer$$

which are both integrated by their respective *GSim integration(array).vi*. The integration parameters are specified to this VI by the *GSim Initialize.vi* in the time interval dt with integration method specified by the user. Finally, the new state vectors $x(t)$ and $x_e(t)$ are plotted as well as fed to their respective shift register for the next iteration. Notice that as the observer states $x_e(t)$ converge to $x(t)$, the terms $-LCx_e(t)$ and $LCx(t)$ cancel each other on the observer differential equation and it becomes the same as the system's equation.

The discrete time version of this simulation, whose block diagram is shown in Figure 10–22, is the *Full State Observer(Discrete Time).vi.* As in the case of the state feedback control, the only changes in this VI to its continuous time counterpart are the change of the *GSim Integrator(Array).vi* blocks for the respective *GSim Unit Delay. vi* blocks and the replacement of the continuous time integration parameters for the discrete time parameters.

Controller Implementations on LabVIEW

In order to implement the control schemes, the VI library *Ipend_RT.llb* was developed. This library contains two VI's, one implements the state feedback control while the other implements the full order observer before applying the state feedback control. The *Full State Feedback(Implementation).vi* performs an approximated state feedback control scheme for the inverted pendulum. The block diagram for this VI is shown in Figure 10–23.

This scheme is an approximation since we are not actually measuring all of the state variables, specifically the velocity variables. These values are obtained by using the *GSim Derivative.vi*, which uses backward differentiation to obtain an approximate of the velocities. The VI calls the *GSim Collect State Vector.vi*, which is a modification of the *GSim Data In.vi* in order to collect two channels. The Data Acquisition(DAQ) was performed using National Instruments AT-MIO-16L-9 DAQ card. Once these values are read, the *Voltage Conditioning-Position and Angle.vi* performs the necessary conversion from voltages into the respective cart position and pendulum angle variables. Once these variables are obtained, they are filtered using the *GSim Discrete Transfer Function.vi* and differentiated using the *GSim Derivative.vi.* After this, the dot product of the resulting state vector

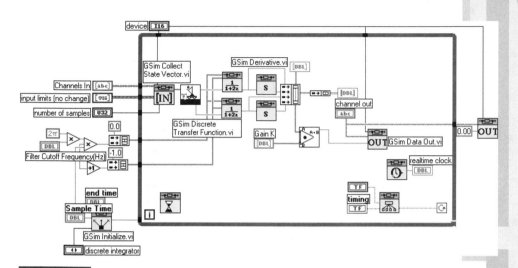

Figure 10–23
Block diagram for implementation of state feedback control

with the gain matrix **K** is obtained and fed to the *GSim Data Out.vi.*

For the implementation of the state feedback using the full order observer, the VI named *Full State Observer.vi*, whose block diagram is shown in Figure 10–24, was developed.

The VI performs the same data acquisition and voltage conversion as the *Full State Feedback(Implementation).vi.* The resulting variables are grouped in vector form to represent the entry and then arranges the vector representing the observer's differential equation as in the observer simulation. Once this vector goes through the *GSim Unit Delay* VI's, the dot product of the gain matrix gain K with the resulting $x_e(k)$ vector is obtained and fed to the *GSim Data Out.vi.*

Conclusions

The implementation of both control strategies yielded outstanding results. The performance of the state feedback control using the full order observer surpassed the state feedback control

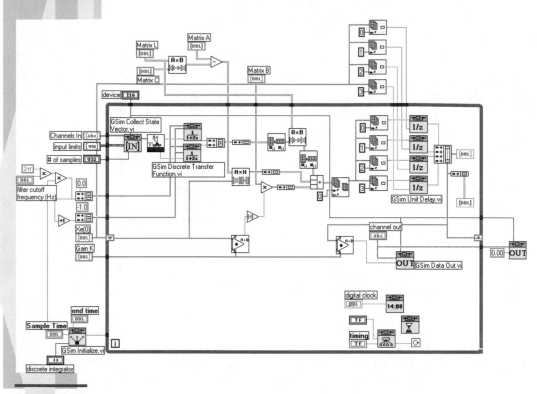

Figure 10–24
Block diagram for the implementation of the full order observer

approximation. This was due to two different factors. The procedure of differentiating incoming signals amplifies any incoming noise that accompanies it. This makes the resulting derivative value less reliable. The other reason is that the observer is more robust to model uncertainties. Therefore, while the observer tries to maintain the estimation error close to zero, it makes up for these model uncertainties.

The simulation of the system in continuous and discrete time was very useful in our design process for the control of the inverted pendulum system. One of the advantages of the simulation, specially in the case of the full order observer, is that the modifications needed to change the VI from a simulation into an implementation was minimal since the structure is basically the same and we only needed to bring the DAQ input and output blocks. The block diagram structure of LabVIEW allows a better visualization and understanding of the control scheme than code written program and with the inclusion of the GSim software, the user can visualize the block diagram in the same way that it is drawn on paper. The GSim software opens the door into allowing easy implementation and simulation of many control strategies.

■ References

[1] Ogata, K., *Modern Control Engineering,* 2nd ed. Englewood Cliffs, N.J.: Prentice Hall, Inc., 1990

[2] Quanser Consulting, *Inverted Pendulum Experiment and Solution,* Quanser Consulting, 1991

[3] Jamshidi, M., Tarokh, M., and Bahram, S., *Computer-Aided Analysis and Design of Linear Control Systems,* Englewood Cliffs, N.J.: Prentice Hall, Inc., 1992

[4] Ogata, K., *Discrete-Time Control Systems,* 2nd ed. Englewood Cliffs, N.J.: Prentice Hall, Inc., 1995

■ Contact Information

Luis V. Meléndez González
Graduate Student-Researcher
Process Control and Instrumentation Laboratory, Electrical Engineering, Depts.,
University of Puerto Rico-Mayaguez Campus, Mayaguez, PR 00681-5000
787-832-4040 x2189
Fax: 787-831-7564
E-Mail: luism@exodo.upr.clu.edu

Overview

In this chapter, you will learn how to use the Digital Filter Design (DFD) Toolkit to design FIR and IIR filters to meet required specifications. You will also see how to use the DFD Toolkit to analyze your filter design in terms of its frequency response, impulse and step responses, and its pole-zero plot.

GOALS

- Review the basics of digital filters.
- Become comfortable with the Digital Filter Design Toolkit.
- Design IIR filters.
- Design FIR filters.
- Use the Digital Filter Design Toolkit to analyze your filter design.

KEY TERMS

- design
- analyze
- filter response
- transition region
- passband
- stopband
- ripple
- filter specifications
- filter coefficients
- IIR filters

Digital Filter Design Toolkit

11

Review of Digital Filters

In Chapter 6 on Digital Filters, you learned about the basic theory behind the operation of digital filters. Below is a brief review of some of the important information you need to know to use the Digital Filter Design (DFD) Toolkit.

■ Filtering

Filtering is one of the most common signal processing techniques and is the process by which the frequency content of a signal is altered. Some of the practical applications of filtering are in the bass and treble controls of your stereo to adjust the frequency response, in the tuning circuits of your radio and television receivers to select a particular channel, in telephone handsets to limit the frequency content of the sound signals to 3 KHz, and many others in the audio, telecommunications, geophysics, and medical fields, such as to alter frequency components (digital equalizer), remove unwanted signals (power line interference) and production of different sound effects (frequency synthesizers).

Why Digital Filters?

Until the advent of the computer age, filtering was in analog form using resistors, inductors, and capacitors. Both the input and the output of the filter were analog signals. Designing analog filters is a specialized task requiring a good mathematical background and proper understanding of the filtering process. However, with the widespread use of computers, digital representation and processing of signals gained immense popularity due to the numerous advantages that digital signals have over their analog counterparts. Because of this, analog filters have gradually been replaced by digital filters. The advantages of digital filters over analog filters are

- They are software programmable, and so are easy to "build" and test.
- They require only the arithmetic operations of multiplication and addition/subtraction and so are easier to implement.
- They are stable (do not change with time or temperature) and predictable.
- They do not drift with temperature or humidity or require precision components.
- They have a superior performance-to-cost ratio.
- They do not suffer from manufacturing variations or aging.

Filter Response Characteristics

The range of frequencies that a filter passes through it is known as the *passband*, whereas the range of frequencies that are attenuated is known as the *stopband*. Between the passband and the stopband is a *transition* region where the gain falls from one (that is, 0 dB in the passband) to zero or a very small value (in the stopband). The passband, stopband, and the transition region for a lowpass filter are shown in Figure 11–1.

Figure 11–1 also shows the *passband ripple*, the *stopband attenuation*, and the *cutoff frequency*, three specifications that are needed in designing digital filters. The passband ripple (in dB) is the maximum deviation in the passband from 0 dB, whereas the stopband attenuation is the minimum attenuation (in dB) in the stopband. In the DFD Toolkit, the passband ripple is also referred to as the *passband response*.

The Digital Filter Design Toolkit

You have already been introduced to the digital filter VIs available in the LabVIEW/BridgeVIEW Analysis Library. Using these VIs is one way to design your digital filters. However, National

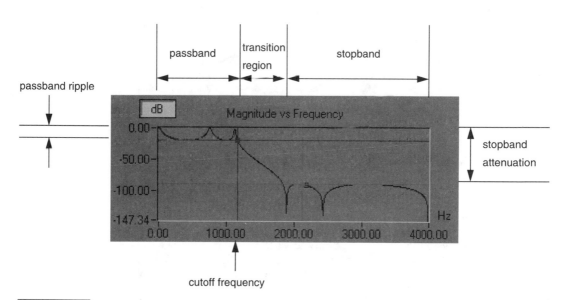

Figure 11–1
Magnitude vs. frequency response of a lowpass digital filter

Instruments also provides the Digital Filter Design (DFD) Toolkit, which is a complete filter design and analysis tool you can use to design digital filters to meet your precise filter specifications. You can graphically design your IIR and FIR filters, interactively review filter responses, save your filter design work, and load your design work from previous sessions. If you have a National Instruments Data Acquisition (DAQ) device, you can perform real-world filter testing from within the DFD application. You can view the time waveforms or the spectra of both the input signal and the filtered output signal while simultaneously redesigning your digital filters.

After you design your digital filter, you can save the filter coefficients to a file on your drive. The filter coefficient files can then be loaded for later implementation by LabVIEW, BridgeVIEW, LabWindows/CVI, or any other application. The diagram in Figure 11–2 shows you the conceptual overview of the DFD Toolkit.

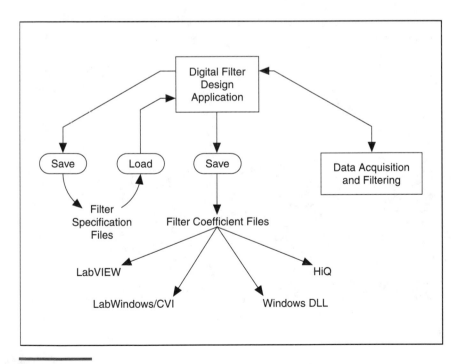

Figure 11–2
Using the Digital Filter Design Toolkit

■ Main Menu

When you launch the DFD application, you get the front panel shown in Figure 11–3, which is referred to as the Main Menu.

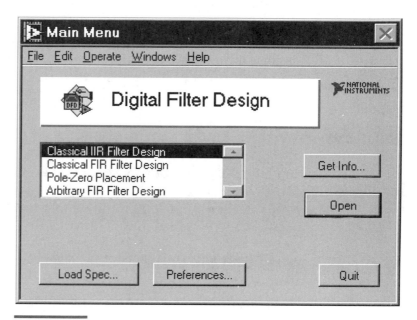

Figure 11–3
Main Menu of the Digital Filter Design Toolkit

■ Design Options

From the Main Menu, you can choose any of the following four methods for designing digital filters:

- *Classical IIR Filter Design*—for designing IIR filters by specifying the frequency response characteristics.
- *Classical FIR Filter Design*—for designing FIR filters by specifying the frequency response characteristics.

- *Pole-Zero Placement*—for designing either IIR or FIR filters by adjusting the location of the poles and zeros (in the z-plane) of the filter transfer function.

- *Arbitrary FIR Filter Design*—for designing FIR filters by specifying the gain of the filter at selected (two or more) frequencies.

If you double-click on one of the four design selections (or single-click on a selection and then click on the **Open** button) in the Main Menu, the DFD application loads and runs the selected design panel, in which you can design your filter.

■ Loading Previously Saved Specifications

You can also load a previously designed filter specification file directly from the Main Menu by clicking on the **Load Spec…** button. You will then be prompted to select the filter specification file that you saved during previous design work.

■ Customizing the DFD Application

By clicking on the **Preferences…** button in the Main Menu, you can edit your DFD application preferences for future design sessions.

The selections in the window in Figure 11–4 tell the DFD application to preload one or more of the filter design panels into memory when it is started. Preloading filter designs increases the time taken for the Main Menu to open. However, when you select a particular design panel from the Main Menu, the corresponding design panel opens almost immediately. If you have a limited amount of memory on your computer, you may want to reconsider how many (if any) of the design panels you preload into memory.

■ Quitting the DFD Application

Choose the **Quit** button to exit the DFD application.

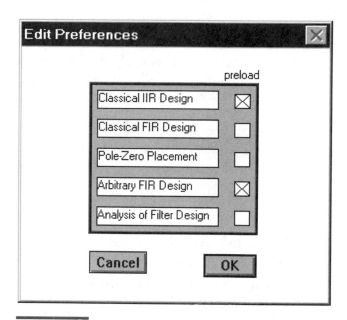

Figure 11–4
The Edit Preferences menu

■ Manipulating the Graphical Display

Each design panel has a graphical display showing you the frequency response of the filter that you are designing. The graphical displays provide you with considerable flexibility in adjusting the way in which you see the magnitude response. Before you move on to the next section (Designing IIR Filters), you should become familiar with the display options. You can see the graphical display by selecting **Classical IIR Filter Design** on the DFD Main Menu and clicking on the **Open** button.

■ Panning and Zooming Options

The graph palette has controls for panning (scrolling the display area of a graph) and for zooming in and out of sections of the graph. A graph with its accompanying graph palette is shown in Figure 11–5.

If you press the x autoscale button, shown at left, the DFD application autoscales the X data of the graph. If you press the y autoscale button, shown at left, the DFD application autoscales the Y data of the graph. If you want the graph to autoscale either of the scales continuously, click on the lock switch, shown at the left, to lock autoscaling on.

The scale format buttons, shown on the left, give you run-time control over the format of the X and Y scale markers, respectively.

You use the remaining three buttons to control the operation mode for the graph.

Normally, you are in standard operate mode, indicated by the plus or crosshatch. In operate mode, you can click in the graph to move cursors around.

The panning tool switches to a mode in which you can scroll the visible data by clicking and dragging sections of the graph.

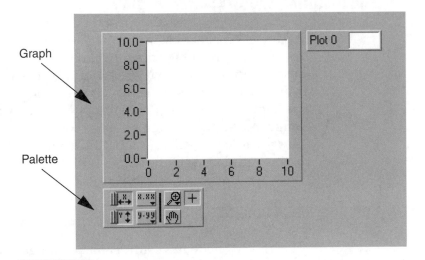

Figure 11–5
Graph with accompanying palette

The zoom tool zooms in on a section of the graph by dragging a selection rectangle around that section. If you click on the zoom tool, you get a pop-up menu you can use to choose some other methods of zooming. This menu is shown in Figure 11–6.

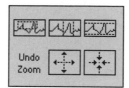

Figure 11–6   Zooming options

A description of other options follows.

 Zoom by rectangle.

 Zoom by rectangle, with zooming restricted to x data (the y scale remains unchanged.

 Zoom by rectangle, with zooming restricted to y data (the x scale remains unchanged).

Undo Zoom   Undo last zoom. Resets the graph to its previous setting.

 Zoom in about a point. If you hold down the mouse on a specific point, the graph continuously zooms in until you release the mouse button.

 Zoom out about a point. If you hold down the mouse on a specific point, the graph continuously zooms out until you release the mouse button.

 For the last two modes, you can zoom in and zoom out about a point. Shift-clicking zooms in the other direction.

■ Graph Cursors

Figure 11–7 is an illustration of a waveform graph showing two cursors and the cursor movement control.

You can move a cursor on a graph or chart by dragging it with the operating tool, or by using the cursor movement control. Clicking the arrows on the cursor movement control causes all cursors selected to move in the specified direction. You select cursors by moving them on the graph with the operating tool.

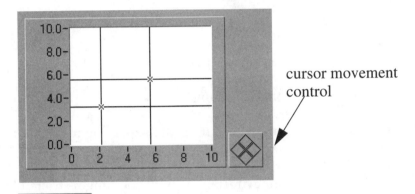

cursor movement control

Figure 11–7
Graph with cursor movement control

Designing IIR Filters

You have seen that IIR filters are digital filters whose impulse response is infinitely long. In practice, the impulse response decays to a very small value in a finite amount of time. The output of an IIR filter depends on both the previous and past inputs and the past outputs. Let $(N_y - 1)$ be the number of past outputs and $(N_x - 1)$ be the number of past inputs. Denoting the inputs to the filter as $x[i]$ and the outputs as $y[i]$, the equation for calculating the current output of an IIR filter can be written as:

$$a_0 y[i] = -a_1 y[i-1] - a_2 y[i-2] - \ldots - a_{N_y - 1} y[i - (N_y - 1)] + $$
$$b_0 x[i] + b_1 x[i-1] + b_2 x[i-2] + \ldots + b_{N_x - 1} x[i - (N_x - 1)]$$

$$y[i] = \frac{1}{a_0} \left(- \sum_{j=1}^{N_y - 1} a_j y[i-j] + \sum_{k=0}^{N_x - 1} b_k x[i-k] \right) \tag{11.1}$$

In Equation 11.1, the b_k are known as the *forward* coefficients and the a_j are known as the *reverse* coefficients. The output sample at the present sample index i is the sum of scaled present and past inputs ($x[i]$ and $x[i-k]$ when $k > 0$) and scaled past outputs ($y[i-j]$ when $j > 0$). Usually N_x is equal to N_y and the *order* of the filter is $N_x - 1$.

■ Implementation of IIR filters

IIR filters implemented in the form given by Equation 11.1 are known as *direct form* IIR filters. Direct form implementations are usually sensitive to errors due to the number of bits used to represent the values of the coefficients (quantization error) and to the precision used in performing the computations. In addition, a filter designed to be stable can become unstable when the number of coefficients (and thus, the *order* of the filter) is increased.

A less sensitive implementation is obtained by breaking up the higher order direct form implementation into an implementation that has several cascaded filter stages, but where each filter in the cascade is of a lower order, as shown in Figure 11–8.

Typically, each lower-order filter stage in the cascade form is a second-order stage. Each second-order stage can be implemented in the *direct form*, where you must maintain two past inputs, ($x[i-1]$ and $x[i-2]$) and two past outputs, ($y[i-1]$ and $y[i-2]$). The output of the filter is calculated using the equation

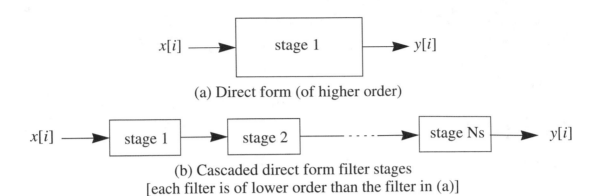

(a) Direct form (of higher order)

(b) Cascaded direct form filter stages
[each filter is of lower order than the filter in (a)]

Figure 11–8
Direct form and cascaded direct form implementation of filters

$$y[i] = b_0x[i] + b_1x[i-1] + b_2x[i-2] - a_1y[i-1] - a_2y[i-2] \tag{11.2}$$

The implementation can also be done in the more efficient *direct form II*, where you maintain two internal states, ($s[i-1]$ and $s[i-2]$). The output of the filter is then calculated as follows:

$$s[i] = x[i] - a_1s[i-1] - a_2s[i-2] \tag{11.3}$$
$$y[i] = b_0s[i] + b_1s[i-1] + b_2s[i-2]$$

The *direct form II* is a more efficient structure because it uses less memory. It needs to store only two past internal states, ($s[i-1]$ and $s[i-2]$), whereas the *direct form* structure needs to store four past values, ($x[i-1]$, $x[i-2]$, $y[i-1]$, and $y[i-2]$).

■ IIR Filter Designs

Depending on whether the ripple in the filter's frequency response lies in the passband and/or the stopband, IIR filters are classified as follows:

- Butterworth: No ripple in either the passband or the stopband
- Chebyshev: Ripple only in the passband

- Inverse Chebyshev: Ripple only in the stopband.
- Elliptic: Ripple in both the passband and the stopband

The advantage of using Butterworth filters is for applications where you want a smooth filter response and no ripple. However, a higher-order Butterworth filter (as compared to Chebyshev, inverse Chebyshev, or elliptic) is generally required for similar filter specifications. This increases the processing time for Butterworth filters.

The advantage of Chebyshev and inverse Chebyshev filters over Butterworth filters is their sharper transition band for the same order filter. On deciding which of these two types of filters to use, the advantage of inverse Chebyshev filters over Chebyshev filters is that they distribute the ripple in the stopband instead of in the passband.

Because elliptic filters distribute the ripple in both the passband and the stopband, they can usually be implemented with the smallest order for similar filter specifications. Hence, they have faster execution speeds than either of the other filters.

■ Applications of IIR Filters

The advantage of IIR filters over FIR filters is that IIR filters usually require fewer filter coefficients to perform similar filtering operations. Thus, they execute much faster and do not require extra memory.

The disadvantage of IIR filters is that they have nonlinear phase characteristics. Hence, if your application requires a linear phase response, then you should use an FIR filter instead. However, for applications where phase information is not necessary, such as in simple signal power monitoring, then IIR filters can be used. Thus, the bandpass filters in real-time octave analyzers are commonly IIR filters, because of their faster speed. Another reason is because it is only necessary to determine the distribution of sound power over several frequency bands, but there is no need to determine the phase of the signal. The applications of such octave analyzers where phase information is not important are vibration tests of aircraft and submarines, testing of appliances, and noise level testing.

■ A Note About the Activities in this Chapter

To perform most of the activities in this chapter, you will need a copy of the demo version of the DFD Toolkit. The demo version can be found on the CD-ROM that comes with this book. Follow the instructions in the preface of this book for directions on how to install the DFD demo version.

Some of the activities in this chapter require the full version of the DFD Toolkit. These activities will be clearly marked by a note placed at the beginning of the activity. You will not be able to perform these activities with the demo version.

Additional functionality available in the full version of the DFD Toolkit, that is not there in the demo version, is

1. Ability to save the coefficients of the designed filter to a file for possible use by other filtering programs.
2. Data acquisition.
3. Capability to design filters using the Classical FIR Filter Design option on the Main Menu.
4. Capability to design filters using the Arbitrary FIR Filter Design option on the Main Menu.

Activity 11-1

Objective: To design an IIR bandpass filter for use in an octave analyzer

1. Launch the **Digital Filter Design Toolkit** application by double-clicking on dfddem32.exe in the `Digital Filter Design Toolkit` folder. The Main Menu panel opens as shown.

■ Front Panel

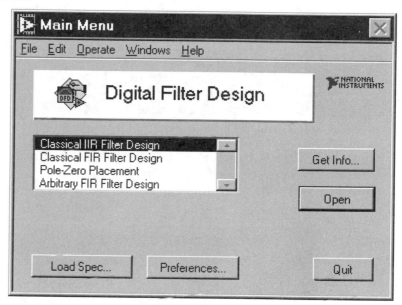

2. In the Main Menu, select **Classical IIR Filter Design** and click on the **Open** button. The design panel of the **Classical IIR Filter Design** opens as shown.

■ Front Panel

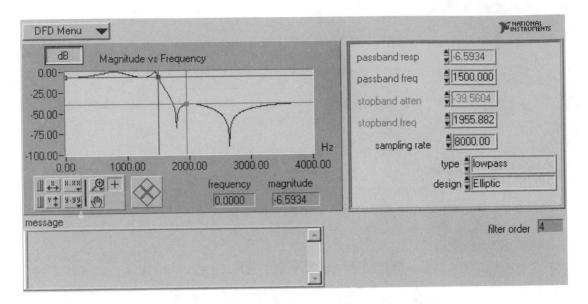

The plot on the left shows the magnitude versus frequency response characteristic of the filter you design. The specifications for your filter can be entered in the text entry portion at the upper-right side of the design panel:

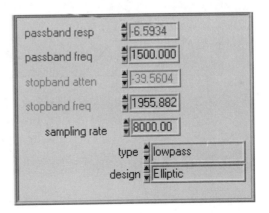

The **passband resp** is the minimum gain allowed in the passband. This is represented by the horizontal blue cursor line in the **Magnitude vs. Frequency** plot. With the reference at 0 dB, it is also the same as the passband ripple.

The **passband freq** determines the frequency edges of the passband. For lowpass and highpass filters, you have only one frequency edge. For bandpass and bandstop filters, you will have two. These frequencies are represented by the vertical blue lines in the **Magnitude vs. Frequency** plot.

The **stopband atten** is the minimum attenuation in the stopband. The horizontal red cursor line represents this attenuation in the **Magnitude vs. Frequency** plot.

The **stopband freq** determines the frequency edges of the stopband. For lowpass and highpass filters, you have only one frequency edge. For bandpass and bandstop filters, you will have two. These frequencies are represented by the vertical red lines in the **Magnitude vs. Frequency** plot.

The **sampling rate** control specifies the sampling rate in samples per second (Hz).

The **type** control specifies one of the four classical filter types:

- Lowpass
- Highpass
- Bandpass
- Bandstop

The **design** control specifies one of the four classical filter design algorithms:

- Butterworth
- Chebyshev
- Inverse Chebyshev
- Elliptic

Below the text entry portion is an indicator showing the order of the designed IIR filter:

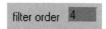

The DFD application automatically estimates the filter order to be the lowest possible order that meets or exceeds the desired filter specifications.

At the bottom left of the Classical IIR Filter Design panel is the **message** window where error messages are displayed:

You will use the Classical IIR Filter Design panel to design an IIR bandpass filter that can be used in an octave analyzer. Octave analyzers are used in applications where you need to determine how the signal power is distributed over a particular frequency range. These applications include the fields of architectural acoustics, noise and vibration tests in aircraft and submarines and testing of household appliances.

An octave analyzer uses bandpass filters to separate the signal power into several frequency bands. The American National Standards Institute requires that these filters adhere to certain specifications. Some specifications for one of these filters are

  fp1 = 890.90 Hz

  fp2 = 1122.46 Hz

  maximum passband ripple ≤ 50 millibels

  fs1 = 120.48

  fs2 = 8300

  stopband attenuation ≥ 65 dB

Because the purpose of the bandpass filter is to determine the level of sound power in a particular frequency band and the phase information in the signal is not being used, it is not necessary for the filter to be linear phase. Hence, you can choose an IIR filter for

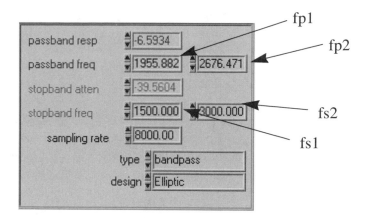

this application. You will use the DFD Toolkit to design the IIR filter to meet these specifications.

3. Change the **type** control in the text entry box to *bandpass*.

The explanation of the controls in the text entry box is the same as before, except note that because you have selected a bandpass filter, there are two controls for the **passband freq** and the **stopband freq**. They are denoted by fp1, fp2, fs1, and fs2, as shown in the following figure, and bear the following relationship:

fs1 < fp1 < fp2 < fs2

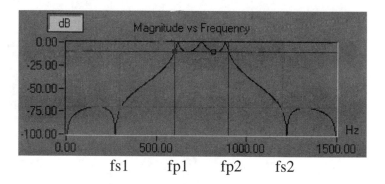

4. Looking at the specifications in step 2, enter the following values in the controls in the text entry box

design	Elliptic
type	bandpass
sampling rate	25600
stopband freq	120.48 and 8300
stopband atten	-65
passband freq	890.90 and 1122.46
passband resp	-0.5

Note that 50 millibels = $\dfrac{50}{1000}$ Bels = $\dfrac{50}{100}$ decibels = 0.5 dB. That is why you entered the passband response as -0.5 dB.

■ Front Panel

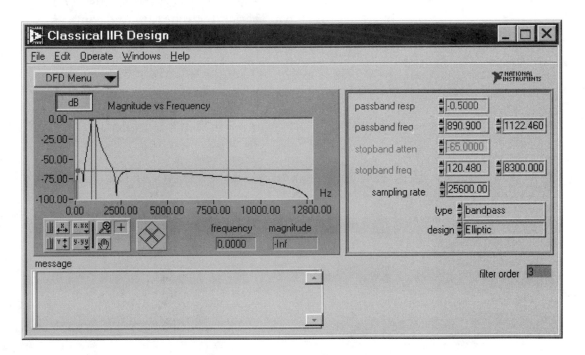

5. The default filter design is Elliptic. Change the filter design one by one to Butterworth, Chebyshev, and Inverse Chebyshev. Note the order of the filter. For the same filter order, which of the four filter designs has the sharpest transition region?

6. You can save the specifications of the filter in a file for later use. From the **DFD Menu,** select **Save Spec.** When asked for the name of the file in which to save the filter specifications, type in `bandpass.iir`. In Activity 11-3, you will load this file and analyze the characteristics of the filter that you have just designed.

7. You can also save the designed filter coefficients in a file for later use with the DFD Toolkit, or with other programs. From the **DFD** menu, select **Save Coeff.** Save the file as a *text* file. When asked for the name of the file in which to save the filter coefficients, type `bpiir.txt`. The format of the text file in which the coefficients are stored is given at the end of this chapter.

This step #7 can only be performed with the full version of the DFD Toolkit.

8. Now that you have saved the filter specifications, and the filter coefficients, you can close the application. Select **File » Close** to close the Classical IIR Filter Design panel. Then select **Quit** in the Main Menu to exit the DFD Toolkit.

■ End of Activity 11-1

Designing FIR Filters

As opposed to IIR filters, whose output depends on both its inputs and outputs, the output of an FIR filter depends only on its inputs. Because the current output is independent of past outputs, its impulse response is of finite length. The output of a general FIR filter is given by

$$y[i] = b_0 x[i] + b_1 x[i-1] + b_2 x[i-2] + \ldots + b_N x[i-N] \qquad (11.4)$$

where N is the order of the filter and b_0, b_1, ..., b_N, are its coefficients.

FIR filters have certain advantages as compared to IIR filters.

- They can achieve linear phase response, and hence they can pass a signal without phase distortion.

- They are always stable. During filter design or development, you do not need to worry about stability concerns.

- FIR filters are simpler and easier to implement.

■ Applications of FIR filters

Many applications require the filters to be linear phase. In this case, you should use FIR filters. However, FIR filters generally need to be of a higher order than IIR filters to achieve similar magnitude response characteristics. So, if linear phase is not necessary, but speed is an important consideration, you can use IIR filters instead.

In applications where a signal needs to be reconstructed after it has been split up into several frequency bands, it is important that the filter be linear phase. In the reconstruction process, the loss of phase could result in the reconstructed signal being quite different from the original one. An example of such an application is in the Wavelet and Filter Bank Toolkit, where you reconstruct the original signal (either a 1D waveform or a 2D image) from the wavelet coefficients. The filters used are FIR filters. The applications are in areas where phase information is an important consideration, such as for noise removal and data compression.

■ Classical FIR Filter Design and Arbitrary FIR Filter Design

Using the DFD Toolkit, you can design FIR filters in the Classical FIR Filter Design panel (shown in Figure 11–9) which is very similar to the Classical IIR Filter Design panel that you have seen.

■ Front Panel

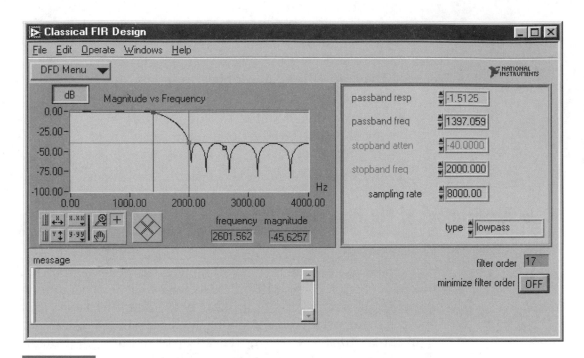

Figure 11–9
Classical FIR Filter Design panel

The panel includes a graphical interface with the **Magnitude vs. Frequency** graph and cursors on the left side, and a text-based interface with digital controls on the right side. The differences are the absence of the **design** control in the text entry box and the addition of the minimize filter order control below the text entry box.

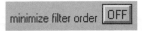

This button controls whether the DFD application minimizes the estimated filter order. If this button is OFF, the DFD application uses a fast formula to estimate the filter order to meet or exceed the desired filter specifications. If this button is ON, the DFD application iteratively adjusts the filter order until it finds the minimum order that meets or exceeds the filter specifications.

The FIR filters that are designed use the Parks-McClellan equiripple FIR filter design algorithm and include the lowpass, highpass, bandpass, and bandstop types. The Parks-McClellan algorithm minimizes the difference between the desired and actual filter response across the entire frequency range.

You can also design FIR filters with an arbitrary frequency response by selecting **Arbitrary FIR Filter Design** in the Main Menu.

"Arbitrary" means that you can specify exactly what the magnitude of the filter response should be at specific frequencies. In the next activity, you will design an FIR filter by specifying an arbitrary frequency response.

Activity 11-2

Objective: To design an FIR filter which filters the data according to A-weighting

The human sense of hearing responds differently to different frequencies and does not perceive sound equally. Certain filters are used to filter the sound applied at their input such that they mimic the human hearing response to audio signals. An application of this is in third-octave analyzers, where, to mimic the response of the human ear, the analyzer output is weighted according to Table 11–1.

This activity requires the full version of the DFD Toolkit.

Table 11-1   *Table showing A-weighting*

Frequency (Hz)	Weighting (dB)	Frequency (Hz)	Weighting (dB)
10	-70.4	500	-3.2
12.5	-63.4	630	-1.9
16	-56.7	800	-0.8
20	-50.5	1000	0
25	-44.7	1250	+0.6
31.5	-39.4	1600	+1.0
40	-34.6	2000	+1.2
50	-30.3	2500	+1.3
63	-26.2	3150	+1.2
80	-22.5	4000	+1.0
100	-19.1	5000	+0.5
125	-16.1	6300	-0.1
160	-13.4	8000	-1.1
200	-10.9	10000	-2.5
250	-8.6	12500	-4.3
315	-6.6		
400	-4.8		

This type of weighting is known as *A-weighting*.

1. Launch the DFD Toolkit.
2. Select **Arbitrary FIR Filter Design** in the **Main Menu**. You will access the following design panel:

■ Front Panel

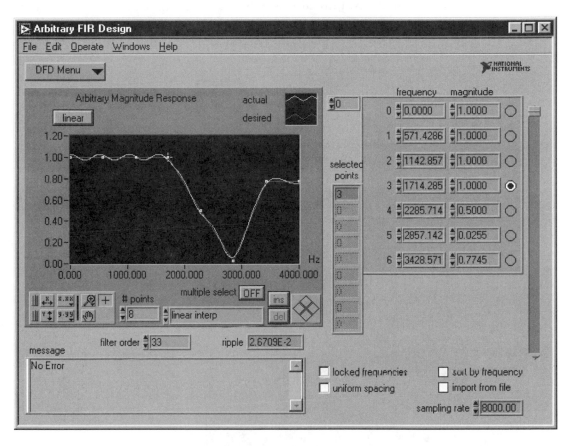

The panel includes a graphical interface with the **Arbitrary Magnitude Response** graph on the left side and a text-based interface with digital controls on the right side. In the array on the right-hand side, you can enter or modify the array magnitude response points (frequency and magnitude). From these points, the DFD application forms a desired magnitude response that covers the entire frequency range from 0.0 to half the sampling rate. The DFD application then takes this desired response, along with the filter order, and uses the Parks-McClellan algorithm to design an optimal equiripple FIR filter.

The graph in Figure 11–10 plots the desired and actual magnitude response of the designed FIR filter.

The *y*-axis is in linear or decibel units, depending on how you set the button in the upper-left corner of the graph. This button controls the display units (linear or decibel) of all the magnitude controls and displays. These controls and displays include the **Arbitrary Magnitude Response** graph (*y*-axis) and the magnitudes in the array of frequency-magnitude points. The *x*-axis is in Hertz. The full scale ranges from 0.0 to Nyquist (sampling rate/2).

| dB |

The array in Figure 11–11 is the array of frequency-magnitude points the DFD application uses to construct the desired filter magnitude response. The DFD application forms the desired filter response by interpolating between these points.

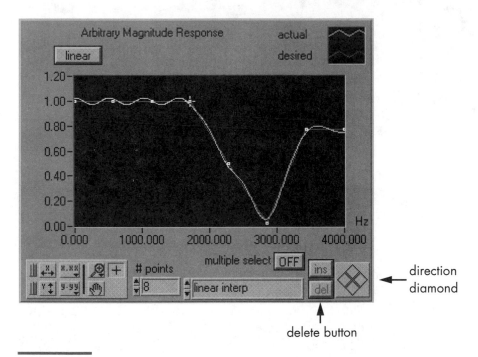

Figure 11–10
Desired and actual magnitude response

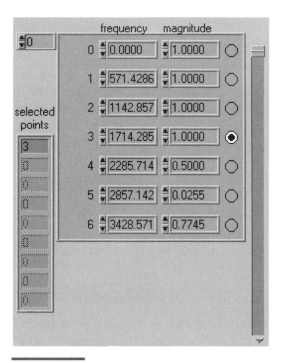

Figure 11–11
Array of frequency-magnitude points

The frequency of each point is in Hertz and the magnitude is in linear or decibel units of gain, depending on the setting of the button in the upper-left corner of the **Arbitrary Magnitude Response** graph.

You can select points in this array by clicking in the circle to the right of each point. You can then delete the selected points by clicking on the delete button (explained later), or move them by clicking on the desired direction diamond in the lower-right corner of the **Arbitrary Magnitude Response** graph.

The **# points** control specifies the total number of frequency-magnitude points the DFD application uses to create the desired filter magnitude response. Reducing this number deletes points from the end of the frequency-magnitude array, while increasing this number inserts the additional number of points to the right of the selected point.

If you want to select more than one frequency-magnitude point on the response graph, you should set the **multiple selection** button to ON. Clicking on a point you already selected removes that point from the selection list:

The interpolation control selects the type of interpolation the DFD application uses to generate the desired response from the array of frequency-magnitude points:

Choose linear interpolation to create "flat" filters (lowpass, highpass, bandpass, and bandstop). Choose spline interpolation to create smoothly varying filters.

To insert a frequency-magnitude point between the selected point and the next point, click on the **ins** button.

If the selected point is the last point in the frequency-magnitude array, the DFD application inserts the new point between the last two points of the array. The DFD application inserts new points halfway along the line connecting the two outer points.

To delete the selected frequency-magnitude points, click on the **del** button. The DFD application deletes all selected points.

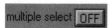

These points are the selected frequency-magnitude points. You can select points on the **Arbitrary Magnitude Response** graph by clicking on the point, or directly from the frequency-magnitude array shown at the right by clicking on the circle to the right of each point:

The **filter order** control specifies the total number of coefficients in the digital FIR filter.

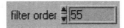

The **ripple** indicator displays the largest absolute error (linear) between the desired and actual filter responses.

The **message** window displays errors that occurred during the FIR design procedure.

The **locked frequencies** box allows you to lock the present frequency values of the frequency-magnitude points. If you click in this box, you can alter only the magnitude or *y*-value of the frequency-magnitude points.

☐ locked frequencies   ☐ sort by frequency
☐ uniform spacing   ☐ import from file

The **uniform spacing** box is used to space the frequency values of the frequency-magnitude points. If you click in this box, the DFD application spaces the frequency-magnitude points uniformly from 0.0 to sampling rate/2, inclusive.

Clicking in the **sort by frequency** box tells the DFD application to sort the frequency-magnitude points in both the response graph and the array according to ascending frequency. The value of each frequency-magnitude point remains unchanged; however, the point order may change.

Clicking on **import from file** enables you to import frequency-magnitude points from a text file.

The **sampling rate** control specifies the sampling rate in samples per second (Hertz).

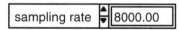

3. Enter these values directly on the front panel and then save them by selecting **Save Specs** from the **DFD Menu**.

When prompted for a file name, type Aweight.fir.

4. With the **multiple selection** control set to OFF, move some of the points on the **Arbitrary Magnitude Response** graph. To do this, move your cursor close to a point until the cursor changes shape, as shown below.

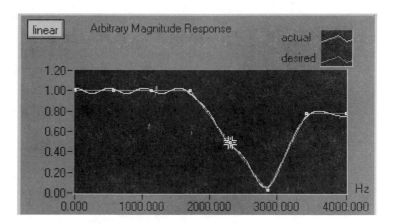

Hold down the left mouse button and move the point. Observe how the response of the filter changes as the point is moved.

5. Choose **locked frequencies** and try to do the same as in the previous step. Now you should be able to change the magnitude of the selected point, but not its frequency.

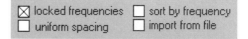

6. Change the methods of interpolation between *linear interpolation* and *spline interpolation*. Observe the difference in the shape of the filter response. Linear interpolation is used to create "flat" filters, such as lowpass, highpass, bandpass, and bandstop filters. Spline interpolation is used to create smoothly varying filters.

7. Close the Arbitrary FIR Filter Design panel by selecting **Close** from the **File** menu.

8. Quit the application by selecting **Quit** from the Main Menu.

■ End of Activity 11-2

Analyzing Your Filter Design

After designing your IIR or FIR filter, you can analyze your
design in several ways. For example, you can see the effect that
the filter has on the amplitude and phase of input signals at different frequencies by observing its magnitude and phase
responses. Several types of analysis methods are available in the
Digital Filter Design Toolkit. These are explained below. You
will first look at the response of the filter to special kinds of
input signals.

■ Impulse Response

The output of the filter when the input is an impulse is known as
the *impulse response* of the filter. An impulse in the digital world
has an amplitude of 1 at index 0 and an amplitude of 0 for all
other indices. An impulse and the impulse response are shown
in Figures 11–12.

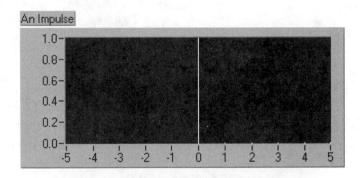

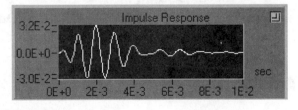

Figure 11–12
Impulse response of a filter

The impulse response has a very special meaning in the case of FIR filters. The impulse response of an FIR filter gives the coefficients of that filter. Furthermore, the number of nonzero terms in the impulse response gives the number of coefficients in the filter. (For an IIR filter, the relationship is much more complicated, and the above discussion does not apply.)

Another use of the impulse response is that, for both FIR and IIR filters, the output of the filter is given by the convolution of the input signal and the impulse response of the filter.

■ Step Response

The output of the filter when the input is a unit step is known as the *step response* of the filter. A unit step in the digital world has an amplitude of 0 for all negative indices, and an amplitude of 1 at index zero and for all positive indices. A unit step and the step response are shown in Figure 11–13.

The step response is important if you will use the filter in a control system. You can then see how the parameters of the con-

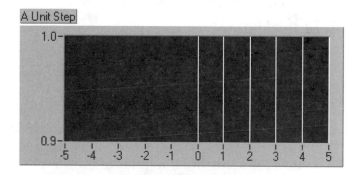

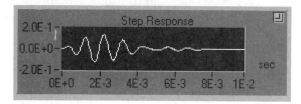

Figure 11–13
Step response of a filter

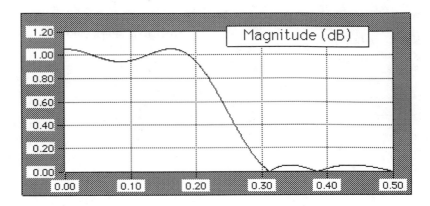

Figure 11–14
Magnitude response of a filter

trol system (such as the rise time, overshoot, etc.) are affected by the filter. The step response also shows you how long the filter will take to respond to a sudden change in the input.

■ Frequency Response (Magnitude Response and Phase Response)

The frequency response shows the effect that the filter has on the amplitude and phase of input signals at different frequencies. Because the filter can affect both the magnitude and the phase of the input signal, the frequency response consists of two parts—the magnitude response and the phase response. An example of these responses is shown in Figures 11–14 and 11–15. The x-axis units are normalized in terms of the sampling frequency.

Not only does the frequency response enlighten you as to the effect that the filter has on signals of specific frequencies, it also allows you to determine what happens to arbitrary signals as they pass through the filter. Because most signals can be expressed as a sum of exponentials (sines and cosines), you can break a signal down into its individual components and determine the effect of the filter on those components.

■ The *Z*-Domain: Transfer Function H(z) and the Pole-Zero Plot

The *transfer function*, $H(z)$, of a digital filter can be expressed as a ratio of polynomials,

$$H(z) = \frac{N(z)}{D(z)}$$

where $N(z)$ is a numerator polynomial, and $D(z)$ is the denominator polynomial. (For an FIR filter, $D(z) = 1$.) $H(z)$ is also known as the *z-transform* of the filter.

The values of z at which $N(z)$ is equal to zero are known as the *zeros* of the filter, because for these values, $H(z)$ is also equal to zero. The values of z at which $D(z)$ is equal to zero are known as the *poles* of the filter, because at these values, $H(z)$ is equal to infinity. A plot of the poles and zeros of the filter is known as the *pole-zero plot*. Because z is a complex number, the pole-zero plot is shown in terms of the real part of z on the *x*-axis and the imaginary part of z on the *y*-axis.

The pole-zero plot is useful in determining the stability of the filter. As long as all the poles of the filter have a magnitude less than one, the filter is stable. If any of the poles of the filter have a magnitude greater than one, the filter will be unstable. This means that the output of the filter will continue to grow indefi-

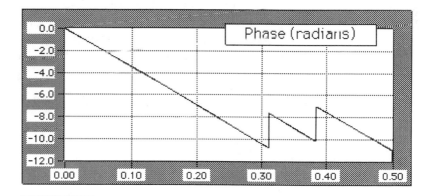

Figure 11–15
Phase response of a filter

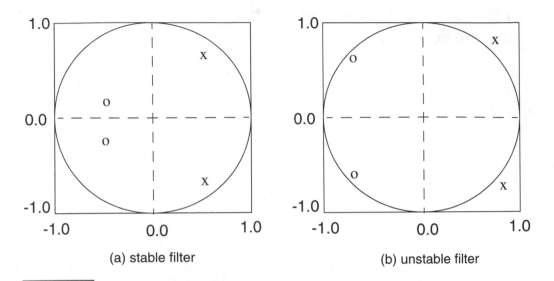

(a) stable filter   (b) unstable filter

Figure 11–16
The location of the poles specifies whether a filter is stable or unstable

nitely even if the input is no longer applied. The values of z for which its magnitude is equal to one are drawn on the pole-zero plot as a circle with its center at the origin and radius equal to one. Thus, so long as the poles of the filter lie inside the circle, the filter will be stable. If even one of the poles lie outside the circle, the filter will be unstable.

The Digital Filter Design Toolkit gives you both the transfer function of the filter and its corresponding pole-zero plot. Figure 11–16 shows the pole-zero plots for both a stable and an unstable filter. Each "o" depicts a zero and each "x" depicts a pole.

Activity 11-3

Objective: To analyze the design of an IIR filter

You will load the filter specifications you saved in Activity 11-1 in the file `bandpass.iir` and see its impulse and step responses, and the pole-zero plot of its transfer function.

1. Open the DFD application by double-clicking on dfddem32.exe. in the **Digital Fitler Design Toolkit** folder.
2. Choose the **Classical IIR Filter Design** panel.
3. When the **Classical IIR Filter Design** panel opens, select **Load Spec** from the **DFD Menu**.

When prompted for the file name, select `band-pass.iir`.

4. From the **DFD** menu, select **Analysis**.

The **Analysis of Filter Design** window opens, as shown below.

■ Front Panel

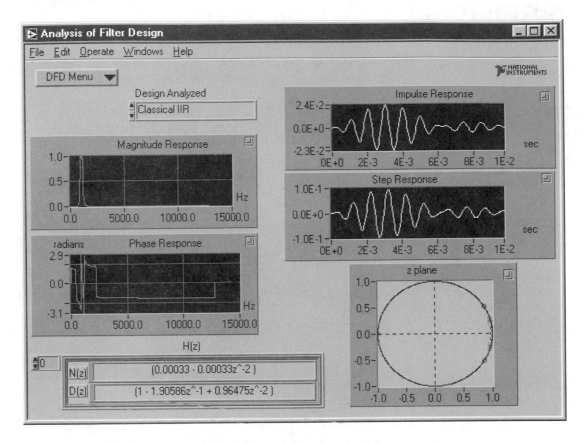

In this panel, you can view the filter magnitude response, phase response, impulse response, step response, and pole-zero plot of the filter you designed in the first activity. You can also view and print full-screen plots of each response. From the full-screen views, you can save the analysis results to text files.

The **DFD Menu** can also be used to load filter designs from previous work, open the **DAQ and Filter** panel, go to the selected filter design panel, or return to the Filter Design Main Menu.

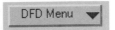

The **Design Analyzed** control selects which filter control to analyze. If you continue to modify the same filter design that is presently being analyzed, the DFD will recompute all filter responses.

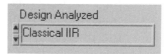

■ Analysis Displays

Each of the five filter plots has a zoom box in the upper-right corner. Clicking in this box brings up a full-screen version of that plot. In the full-screen versions of these plots, you can change the units from linear to decibel (magnitude response), from radians to degrees (phase response) or from seconds to samples (impulse and step responses). From each full-screen view, you can save the response data to text files.

zoom box

■ Magnitude Response

5. The magnitude response is the magnitude of the filter's response $H(f)$ as frequency varies from zero to half the sampling rate. Look at the magnitude response of the designed filter. You can see that it is indeed a bandpass filter.

■ Phase Response

6. The phase response is the phase of the filter's response $H(f)$ as frequency varies from zero to half the sampling rate. The following figure illustrates the phase response of the selected filter design. Note that the phase is displayed in radians. You can obtain a display in degrees by clicking on the zoom box on the top right hand side of the **Phase Response** plot. A new window will appear, which will give you the option of choosing between the appropriate display units.

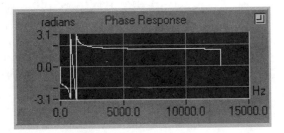

■ Impulse Response

7. The impulse response of a digital filter is the filter's output when the input is a unit sample sequence (1, 0, 0, ...). The input before the unity sample is also zero. The following figure shows the impulse response of the selected filter design.

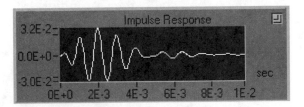

Observe that although it is an IIR filter, the impulse response decays toward zero after a finite amount of time.

■ Step Response

8. The step response of a digital filter is the filter's output when the input is a unit step sequence (1, 1, 1, …). The input samples before the step sequence are defined as zero. The following figure shows the step response of the designed filter.

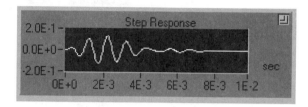

■ Z-Plane Plot

9. The following figure illustrates the z-plane plot of the filter poles and zeros.

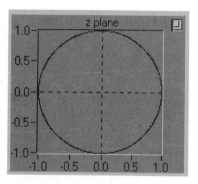

Each pole is represented by a red "x." Each zero is represented by a blue "o." Compare this with the **Magnitude Response** of the filter and observe that the zeros at 0 and π in the pole-zero plot correspond to frequencies 0 and Nyquist ($f_s/2$). The location of the poles is about 1000 Hz, which is why there is a peak in the magnitude response around 1000 Hz.

■ H(z) for IIR Filters

10. $H(z)$ is the z-transform of the designed digital filter.

Previously, you saw that the IIR filters are implemented as cascaded second-order stages. For an IIR filter, $H(z)$ can be represented by a product of fractions of second-order z polynomials.

$$H(z) = \prod_{k=1}^{N_s} \frac{N_k(z)}{D_k(z)}$$

$N_k(z)$ = numerator for stage k

$D_k(z)$ = denominator for stage k

N_s = number of second-order stages

You can view the $N(z)$ and $D(z)$ polynomials for other stages by incrementing the index shown in the upper-left corner of the $H(z)$ display.

11. Close the panel by selecting **Close** from the **File** menu. Quit the DFD application.

■ End of Activity 11-3

Format of Filter Coefficient Text Files

When you save your filter coefficients to a text file, the DFD application generates a readable text file containing all the information you need to implement the designed FIR or IIR digital filter. This section details the format for both FIR and IIR filter coefficient files.

■ FIR Coefficient File Format

Table 11–2 gives an example FIR coefficient text file and description.

Table 11-2 *Format of FIR coefficients saved in a text file*

Coefficient File Example	Description
FIR Filter Coefficients	type of file label
Sampling Rate	sampling rate label
8.000000E+3	sampling rate in Hz
N	filter order label
22	filter order
$h[0...21]$	coefficients label
6.350871E-3	1st coefficient, $h[0]$
-8.833535E-3	2nd coefficient, $h[1]$
-2.847674E-2	.
4.626607E-2	.
4.103986E-2	.
-1.114579E-1	
-1.412791E-2	
1.810791E-1	

(Continued)

Table 11-2 *Format of FIR coefficients saved in a text file*

Coefficient File Example	Description
-5.984635E-2	
-2.002337E-1	
1.516199E-1	
1.516199E-1	
-2.002337E-1	
-5.984635E-2	
1.810791E-1	
-1.412791E-2	
-1.114579E-1	
4.103986E-2	
4.626607E-2	.
-2.847674E-2	.
-8.833535E-3	.
6.350871E-3	last coefficient, $h[N-1]$

You can implement the FIR filter using Equation 11.4 directly.

■ IIR Coefficient File Format

IIR coefficient files are slightly more complex than FIR coefficient files. IIR filters are usually described by two sets of coefficients, *a* and *b* coefficients. There are a total of $M*S$ *a* coefficients and $(M+1)*S$ *b* coefficients, where M is the stage order (usually 2) and S is the number of stages. An IIR filter with three second-order stages has two *a* coefficients per stage for a total of six *a* coefficients, and three *b* coefficients per stage for a total of nine *b* coefficients.

Table 11–3 gives an example IIR coefficient text file and description.

Table 11-3 *Formats of IIR coefficients saved in a text file*

Coefficient File Example	Description
IIR Filter Coefficients	coefficient type label
Sampling Rate	sampling rate label
8.000000E+3	sampling rate in Hz
Stage Order	stage order label
2	order of each stage
Number of Stages	number of stages label
3	number of stages
a Coefficients	a coefficients label
6	number of a coefficients
3.801467E-1	a_1 for stage 1
8.754090E-1	a_2 for stage 1
-1.021050E-1	a_1 for stage 2
9.492741E-1	a_2 for stage 2
8.460304E-1	a_1 for stage 3
9.540986E-1	a_2 for stage 3
b Coefficients	b coefficients label
9	number of b coefficients
1.514603E-2	b_0 for stage 1
0.000000E+0	b_1 for stage 1
1.514603E-2	b_2 for stage 1
1.000000E+0	b_0 for stage 2
6.618322E-1	b_1 for stage 2
1.000000E+0	b_2 for stage 2
1.000000E+0	b_0 for stage 3
1.276187E+0	b_1 for stage 3
1.000000E+0	b_2 for stage 3

You can implement the IIR filter in cascade stages by using Equation 11.2 (maintaining two past inputs and two past outputs for each stage), or by using the direct form II equations (maintaining two past internal states), as in Equation 11.3.

Wrap It Up!

- FIR filters are used for applications where you need a linear phase response, such as for applications that require reconstructing the original waveform after filtering, noise removal, and data compression. For applications where phase is not an important consideration (such as for simple signal power monitoring) and where faster speeds are necessary, you can use IIR filters.

- Using the DFD Toolkit you can interactively design both FIR and IIR filters. The design can be done either by specifying the filter parameters (classical IIR design and classical FIR design), deciding the location of the poles and zeros in the z-plane (pole-zero placement), or arbitrarily specifying the magnitude response characteristics (arbitrary FIR design) of the filter.

- After you have designed a filter using the DFD toolkit, you can analyze your filter in terms of its magnitude and phase responses, impulse and step responses, and the pole-zero plot. You can also save the filter coefficients for use in other applications.

■ Review Questions

1. What are the advantages of digital filters over analog filters?
2. Name any application where one would actually use an analog filter instead of a digital filter.
3. Why would one want to implement a higher order filter as cascaded stages of lower order filters?
4. Name an application where you would use

   a) an IIR filter instead of an FIR filter
   b) an FIR filter instead of an IIR filter

5. Why does one usually see the magnitude response of a filter in dB?

6. Why would one want to determine the impulse response or the step response of a filter?

7. True or False?

   a) Zeros outside the unit circle in the z-plane indicate an unstable filter.

   b) Digital filters are commonly used to prevent aliasing.

   c) For similar filter specifications, elliptic filters usually have the lowest order.

■ Additional Activities

1. Using the DFD Toolkit, design an IIR Filter to remove the 60 Hz (or 50 Hz) power line pickup. Keep the stopband attenuation to –50 dB, and the passband ripple to –0.5dB. Decide on suitable values of the sampling rate, and of the passband and stopband frequencies. Which filter design (Butterworth, Chebyshev, Inverse Chebyshev or Elliptic) gives you the lowest filter order? The highest?

2. In temperature control applications, measurements are usually taken at intervals of 1 Hz or less. These measurements are many times corrupted by noise generated by the surroundings. Design a lowpass filter to pass the measured temperature, but to remove the noise from the surroundings. Choose a sampling frequency of 1 Hz, passband and stopband frequencies of 0.1 and 0.15 Hz respectively, and a stopband attenuation of –40 dB. The passband ripple should be less than – 0.1 dB.

   a) Narrow the transition region by changing the stopband frequency to 0.135 Hz. What happens to the filter order?

   b) Reduce the stopband attenuation to –30 dB. What happens to the filter order?

3. In telephone transmissions, there are several types of signals that are transmitted down a telephone line. These are digital (ON-OFF) signals, analog tone signals (dial tone, busy tone), and analog voice signals. It has been found that for intelligible speech, most of the energy lies in the range of 200 Hz–4000 Hz. Thus, the pass band of voice signals is limited to the range of 300 Hz to 3000 Hz.

   a) Using the **Classical IIR Design** panel of the DFD Toolkit, design a bandpass filter for a **sampling rate** of 8000 Hz, **passband frequencies** of 300 Hz and 3000 Hz, and **stopband frequencies** of 200 Hz and 3500 Hz. Let the **stopband attenuation** be –40 dB, and the **passband ripple** be –0.2 dB. Choose different filter designs (Butterworth, Chebyshev, Inverse Chebyshev, and Elliptic) and note the corresponding filter orders.

   b) Choose the elliptic filter design, and increase the attenuation (keeping all other specifications same as in (a) above) till the order of the elliptic filter equals the order of the Butterworth filter in (a). What is the level of attenuation for the elliptic filter?

A Virtual Sound-Level Meter

Dean E. Capone
Applied Research Laboratory, Pennsylvania State University

The Challenge Creating a "virtual" sound-level meter for use in teaching fundamental concepts of sound-level meter operation and measurement techniques to students in acoustics and noise control engineers.

The Solution Using LabVIEW's signal generation and analysis tools to create an interactive sound-level meter with all the functionality of a commercially available sound-level meter. The front panel of the sound-level meter is shown in the figure below.

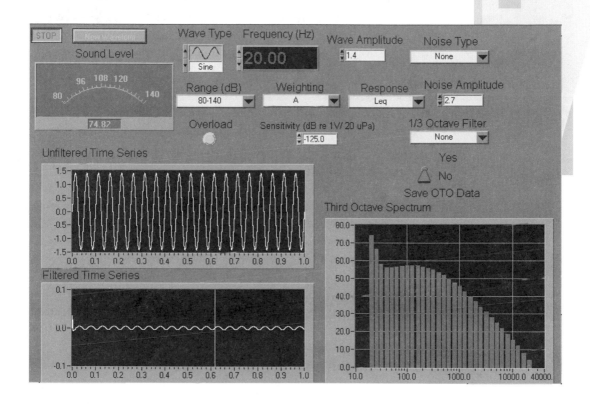

Introduction

One of the primary tools for practicing acoustical or noise control engineers is a Sound-Level Meter (SLM). A SLM is typically used as a field tool to measure the overall sound level, or spectral character, of noise. National, state, and local regulations govern such things as noise levels for aircraft and exposure levels for personnel in industrial environments. In most cases, the allowable noise levels are specified as quantities that are measured using a SLM. In order for the operator of a SLM to correctly measure noise levels, the concept of operation and the limitations of the equipment must be thoroughly understood. Using a Virtual SLM (VSLM), individuals can learn the fundamental operating characteristics and limitations of a typical SLM in a controlled environment.

The Software

The starting point for the meter design was ANSI standards S1.4, Specification for Sound-Level Meters, and S1.11, Specification for Octave-Band and Fractional Octave-Band Analog and Digital Filters. Although the sound-level meter was designed around the appropriate ANSI standards, the primary goal of the software was to be used as a learning tool, not a precision measurement device.

Figure 11–17 shows a block diagram for a sound-level meter.

At a minimum, a SLM consists of a microphone, preamplifier, weighting filter, range control, processing circuit, and an analog or digital display. Many sound-level meters, including the virtual sound-level meter, also have the capability to measure, display, and record one-third-octave[1] Sound Pressure Levels

---

1. An octave is a doubling in frequency, for example from 100 to 200 Hz. Octave filters are termed constant percentage bandwidth filters; the bandwidth filter is always 70.7% of the center frequency. One-third-octave frequencies are obtained by dividing each octave band into three subsections with a constant percentage bandwidth of 23.1%. The one-third-octave-band center frequencies from 100 to 200 Hz are 100, 125, 160, and 200 Hz.

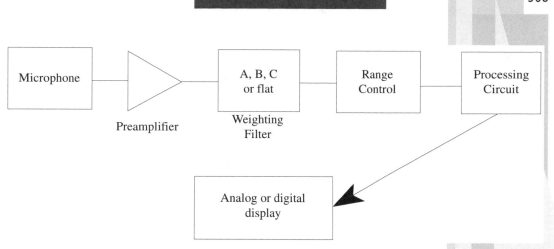

Figure 11–17
Block diagram of a sound-level meter

(SPLs). For the design of the meter, each block in Figure 11–17 had to be emulated using LabVIEW signal generation or analysis tools. The resultant program was implemented using the flowchart in Figure 11–18.

Considering the first two components, a microphone and preamplifier, it was decided to generate the signals internally for two reasons. First, this was easy to do with existing signal generation VIs. The signals are generated using a slightly modified version of the **Function Generator Example** VI. (The **Function Generator Example** VI can be found in the **Examples >> Analysis >> sigxmpl.llb** library in the LabVIEW folder on your computer.) The second, and more important reason, is that deterministic signals, such as sine, square, and triangle waves, have well-defined characteristics, which are used to demonstrate particular points. When a Fourier transform is performed on a sine wave, only the fundamental frequency of the wave will be present. If a Fourier transform is performed on a triangle wave, the fundamental frequency and the odd harmonics are present, while a sawtooth wave generates the fundamental frequency along with all the harmonics. The different spectral characteristic of these waves is visible in the resultant one-third-octave spectrum and the measured sound pressure levels. To illustrate how background noise

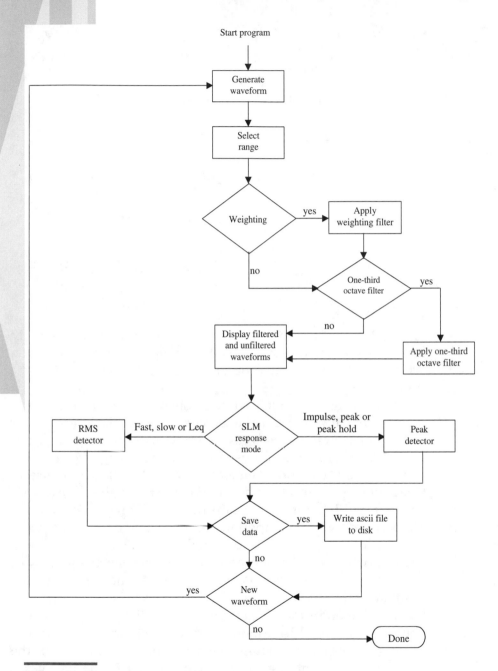

Figure 11–18
Flowchart of virtual sound-level meter program

can mask a signal, the capability to generate random noise was added to the **Function Generator Example** VI. The random noise, uniform white noise, is added to the deterministic portion of the signal and the combined signals are then analyzed. Once generated, the time series data are displayed on the front panel.

The next portion of the program, and the first function of a real sound-level meter, is to condition the incoming signal. The human ear is not equally sensitive to all frequencies. In an effort to mimic the response of the human ear, three weighting networks, A, B, and C were developed. Figure 11–19 shows a plot of the three weighting networks as a function of frequency. The markers on the curves in Figure 11–19 are located at the one-

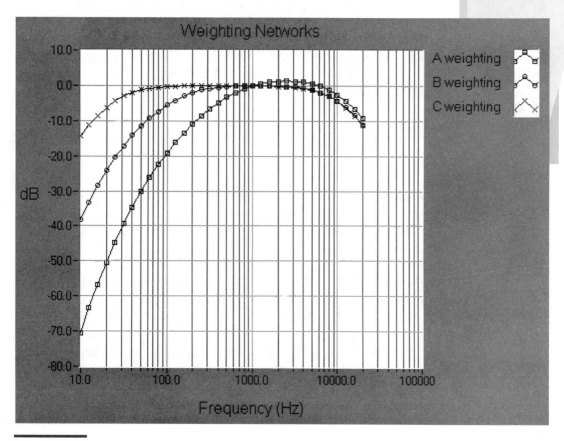

Figure 11–19
A, B, and C weighting curves as a function of frequency

third-octave center frequencies. As an example, it can been seen in Figure 11–19 using the A-weighting scale, the level at 20 Hz is reduced in magnitude by 50.5 dB relative to the level at 1000 Hz. Today the majority of noise measurements made, reported in the literature and used in noise regulations, specify the A-weighting scale. The resultant overall sound level, if for example A-weighting was applied, is provided as a value in dBA. The weighting functions are specified, in ANSI Specification S1.4, in terms of pole-zero configurations in the complex frequency plane. The half-power (3-dB down) points for the weighting networks are also provided. The 3-dB down points were used as the high- and low-cutoff frequencies for a Butterworth first-order filter.

The capability to apply individual one-third-octave filters to a measured signal is often desired when performing measurements with a SLM. A spectrum as a function of frequency is provided if one-third-octave filtering is incorporated in the SLM. The high and low cutoffs for one-third-octave filters can be computed using

$$f_l = 2^{-1/6} f_c$$

$$f_h = 2^{+1/6} f_c$$

where f_c is the center frequency of the one-third-octave band. In the VSLM the one-third-octave filters are computed using these cutoffs for center frequencies from 20–20,000 Hz. Butterworth filters were chosen originally in a 1966 ANSI standard, presumably due to the maximally flat filter characteristic. A filter order of at least three is recommended, with higher-order filters suggested if the spectral slope of the data you are measuring is greater than –30 dB. Since the data generated and analyzed by the VSLM do not have a steep spectral slope, a filter order of three was deemed sufficient. Once the time series data has been filtered, it is also displayed on the front panel of the VSLM. Having both the filtered and unfiltered data displayed allows the user to see the effects weighting and filtering have on the time series.

Once the signal is conditioned, the processing circuit portion of the VLSM computes the sound level. The calculation of the sound levels is comprised of two stages: the measurement of the magnitude for a specified period of time and the exponential time average of successive samples. The incoming time series, which has been filtered in the time domain according to the user's instructions, is squared and passed to one of two types of detectors. The detector depends on the measurement mode, which the user selects on the SLM.

Typical measurement modes available on SLMs are fast, slow, impulse, peak, and peak hold. The original specifications for an analog sound-level meter's response modes, and hence the origin of the terminology, were based on the time constant of the detection circuit. The slow mode has a time constant of 1000 milliseconds, the fast mode has a time constant of 125 milliseconds, and the impulse mode has a time constant of 35 milliseconds. The meter response mode selected usually depends on one or both of the following: the character of the sound to be measured and the specification in noise regulations. For sounds which change rapidly with time, the fast or impulse modes may be appropriate, while for sounds that change slowly with time, the slow mode is appropriate. The peak function will measure the highest instantaneous sound pressure during a given period of time, while the peak hold will measure the peak level and retain it on the display, or in memory, until cleared.

The output of the detector, the peak or RMS magnitude of the sound level, is then exponentially averaged. A digital representation of the exponential time average uses a running two-point linear average of the exponentially weighted sound pressure levels. The exponential weighting factor is determined by the response mode of the SLM. Once the exponential time average is computed, the final sound pressure level is calculated by taking 10 times the logarithm to the base 10 and applying the microphone sensitivity. The entire one-third-octave spectrum was also computed by successive application of digital filters to the time series data. It is interesting to note that although a one-third-octave spectrum is computed, it is done completely in the time domain, with RMS or peak detectors. No Fourier transforms were necessary to provide the spectrum.

Results

The performance of the signal processing of the VSLM was validated using test cases with known results. The attenuation of the weighting networks in the frequency domain is known for each one-third-octave center frequency. Using the weighting filter sub-VI, a pure sine wave of 1Vrms, 0 dBV, was input as the reference level. The attenuation provided by the filter for each one-third-octave center frequency was compared to the specified values. For all three weighting networks the performance of the VSLM filters were within ±0.5 dB.

The one-third-octave filtering portion of the program was checked using white noise as the input source. White noise, when measured using a one-third-octave analyzer, has an increase of 1 dB per one-third-octave band. This can be verified with the following simple exercise.

> Assume you have white noise and the magnitude is constant for all frequency with a value of 0.5 Vrms, with a frequency resolution of 1 Hz. For a center frequency of 1000 Hz compute the upper- and lower-cutoff frequencies of the one-third-octave band. Assuming these cutoff frequencies represent brick wall filters, compute the RMS value of the white noise in the 1000-Hz one-third-octave band and convert to dB. Now repeat the same steps for the 1250-Hz one-third-octave band and compare your dB level to that for the 1000-Hz band. Why is the level higher for the 1250-Hz band than for the 1000-Hz band?

The VLSM correctly computed the one-third-octave spectrum of the white noise, with a spectral slope of approximately 1 dB per one-third-octave band.

Future Work

At the present time the VSLM uses only time series data generated by the internal function generator as input. Future development will involve reading data from an existing file, using an A/D board inside the computer to perform real-time sound-level measurements and a sound card to record and analyze.WAV files. Also as actual sounds are recorded, the operator will be able to listen to the sounds before and after filtering. Once the capability to read and analyze .WAV files is added, any computer with a sound card becomes a VSLM.

Using the signal processing VIs readily available in LabVIEW, individuals can explore the world around them from a different perspective. The tools are available to observe what various sounds look like in the time and frequency domains. One can also listen to the difference signal conditioning can make in the way a sound is perceived. The Acoustical Society of America, composed of over 7000 members, is the primary scientific body for the study of acoustics in America. For more information on the society and the uses of signal processing in acoustics, visit their Web site at http://asa.aip.org/.

■ Contact Information

Dean E. Capone
Applied Research Laboratory
Pennsylvania State University
814-863-9893
Fax: 814-863-5578
Email: dec@wt.arl.psu.edu

Overview

The G Math Toolkit offers a new paradigm for mathematics, numerical recipes in G, *with hundreds of math VIs for solving differential equations, optimization, root finding, and other mathematical problems. All VIs in the G Math Toolkit are written in G, so you can quickly modify them for your custom applications. A main feature of the G Math Toolkit is that it adds to LabVIEW and BridgeVIEW the ability to enter complicated formulas directly onto the front panel of a VI.*

This toolkit is intended for use by scientists, engineers, mathematicians, and anyone else needing to solve mathematical problems in a simple, quick, and efficient manner. It can also be used as an educational aid by those interested in expanding their knowledge of mathematics.

GOALS

- Become familiar with the organization of the G Math Toolkit.
- Learn about the different types of Parser VIs and how to use them to parse formulas entered directly on the front panel.
- Solve differential equations using the differential equation VIs.

KEY TERMS

- differential equations
- parser
- visualization
- functions
- Euler
- Runge-Kutta
- Cash Karp

G Math Toolkit

12

Organization of the G Math Toolkit

The G Math Toolkit consists of nine libraries, each specifically suited to solve problems in a particular area of mathematics. These libraries are

 `Parser.llb`: Consists of the VIs that act as an interface between the end user and the programming system. These VIs parse the user-given formula and convert it to a form that can be used for evaluating the results.

 `Visualiz.llb`: These are the data visualization VIs for plotting and visualizing data in both 2D and 3D. They include advanced methods such as animation, contour plots, and surface cuts.

 `Ode.llb`: The VIs in this library solve ordinary differential equations, both numerically and symbolically.

`Zero.llb`: Used for finding the zeros of 1D or nD, linear or nonlinear functions (or system of functions).

 `Opti.llb`: The optimization VIs that determine local minima and maxima of real 1D or nD functions. You can choose between optimization algorithms based on derivatives of the function and others working without these derivatives.

 `1Dexplo.llb`: Contains VIs that allow the study of real-valued 1D functions, with and without additional parameters, given in symbolic form.

 `2Dexplo.llb`: A collection of VIs that allow the study of real-valued 2D functions given in symbolic form, where parameterization is allowed. Extrema (minima and maxima) and partial derivatives can be numerically calculated.

 `Function.llb`: These VIs evaluate some common mathematical functions.

 `Trans.llb`: A group of VIs that implement some transforms commonly used in mathematics and signal processing.

The G Math libraries consist of more than 100 VIs you can use for solving your mathematics problems. In the next section, you will concentrate on learning more about the Parser VIs and use some of them to build a simple arbitrary waveform generator. Thus, the user can enter a formula on the front panel and the arbitrary waveform generator will generate and plot the corresponding signal. The importance of the VIs in the Parser Library is in enabling users to enter formulas directly on the front panel.

Parser VIs

The VIs in the Parser Library act as the interface between the user and the VIs in the other libraries. The formulas entered on the front panel can have any number of variables. The formulas are first parsed to determine the variables and the values to be assigned to them. They are then evaluated to a number by substituting numeric values for the variables.

■ Direct and Indirect Forms

Because there are basically two steps involved in this process (parsing and then evaluation), there are two forms of parser VIs—the *direct* form and the *indirect* form. In the direct form, both the parsing and evaluation are done in the same VI. The **Eval Formula Node** VI is one of the direct form Parser VIs. The connections to this VI are shown in Figure 12–1.

In the indirect form, the parsing and evaluation are done in separate VIs. The indirect forms split the VI in two subVIs, as shown in Figure 12–2. You can use the indirect form in larger applications, where a two-step process (parsing and then evaluating) is more efficient.

Table 12–1 summarizes the different types of Parser VIs.

The front panel of each Parser VI has an example that shows how to enter values in the control inputs.

The direct form of Parser VIs is more widely used than the indirect form. The flowchart in Figure 12–3 guides you through the selection of a specific Parser VI of the direct form.

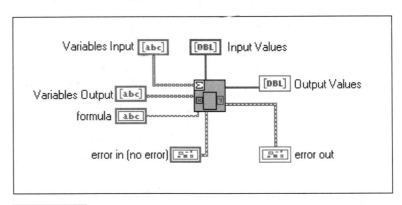

Figure 12–1
Direct form of Parser (Parsing and evaluation in the same VI)

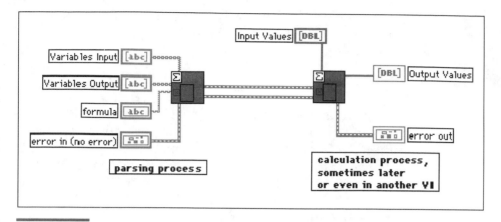

Figure 12–2
Indirect form of Parser (Parsing and evaluation in separate VIs)

Table 12–1 *Direct and indirect forms of Parser VIs. The first column specifies the number of variables, the second column gives the name of the VI, and the third column gives its desciption.*

Direct Form (Parsing and evaluation done in one VI)			Example
No Variables	Eval Formula String	Evaluates RHS of a formula without variables.	$\sin(1.2) + 5$
Single Variable	Eval Single-Variable Scalar	Evaluates RHS of a formula of one variable at one specified point.	$\cos(x)$ at $x = 3.142$
	Eval Single-Variable Array	Evaluates RHS of a formula of one variable at several specified points.	$\cos(x)$ at $x = 0, 0.1, 0.2, \ldots$
Many Variables	Eval Formula Node	Evaluates both sides of formula(s) with several variables.	$x = a + b$ at $a = 1, b = 2$
	Eval Multi-Variable Scalar	Evaluates RHS of a formula with several variables at one specified point.	$\sin(x) + \cos(y) + z$ at $x = 1.0, y = 2.5,$ $z = 3.1$

(Continued)

Table 12–1 Direct and indirect forms of Parser VIs. The first column specifies the number of variables, the second column gives the name of the VI, and the third column gives its desciption.

Direct Form (Parsing and evaluation done in one VI)			Example
	Eval Multi-Variable Array	Evaluates RHS of a formula with several variables at several specified points.	$\sin(x) + \cos(y) + z$ at $x = 1$, $y = 2$, $z = 3$ and $x = 3$, $y = 10$, $z = 1$ and $x = 0.1$, $y = -2$, $z = 0.0$
Indirect Form (Parsing and evaluation in separate VIs)			
Many Variables (The combination evaluates the RHS of a formula of several variables at the specified point.)	Parse Formula String	Parses the RHS of a formula to determine input variables and the operations performed on them.	$x = \sin(z) + 7 * y$
	Eval Parsed Formula String	Evaluates the parsed formula with the specified values for the input variables.	
Many Variables (The combination evaluates both sides of formula(s) of several variables at the specified point.)	Parse Formula Node	Parses both sides of the formula(s) to determine the input and output variables and the operations to be performed on them.	$x = a + b$ $y = a * b$
	Eval Parsed Formula Node	Evaluates the parsed formula(s) with the specified values for the input variables.	
Others			
	Substitute Variables	Substitutes specified formulas for variables in the main formula.	

■ Comparison with Formula Node

A parser VI scans an input string and interprets this string as a collection of formulas. Then, the Parser VI replaces the formulas with numeric calculations and outputs the results. The Parser VI routines deal only with real numbers. There are some differences between the parser in the G Math Toolkit and the Formula Node

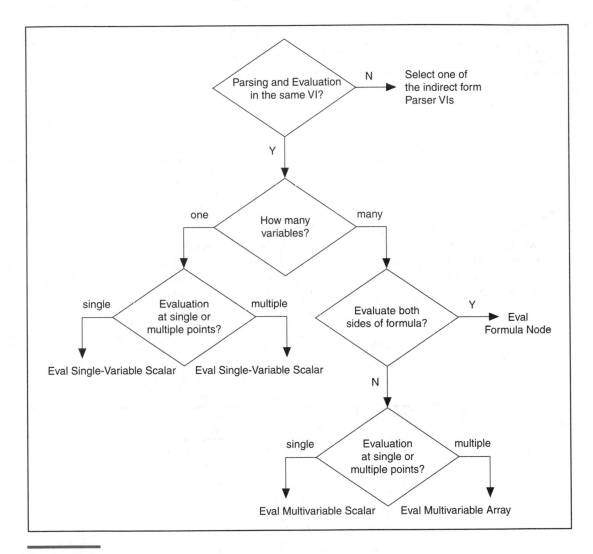

Figure 12–3
Choosing a Parser VI

found in the original LabVIEW (or BridgeVIEW) package. Table 12–2 outlines these differences.

Table 12–2 Differences between the Parser VIs and the Formula Node

Meaning	Formula Node	Parser VI Routines (G Math Toolkit)
Variables	No restrictions	Only $a, a_0, ..., a_9, ...$ $z, z_0, ..., z_9$, are valid
Binary functions	Max, min, mod, rem	Not available
More complex math functions	Not available	Gamma, ci, si, spike, step, square
Assignment	=	Not available
Logical, conditional, inequality, equality	?:, | |, &&, !=, ==, <, >, <=, >=	Not available
π	pi	$pi(1)= \pi, pi(2)= 2\pi$

The precedence of operators is the same for the G Math Toolkit Parser VIs and for the formula nodes in LabVIEW and Bridge-VIEW.

■ Error Structure

The Parser VIs use the following error handling structure. This structure consists of a Boolean **status** button, a signed 32-bit integer numeric **code** indicator, and a string **source** indicator. These error handler components are explained below.

> **Status** is TRUE if an error occurred. If *status* is TRUE, the VI does not perform any operations.

> **Code** is the error code number identifying the error.

> **Source** explains the error in more detail.

The default status of the **error in** structure is FALSE (no error), indicated by an error code of 0.

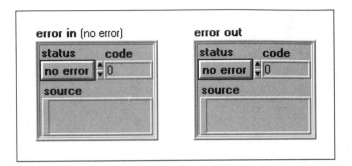

With this structure, an application can programmatically decide the accuracy of formulas and control the data flow in case of errors. The application uses the **source** field of the error handling structure as a storage for a wrong formula input. This field displays limited error descriptions if an error is detected in your program. See the error codes in Appendix B, Table B-4, for the error codes and the error messages of the Parser VI routines.

■ Functions Available for Use with Parser VIs

Most of the functions that you can use in the Formula Node can also be used in the Parser VIs. However, there are some differences. For a complete list of functions that you can use with the Parser VIs, refer to Appendix B, Table B-3.

The Parser VIs are extremely powerful and can be used in a wide variety of applications. The first few activities in this chapter will teach you how to build an arbitrary waveform generator that uses several of the Parser routines and will help you to understand their functionality.

■ A note about the activities in this chapter.

The GMath VIs included on the accompanying CD are a subset of the VIs that come with the complete GMath Toolkit. Follow

the instructions in the preface of this book for directions on how
to install the GMath VIs that you can use with all of the activities
in this chapter. Even though the complete GMath Toolkit consists of nine different libraries, the VIs that you need for this
chapter are all grouped together in one library, as shown in the
figure below.

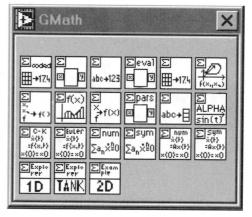

Figure 12–4
The GMath VIs

Activity 12-1

Objective:To build an arbitrary waveform generator using the Eval Single-Variable Array VI

In this activity, you will build an arbitrary waveform generator using the Parser VIs. You will enter the formula for the waveform on the front panel and see the output waveform on a graph indicator.

■ Selection of Direct or Indirect Form Parser VI

First, you must decide whether to use the direct or the indirect form of the Parser VIs. Because you are interested only in displaying the result of the evaluation on a graph and are not concerned about the results of parsing, you will use the direct form of the Parser VIs.

■ Selection of Parser VI

Looking at Table 12–1, depending on the number of variables in your formula, you need to choose the appropriate VI from several available choices of the direct form. Because you will plot a 1D function, you need a VI that can handle functions of at least one variable. And because you want to evaluate that function at more than one point, choose the **Eval Single-Variable Array** VI (**Functions » G Math** subpalette). The inputs and outputs of this VI are shown below:

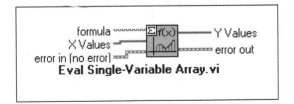

formula: A control for specifying your mathematical formula consisting of one variable.

X Values: Values of the variable at which the formula is evaluated.

Y Values: Numeric results of the evaluation of the formula in **formula** at the values specified in **X Values**.

You need to specify at least the inputs **formula** and **X Values**. For the **formula** input, you will enter a control on the front panel. On this control, you can type in your input formula. For the **X Values**, you will build a For Loop on the block diagram that will generate 1000 points at which to evaluate the equation you will type in the formula control. The corresponding front panel and block diagram are shown below. Note that the For Loop generates **X Values** from 0 to 9.99 in increments of 0.01.

1. Build the VI front panel and block diagram shown below.

■ Front Panel

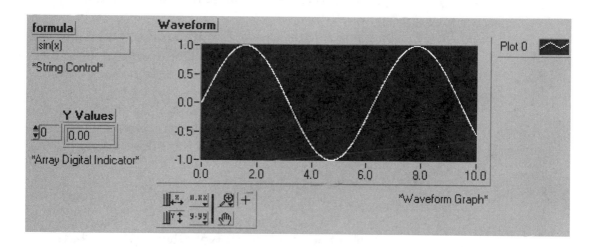

The **formula** string control is where you enter your formula.

The **Y Values** indicator shows the result of the calculation of the formula at specified discrete points.

The **Waveform** graph shows you a plot of the function you typed in the formula control.

■ Block Diagram

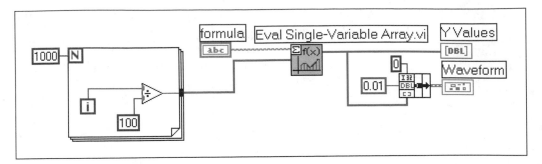

 Eval Single-Variable Array VI (**G Math** subpalette) evaluates a function of a single variable at multiple points.

The **For Loop** generates the discrete set of points at which the formula in the **formula** control will be evaluated. These points range from 0 to 9.99 in increments of 0.01.

2. After building the above VI, return to the front panel.

3. Type in the equation sin(x) in the **formula** control and run the VI. See the result displayed on the graph indicator.

4. Note that sin(x) has only one variable, called x. You could also call the variable y or z or anything else. The first row in Table 12-2 lists the rules for variable names.

5. Change the equation to sin(x) + cos(x/2) and run the VI. Observe the plot.

6. Change the equation to step(y-1) + sinc(y-2) + sin(y*10) and run the VI. Note that now you have changed the variable to y.

7. Type in any other equation in the **formula** control and run the VI. Appendix B, Table B-3, contains a list of available functions you can use in the equation.

If you get an error, refer to Appendix B (Tables B-2 and B-4) for a list of error codes.

8. Save the VI as **EvalSVA.vi** in the library `Activi-ties.llb`.

9. Close the VI. Save any changes.

You have now seen how easy it is to build the waveform generator where the user can specify the formula for an arbitrary waveform on the front panel. In the next activity, you will modify this waveform generator to have the functionality of easily specifying even more complex waveforms.

■ End of Activity 12-1

Activity 12-2

Objective: To generate even more complex waveforms by incorporating the Substitute Variables VI in the function generator

Another useful VI in the parser library is the **Substitute Variables** VI (**G Math** subpalette). The connections to this VI are shown in the figure below. It is used to substitute formulas for parameters that are already defined in formulas.

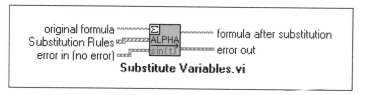

original formula: The main formula you type in.

Substitution Rules: Specify the substitutions to be made for the parameters in **original formula**.

formula after substitution: The resulting formula after the parameter substitutions specified in **Substitution Rules**.

You will now modify the waveform generator you built in the previous exercise to use the **Substitute Variables** VI. Suppose you want to generate a "generic" waveform that is of the form $\sin(A) + \cos(B)$, where A and B can themselves be functions like $\sin(x)$, $\cos(x)$, square(x), sinc(x), $\ln(x)$, and so on. But you want to use different functions for A and B each time you run the VI.

This is accomplished as shown in the block diagram and front panel below.

■ Block Diagram

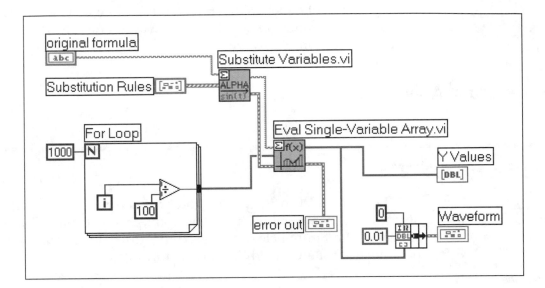

 Substitute Variables VI (**G Math** subpalette) substitutes for parameters defined in formulas. The substitution is done according to the rules specified in the **Substitution Rules** cluster array.

 Eval Single-Variable Array VI (**G Math** subpalette) evaluates a function of one variable at specified multiple points.

The **For Loop** generates a thousand points, ranging from 0 to 9.99 in steps of 0.01, at which to evaluate the user-specified formula.

■ Front Panel

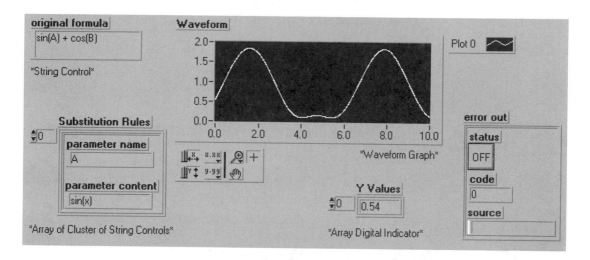

The **original formula** string control is where you enter your formula of one variable. The formula may have one or more parameters.

Each element (cluster) in the **Substitution Rules** array specifies a parameter and its corresponding substitution.

Y Values is an array digital indicator that shows the result of calculating the formula at specified points.

1. Build the VI as shown in the front panel and block diagram.

Build the block diagram first, pop up to create the controls and indicators, and then add the waveform graph.

The **formula after substitution** output of the **Substitute Variables** VI goes to the **formula** input of the **Eval Single-Variable Array** VI.

2. In the **original formula** control, type in the formula *sin(A) + cos(B)*. This will be the "generic" formula. In the **Substitution Rules** control, you will provide the

user with the capability of specifying the functions to be substituted for *A* and *B*.

3. In element 0 of the **Substitution Rules** control, type in

parameter name: A

parameter content: sin(x)

In element 1, type in

parameter name: B

parameter content: cos(x)

Note that this is the same as if you had typed sin(sin(x)) + cos(cos(x)) in the **formula** control in Activity 12-1.

4. Run the VI and see the waveform on the graph display.

5. Change the values in the **Substitution Rules** control to

element 0   **parameter name:**   A

                      parameter content: sin(x)

element 1   **parameter name:**   B

                      parameter content: square(x)

Run the VI. The resulting equation is the same as if you had typed sin(sin(x)) + cos(square(x)) in the formula control in Activity 12-1.

6. Now change the control to

element 0   **parameter name:**   A

                      parameter content: step(x-5)

element 1   **parameter name:**   B

                      parameter content: square(x)

Run the VI.

With this simple VI that you have built, you now can generate any type of waveform that you so desire, with the functions given in Appendix B.

7. Save the VI as **ESVA_SV.vi** in the library `Activi-`
`ties.llb`.
8. Close the VI.

■ End of Activity 12-2

Solving Differential Equations

You can use G Math to solve both ordinary or partial, linear or nonlinear, and either first- or higher-order differential equations. G Math has seven VIs for solving various types of differential equations. These VIs are listed below, classified according to the order of the differential equation they can solve.

The order of a differential equation is the order of the highest derivative in the differential equation.

The first five VIs can solve a set (one or more) of first-order ordinary differential equations (ODE's). The next two VIs are for solving higher-order ordinary differential equations. Note that you can convert a higher-order differential equation into a set of first-order differential equations (later in this chapter, you will see an example of how to do this).

VIs for solving a set (one or more) of first-order ordinary differential equations:

 ODE Cash Karp 5th order for solving differential equations using the Cash Karp method.

 ODE Euler Method for solving differential equations using the Euler method.

ODE Runge Kutta 4th order for solving differential equations using the Runge-Kutta method.

 ODE Linear System Numeric for numerical solution of a linear system of differential equations.

 ODE Linear System Symbolic for symbolic solution of a linear system of differential equations.

Out of these, the first three are for solving *nonhomogeneous* (right side ≠ 0) differential equations, whereas the last two are for *homogeneous* (right side = 0) differential equations.

VIs for solving higher-order ordinary differential equations:

 ODE Linear *n*th order Numeric for numeric solution of a linear system of *n*th order differential equations.

 ODE Linear *n*th order Symbolic for symbolic solution of a linear system of *n*th order differential equations.

Both these VIs are for solving *homogeneous* differential equations.

■ Solving Nonhomogeneous Differential Equations

From the previous discussion, you see that there are three VIs available for solving nonhomogeneous differential equations. They are

- ■ **ODE Cash Karp 5th order** VI
- ■ **ODE Euler Method** VI
- ■ **ODE Runge Kutta 4th order** VI

Each VI employs a different method for solving the differential equations. Each method uses a parameter known as the *step size* (denoted by h) that determines the spacing between points at which the solution is evaluated. This step size is a constant for the last two VIs (which employ the Euler and Runge-Kutta methods), whereas it is variable (it automatically adapts itself to the solution) for the first VI (which employs the Cash Karp method). The Euler method is the simplest method, but the Runge-Kutta method gives a more accurate solution.

The question arises as to which of these three VIs you should choose for your application. The general guidelines are

- Select the **ODE Euler Method** VI for very simple ODEs.

- For all other cases, choose the **ODE Runge-Kutta 4th Order** VI or the **ODE Cash Karp 5th Order** VI.

  — If you need equidistant points (that is, constant *h*—for example, for robot control applications), choose the **ODE Runge-Kutta 4th Order** VI.
  — For more complicated problems, where a variable step rate is needed, choose the **ODE Cash Karp 5th Order** VI. As an example, this VI is preferred for studying the motion of three bodies when their paths are very close to each other.

■ A General Class of Second-Order Differential Equations

As an example, a general class of second-order differential equations is described by the following initial value problem:

$$a\frac{d^2y}{dt} + b\frac{dy}{dt} + cy = g(t) \tag{12.1}$$

with

$$y(0) = y_0 \text{ and } \frac{dy(0)}{dt} = \frac{dy_0}{dt}$$

where $y(0)$ is the value of y at $t = 0$, $dy(0)/dt$ is the value of dy/dt at $z=0$, and $g(t)$ is known as the *forcing function*. $y(0)$ and $dy(0)/dt$ are known as the *Initial Conditions* (ICs). Note that because the right side of Equation 12.1 is not equal to zero, it is a *nonhomogeneous* differential equation. In the absence of a forcing function ($g(t) = 0$), it would be a *homogeneous* differential equation.

Some of the practical applications of this class of differential equations are for modeling second-order systems such as:

1. The motion of a mass on a vibrating spring:

$$m\frac{d^2y}{dt^2} + c\frac{dy}{dt} + ky = F(t)$$

where m is the mass, c is the damping coefficient, k is the spring constant, and $F(t)$ is the applied force.

2. Flow of an electrical current in a series circuit:

$$L\frac{d^2Q}{dt^2} + R\frac{dQ}{dt} + \frac{Q}{C} = \frac{dE(t)}{dt}$$

where R, L, and C are the resistance, inductance, and capacitance, respectively, in the circuit, Q is the charge flowing in the series circuit, and $E(t)$ is the applied voltage.

3. The motion of an oscillating pendulum:

$$\frac{d^2\theta}{dt^2} + \frac{c}{ml}\frac{d\theta}{dt} + \frac{g}{l}\sin(\theta) = 0$$

where m is the mass of the pendulum, l is the length of the rod, g is the acceleration due to gravity, and θ is the angle between the rod and a vertical line passing through the point where the rod is fixed (that is, the equilibrium position).

To solve equations of the type of Equation 12.1 using the **ODE Euler Method** VI, the **ODE Runge-Kutta 4th Order** VI, or the **ODE Cash Karp 5th Order** VI, you first need to convert them into a set of first-order differential equations. This is achieved by making the substitution $x_1 = y$ and $x_2 = dy/dt$. Substituting these in Equation 12.1, you get

$$a\frac{dx_2}{dx} + bx_2 + cx_1 = g(t)$$

$$\frac{dx_2}{dt} = \frac{g(t) - cx_1 - bx_2}{a}$$

Thus, now you have converted the second-order differential equation given by Equation 12.1 to the following equivalent set of first-order differential equations:

$$\frac{dx_1}{dt} = x_2 \tag{12.2}$$

$$\frac{dx_2}{dt} = \frac{g(t) - cx_1 - bx_2}{a} \tag{12.3}$$

You will use G Math to solve this set of equations in the next activity.

Activity 12-3

Objective: To build a general G Math VI for applications whose solutions are second-order differential equations

In this activity, you will build a VI that will solve the general second-order differential equation of the type given in Equation 12.1. This equation is also given below:

$$a\frac{d^2y}{dt^2} + b\frac{dy}{dt} + cy = g(t)$$

with

$$y(0) = y_0 \text{ and } \frac{dy(0)}{dt} = \frac{dy_0}{dt}$$

You will see how incorporation of the **Substitute Variables** VI enables you to solve problems related to a wide variety of applications.

Because you want to solve the above equation, which is a non-homogeneous equation, you are faced with three choices of VIs:

- **ODE Cash Karp 5th Order**
- **ODE Euler Method**
- **ODE Runge Kutta 4th Order**

Because this is a simple second-order differential equation, you can actually select any of these VIs. Choose the **ODE Euler Method** VI, which has the following inputs and outputs:

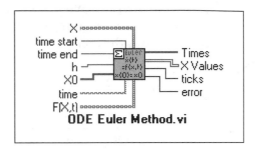

ODE Euler Method.vi

X: An array of strings listing the dependent variables.

time start: The point in time at which to start the calculations.

time end: The point in time at which to end the calculations.

h: Time increment (step rate) at which to perform the calculations.

X0: The initial conditions (IC). There is a one-to-one relationship between the components of **X** and those of **X0**. That is, the first value in **X0** is the IC of the first variable listed in **X**, the second value in **X0** is the IC of the second variable listed in **X**, and so on.

time: A string that defines what the independent variable is, usually time (*t*).

F(X,t): The right sides of the set of first-order differential equations. There is a one-to-one relationship between the elements of **F(X,t)** and **X**. This will be explained in more detail later.

F(X,t) requires its input as a set of first-order differential equations. Therefore, to solve Equation 12.1 with this VI, you need to use the equivalent form given by Equations 12.2 and 12.3.

ticks: The time in milliseconds required for the calculations.

Times: The time instants at which the solution of the differential equations are evaluated. For the **ODE Euler Method** VI and the **ODE Runge Kutta 4th Order** VI, these instants begin at **time start** with increments of **h**, until **time end**.

X Values: Contains the solution for each of the dependent variables of the differential equations at each time instant in **Times**. The first column is the solution for the first variable in **X**, the second column is the solution for the second variable in **X**, and so on.

error: Contains an error code in case of any error. Refer to Appendix B for the list of errors and the corresponding error codes.

1. Build the VI whose front panel and block diagram are shown below.

*An easy way to build the VI is to build the block diagram first, pop up on the terminals of the **ODE Euler Method** VI, and select **Create Control** for the inputs and **Create Indicator** for the outputs. The controls and indicators automatically appear in the correct form on the front panel. Finally, you can add the Waveform graph.*

*For the **Index Array** function, to slice out arrays (rows or columns) popup on the corresponding terminal and select **Disable Indexing**.*

■ Front Panel

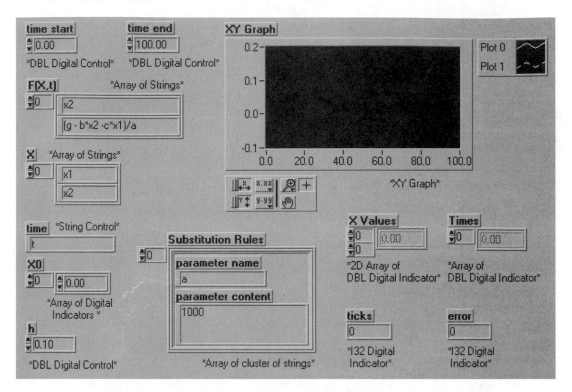

The **time start** and the **time end** controls contain the starting and ending time instants.

The differential equations are typed in the **F(x,t)** control.

The dependent variables are entered into the **X** control.

The independent variable is entered into the **time** control.

X0 contains the initial values (initial conditions) of the dependent variables.

h contains the step size.

Substitution Rules contains the parameters and the corresponding substitution.

There is a one-to-one correspondence between the elements of X, XO, and F(x,t). The first element in XO corresponds to the initial value of the first element in X, the second element in XO corresponds to the initial value of the second element in X, and so on. Also, the first element in F(x,t) corresponds to the derivative of the first element in X, the second element in F(x,t) corresponds to the derivative of the first element in X, and so on.

■ Block Diagram

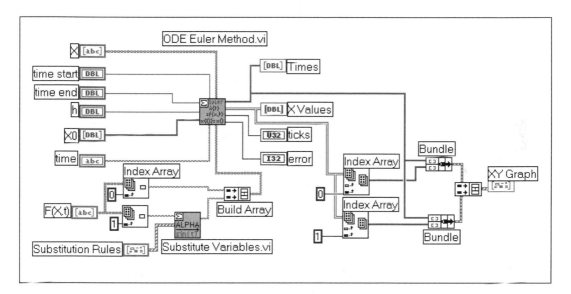

 ODE Euler Method VI (**Functions » G Math)** subpalette solves a set of differential equations using the Euler method.

 Substitute Variables VI (**Functions » G Math** subpalette) substitutes the specified values for the corresponding parameters.

2. Enter the right side of Equations 12.2 and 12.3, respectively, in the first two elements of **F(X,t)**. Thus, you will enter x_2 and $(g - b{*}x_2 - c{*}x_1)/a$.

Equation 12.2 in the second element of **F(X,t)** contains the variables a, b, c, and g. The values of these variables will change depending on the application. Hence, in the block diagram, you use the **Substitute Variables** VI to assign different values to these variables. You can enter these values in the **Substitution Rules** control on the front panel.

3. **X** contains the names of the dependent variables, which in this case are entered as x_1 and x_2.

4. **time** contains the name of the independent variable, which in this case is t.

   X0 will contain the initial conditions for x_1 and x_2. The first value entered in **X0** will correspond to the first variable entered in **X**. The second value entered in **X0** will correspond to the second variable entered in **X**. The values of the ICs will vary depending on the application.

5. Save the VI as **ODE_2nd.vi** in the library `Activi-ties.llb`.

   Now you have a VI ready for solving any second-order nonhomogeneous differential equation. All that remains is to enter the corresponding values of a, b, c, and g in the **Substitution Rules** control, and the initial conditions of the variables **X0** control. Then you can specify the time interval (between **time start** and **time end**, in increments of **h**) for which you want the solution and run the VI. You will continue this procedure in the next activity.

■ End of Activity 12-3

Activity 12-4

Objective: To design the suspension system of an automobile

You will now use the VI you built in the previous activity to solve a typical mechanical modeling problem. In particular, you will design the suspension system of an automobile to have a suitable response when the automobile goes over a pothole, or a speed bump.

You can model the suspension system by the nonhomogeneous differential equation given in Equation 12.1 (reproduced below),

$$a\frac{d^2y}{dt^2} + b\frac{dy}{dt} + cy = g(t)$$

with $y(0) = y_0$ and $\dfrac{dy(0)}{dt} = \dfrac{dy_0}{dt}$

where

> $y(t)$ is the position of the automobile. $y(t) = 0$ indicates that the vehicle is in a balanced or a stable position.

> dy/dt is the up and down speed of the automobile.

> c is the coil constant of the spring used in the suspension system. It is proportional to the displacement of the spring from its equilibrium position. A larger value of c indicates a strong spring coil and thus a tough suspension system. c is known as the *spring constant*.

> b is a resistance coefficient that is proportional to the up and down speed of the automobile. A larger value of b indicates a stronger resistance to the up and down movement of the automobile, and thus a smoother suspension system. b is known as the *shock constant*.

> a is the weight of the vehicle.

> $g(t)$ is the external force that causes the suspension system of the automobile to deviate from its equilibrium position. Some possible causes may include a speed bump, a pothole, or a stone in the middle of the road.

1. Open the VI developed in the previous activity, and enter the following values in the controls on the front panel:

    time start: 0

    time end: 100

    h: 0.1

    Substitution Rules

parameter name:	a	b	c	g
parameter content:	1000	125	1000	0
X0:		0	0	

    The initial conditions of 0 specified in X0 indicate that the vehicle is initially in a stable equilibrium position. $g = 0$ means that there is no disturbing force.

2. Run the VI and note the waveforms on the XY Graph. What do you see? Explain the results.

3. Now assume that after driving for 10 seconds, the driver of the automobile drove on to the sidewalk. This action can be modeled by substituting a step function for g. In the **Substitution Rules** control, change the value of g from **0** to **100*step(t-10)** and run the VI.

    Notice that until t = 10, the values of x_1 and x_2 are equal to zero because no force (g) has been applied until then. At $t = 0$, a step function of magnitude 100 is applied and both the position and the up and down speed of the automobile change. However, as time passes, the oscillations subside and both values tend to settle down to an equilibrium position. You can verify this by entering 400 in the **time end** control and running the VI.

4. A speed bump can be approximately modeled by the following value for g:

    10*sin(t)*(step(t) - step(t-3.142))

The multiplying constant (in this case 10) controls the height of the speed bump, while the remaining term models one "bump."

How the above formula corresponds to a speed bump is shown in the figure below. The upper plot is a plot of one cycle of sin(t). The middle plot is a plot of step(t) - step(t-3.142), whereas the lower plot is the multiplication of the top two plots resulting in sin(t)*(step(t) - step((t-3.142)).

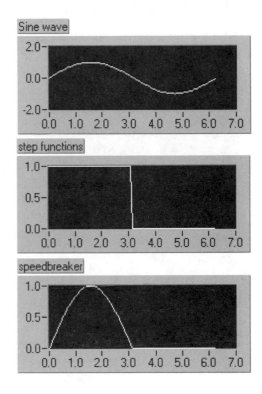

Enter the above value of g and run the VI.

What do you see? Explain the results.

5. A pothole can be modeled in a similar fashion, but with the opposite sign: -10*sin(t)*(step(t) - step(t-3.142))

Enter the above value of g and run the VI. What change do you notice as compared with the results of step 4?

6. Experiment with the value of c so that the oscillations reduce to zero in less than 80 seconds.

Reduce the value of c to a lower value (for example, 250)

7. Keeping the value of c at 1000, change the value of b so that the oscillations reduce to zero in less than 80 seconds.

Increase the value of b to a higher value (for example, 175)

Thus, you see that by proper selection of b and/or c, you can design the suspension system of an automobile to behave in a certain way (that is, to have a certain response). The selection of b and/or c could correspond to a particular choice of material to be used in building the suspension system.

8. Close the VI when you are finished.

■ End of Activity 12-4

Activity 12-5

Objective: To see some of the capabilities of the G Math Toolkit for 1D functions

1. Open the **1D Explorer Example** VI from the library `vi.lib » addons » gmath.llb` found in your LabVIEW folder. This VI shows a good overview of some of the mathematical functionalities of G Math.

■ Front Panel

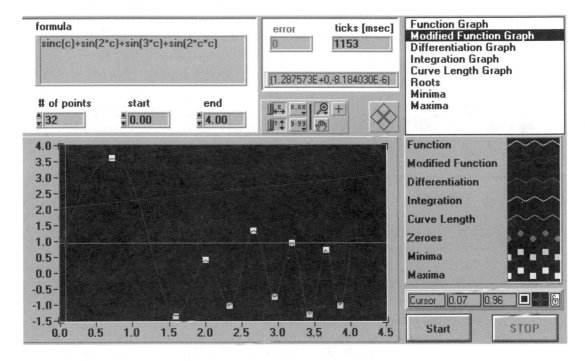

2. The default formula in the **Formula** control is sinc(c)+sin(2*c)+sin(3*c)+sin(2*c*c). Choose the following options by holding down <Shift> and clicking the left mouse button on the following selections:

   ■ Modified Function Graph

   ■ Integration Graph

   ■ Roots

   ■ Maxima

3. Run the VI and click the **Start** button on the bottom right of the panel. Plots of the above selections are displayed. Note that you can enter the start and end points directly on the front panel through the **start** and **end** controls.

4. Line up the cursor on a maxima and see its value.

5. Enter a new formula sin(exp(x)) in the **Formula** control and click on the **Start** button.

6. Zoom in on a maxima using the graph zoom option.

7. Experiment by typing in other formulas in the formula control.

8. When you finish, you can stop the VI by clicking on the **STOP** button at the bottom right of the panel.

9. Close the VI. Do not save any changes.

■ End of Activity 12-5

Activity 12-6

Objective: To simulate the tank flow problem

This activity is a practical application that combines entering formulas on the front panel, solving an ordinary differential equation, and visualizing a process to simulate a tank inflow and outflow process.

Consider a cylindrical tank of constant cross section A cm^2. Water is pumped into the tank from the top at a constant rate $f_i(t)$ cm^3/s. Water flows out of the bottom of the tank by a valve of area $a(t)$ cm^2. Note that both the input and output flow rates are functions of time. You will observe how the height of the water, $h(t)$, in the tank varies with time.

The solution to this problem is a first-order differential equation given by

$$\frac{dh(t)}{dt} = -\sqrt{2gh(t)} - \frac{a(t)}{A} + \frac{f_i(t)}{A}$$

where g is the acceleration due to gravity, equal to 980 cm/s^2.

The input flow rate to the tank, as well as the area of the outflow valve, can be modeled as equations on the front panel. Select the following equations for these values:

Input flow, $f_i(t)$:   340*square(t)

Area of valve, $a(t)$, for the outflow: sin(t) + 1

■ Front Panel

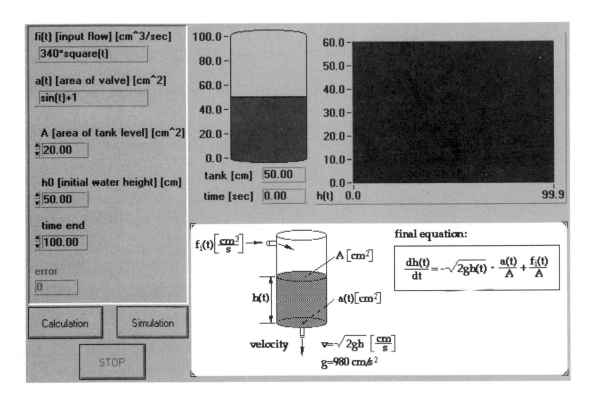

1. Open the **Process Control Explorer Example** VI from the library vi.lib » addons » gmath.llb. The explanation of the front panel is as already described above.

2. Run the VI and select the **Calculation** button. This button will flash while the calculation is in progress.

In this example, all variables in the differential equation can be controlled from the front panel.

3. Once the calculation is over, select the **Simulation** button. The simulated system is then graphed.

4. Change the input flow rate and area of tank to:

   $f_i(t)$, input flow rate: 340*sin(t)

   $a(t)$, area of valve: 0.01

5. Select the **Calculation** button and then the **Simulation** button. Notice that the tank is now gradually emptying.

6. Select the **STOP** button.

7. Close the VI. Do not save any changes.

■ End of Activity 12-6

Activity 12-7

*Objective: To see
an example of the G
Math Toolkit for
data visualization*

■ Front Panel

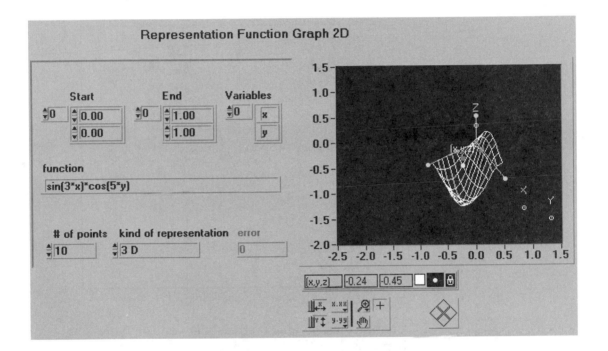

1. Open and run the **Representation Function Graph 2D** VI from the `vi.lib » addons » gmath.llb` library.

   When the interactive subVI comes up, you can look at the plot from different viewpoints by controlling the values of *psi*, *phi*, and *r* with the help of the slider controls. *Psi* and *phi* control the angles with respect to the *xz* and *xy* planes, respectively, where *r*

The default function value is sin(3*x)*cos(5*y). This is a function of two variables, x and y. The range of values to be plotted is indicated on the Start and End controls on the front panel. The default values are 0.0 and 1.0 for both the x and y variables.

controls the distance from the origin. How these controls correspond to different viewpoints is shown in the figure below:

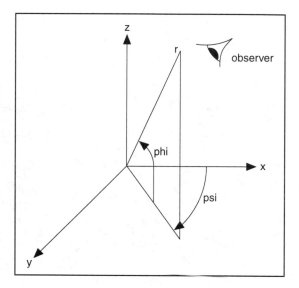

2. Change the *psi* value on the slider control. Note what happens.
3. Change the *phi* value on the slider control. Note what happens.
4. To return to the main VI, select the **STOP** button on the front panel.
5. When you are done, close the VI. Do not save any changes.

■ End of Activity 12-7

Wrap It Up!

- The GMath Toolkits offers a new paradigm for solving your mathematics problem in G.

- An extremely important feature is the ability to enter formulas directly on the front panel of your VI.

- There are nine GMath libraries that can be used for applications that require optimization, solving differential equations, signal processing, and much more.

- In this chapter, you built a simple arbitrary waveform generator using the VIs in the parser library, designed the suspension system of an automobile using the differential equation VIs, and experimented with other VIs that showed some of the more advanced capabilities of GMath.

■ Review Questions

1. Which GMath VI would you use for

   a) evaluating the formula $\sin(x) + 3x$ at multiple values of x?

   b) evaluating the formula $\dfrac{\cos(x)}{a + 3b - \sin(y)}$ at one specified value of x, a, b, and y?

   c) evaluating the formula $z = p/q$ at $p = 3$ and $q = 5.2$?

2. How would you enter the number "3π" in a parser VI?

3. Which of the differential equation VIs would you use to solve nonhomogeneous differential equations? When would you prefer one over another?

■ Additional Activities

1. A simple model that describes the interaction between two species, one of which (the predator) eats the other species (the prey) is given by the Predator-Prey equations:

$$\frac{dx}{dt} = ax - bxy$$

$$\frac{dy}{dt} = -cy + dxy$$

where x: population of the prey

         y: population of the predator

         a, b, c, d: positive constants

Examples of such species are foxes (predators) and rabbits (prey) or lions (predators) and deer (prey).

Build a VI that solves the Predator-Prey equations for values of the constants a, b, c and d which can be specified on the front panel. The VI should return an error if a, b, c, or d are not positive.

2. Calculus is widely used in the field of Economics. Economists use the term **marginal analysis** when they apply derivatives to economic problems. As an example, if the cost of producing x units of a particular item is $C(x)$, then the derivative $\frac{dC(x)}{dx}$ is known as the **marginal cost.**

Suppose that the cost of a weekly production of x bales of cotton is

$$C(x) = \frac{x^2}{25} + 2x + 156$$

Build a VI that calculates the cost, and the marginal cost, where the number of bales of cotton is specified on the front panel.

Using LabVIEW and the G Math Toolkit to Develop a User-Configurable Telemetry Program

Michael Fortenberry
Aerospace Engineer
National Scientific Balloon Facility/Physical Science Laboratory

The Challenge Developing a user-configurable program to provide telemetry display and command capability for long-duration high-altitude scientific balloon flights using multiple serial links.

The Solution LabVIEW and the G Math Toolkit provided the capability to make the entire system user-configurable and provide compatibility with existing software. LabVIEW also allowed extensive code reuse from previously developed systems which shortened the development time.

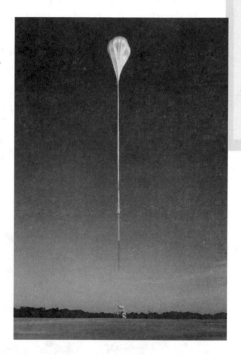

Figure 12–5
Balloon launch. (Launch vehicle is holding science payload.)

Introduction

Balloons have been used for decades to conduct scientific studies. NASA balloons (Figures 12–5 and 12–6) are made of a thin 0.002-centimeter (0.8-mil) polyethylene material, about the same thickness as ordinary sandwich wrap. The volume of the balloons can range up to 40 million cubic feet. The balloon system includes a balloon, a parachute, and the payload which contains the instruments to conduct the experiment. Scientific balloons can carry a payload weighing as much as 8000 pounds (3630 kilograms). They can fly to an altitude of 26 miles (42 kilometers), with flights lasting an average of 12 to 24 hours. Some special-purpose long-duration balloon flights have lasted more than two weeks.

The National Scientific Balloon Facility (NSBF), a NASA facility operated by the Physical Science Laboratory of New Mexico State University, provides high-altitude scientific ballooning services to the scientific community. Part of these services include providing a telemetry/command link between the balloon systems and the operators. For long-duration balloon flights, there are Line-Of-Sight (LOS) links using radio communications and Over-The-Horizon (OTH) links which use satellites to relay the data as shown in Figure 12–7.

Figure 12–6
Balloon at 120,000 feet as seen through telescope

A need was identified for a windows-based telemetry/command program which would run on a PC mounted in an aircraft. Since there was already an ADA program running on a UNIX platform available for ground use, the setup files for the two systems needed to be compatible. The NSBF had successfully used LabVIEW to create telemetry/command software for other systems, so the decision was made to leverage these previous efforts and use LabVIEW to develop the new software.

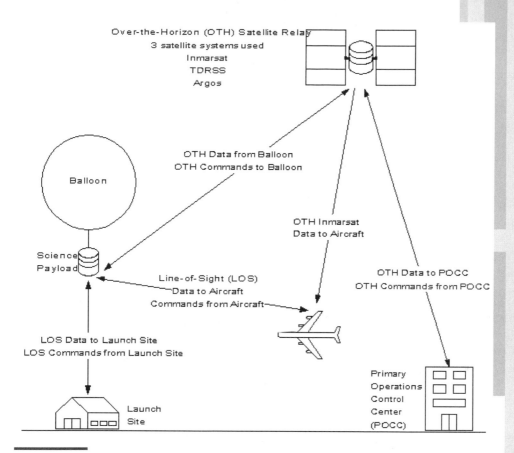

Figure 12–7
Diagram of telemetry/command links for long-duration flights

Hardware Requirements

The aircraft telemetry command system consists of the equipment required to support one LOS link and one OTH link as shown in Figure 12–8. Line-of-sight data from analog and digital sources along with four streams of packetized RS-232 data from balloon-borne flight computers are encoded on the balloon with a Pulse Code Modulation (PCM) encoder built by PSL. This data sequence is received on the aircraft using an L-band receiver. The telemetry signal from the receiver is then fed to a computer which contains a bit synchronizer board and PCM decommutation board built by PSL. The bit synchronizer and decommutation board are used to extract the analog and digital data from the telemetry stream. This high-rate data sequence is displayed on the computer using software developed using Labwindows/CVI. The decommutation board is also able to extract the packetized RS-232 data sequence and forward it to the telemetry/command computer. The LOS uplink is an RS-232 modulated UHF transmitter.

The OTH system consists of an aircraft-rated Inmarsat transceiver. Data from balloon-borne flight computers is routed to the aircraft transceiver through the Inmarsat satellite system and stored onboard until retrieved using an RS-232 link. While the Inmarsat transceiver can be used for sending commands to the balloon, this feature has not yet been implemented.

The telemetry/command computer is connected through three RS-232 serial links to the LOS uplink/downlink and the OTH downlink. There is an additional capability to use two additional RS-232 links to provide science groups with science data and command capability in the aircraft.

Software Requirements

Because of the reconfigurability of the flight hardware, a requirement was imposed on the software to allow the telemetry displays to be user configurable. The software was also required to allow the user to custom-scale analog parameters. These requirements give the user the power to customize telemetry displays and display the parameters in a way that is appropriate and meaningful. An additional requirement was to maintain compatibility of setup files with the existing UNIX-based software.

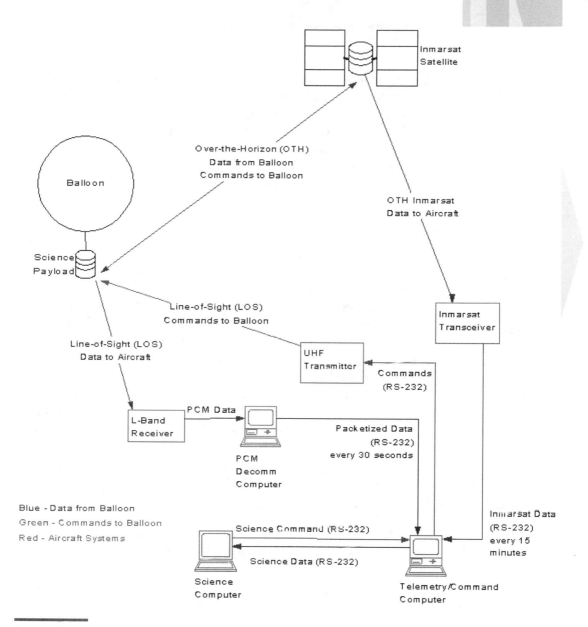

Figure 12–8
Diagram of aircraft telemetry system

System Design

The telemetry/command software uses the natural ability of LabVIEW to handle multiple tasks in parallel. Each module (telemetry retrieval, telemetry display, NSBF commanding, science commanding, etc.) of the program is a separate task/VI and runs in parallel with the other tasks/VIs. The only interactions between the modules are a few global variables to pass data and program control parameters.

■ Data Retrieval

Telemetry data retrieval and conversion to raw data are accomplished using different techniques for the LOS and OTH links. The LOS data arrive at the telemetry/command PC asynchronously approximately every 30 seconds. The OTH data are retrieved by querying the Inmarsat transceiver and requesting the most recent data. New OTH data arrive every 15 minutes.

Because of limitations in the amount of data that can be relayed to the aircraft, the data are formatted in packets which are packed as tightly as possible on irregular bit boundaries. Each packet of data has a unique header to identify the source of the data and a checksum to guarantee data validity.

The incoming data packets arrive as strings from the serial ports. These strings are time-tagged, logged to disk, and are then converted into an array of bits. The information from a packet setup file is then used to extract an array of data in raw counts from the bit array. This process is executed in two parallel loops, one for the LOS data and another for the OTH data.

■ Telemetry Display and Conversion of Data

The telemetry display (Figure 12–9) is created using the stored arrays of raw data and 12 data screen setup files. The telemetry display is presented as a table with three columns: label, value, and either units or Boolean text. The setup files are used to format the table display and contain a label string, the location of the raw data in the array, and a format option for each value. Currently, there are three options for data display: Boolean with text; scaling using a custom equation; and raw counts. The user is able to select which data screen is displayed using function keys and the user can also select which set of data, LOS or OTH, is displayed.

The most difficult requirements to accommodate were the need to custom-scale analog parameters and the need to maintain compatibility with existing setup files. Some of the scaling equations are extremely complex as shown in the following example:

```
s1 = (A * 0.01221) - (B * 0.01221)
s2 = A * 0.357627 + 6.89
s3 = (A * 0.007358) * (B * 0.001604)
s4 = A * 0.345764 + 8.12
s5 =  1.0/((0.2505*(10.0^-2.0))+((0.238*(10.0^-
       3.0))*(ln((((100000.0*A)/4095.0)/(1.0-(A/
       4095.0))))/1000.0)))+((0.313*(10.0^-
       5.0))*((ln((((100000.0*A)/4095.0)/(1.0-(A/
       4095.0))))/1000.0))^2.0))+((0.764*(10.0^-
       7.0))*((ln((((100000.0*A)/4095.0)/(1.0-(A/
       4095.0))))/1000.0))^3.0))))-273.15
```

The LabVIEW G Math Toolkit gave us the capability of easily reading these complex equations from a file and using them to scale the raw telemetry data. The indirect form of the Parser VIs were used in this application. At program initiation, the equations are read from the file and then parsed. The results of the parsing process are stored in an array for use during construction of the telemetry display.

After program initiation is complete, the parsed equations are used to evaluate the raw telemetry data based on the contents of the screen setup files described above. If the screen setup file indicates that a value needs to be scaled, the appropriate equation and raw data are evaluated and inserted into the telemetry display table as it is being constructed.

This ability to store equations in an ASCII file and then call them programmatically is very powerful. It allows custom-scaling without any code modification, which is very important for source code control and allows the end user ultimate control over how the data sequence is scaled.

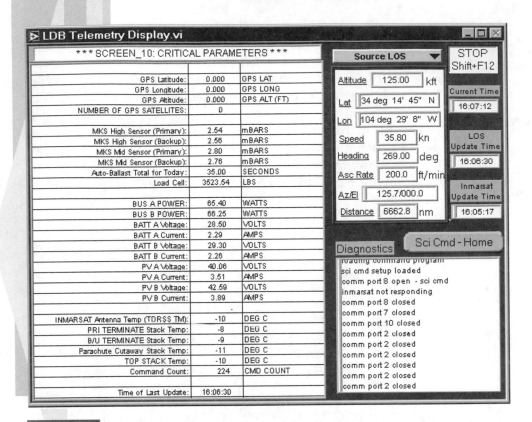

Figure 12–9
Example of user-configurable telemetry display screen

Postflight Processing of Log Files

For our application, there is an added benefit to this implementation with regard to postflight data processing. The data are logged as time-tagged strings of packed data to reduce the amount of disk I/O required during the flight. After the flight, these compressed log files are converted into a spreadsheet file using a slightly modified version of the process described above to construct a telemetry table display. The user has complete control over the content of the spreadsheet file, since the screen setup files and the equation are used to control the content.

Commanding and Science Data

Commands from science groups and the balloon operator are sent to the balloon using a UHF LOS transmitter which has a built-in RS-232 modem for modulation. The list of available commands is completely user-configurable using ASCII setup files (Figure 12–10).

Science groups interface with the telemetry/command PC via two RS-232 connections. One of the connections is for commanding and the other is for telemetry. It is through these connections that scientists can initiate a command to their payload and receive telemetry data from their payload.

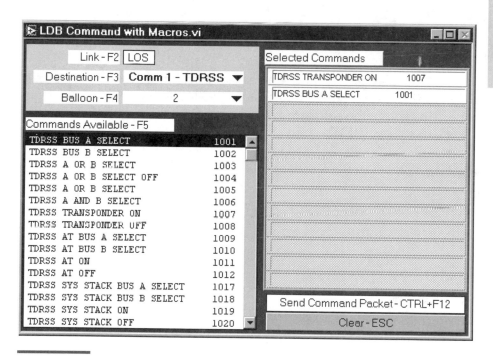

Figure 12–10
Example of command selection window

Conclusions

The telemetry/command software has been used successfully for several flights. During a circumpolar flight in the summer of 1997, the aircraft system allowed the payload engineers to monitor the balloon continuously as they traveled from the launch site in Fairbanks, Alaska, to the primary operations control center in Palestine, Texas.

The use of LabVIEW and the G Math Toolkit allowed the software development to proceed quickly. Some of the primary factors in the success of this project were the ability of LabVIEW to run tasks in parallel, the ability to easily reuse LabVIEW code developed for other systems, and the utility of the G Math Toolkit in using equations stored in an ASCII file.

Future Work

The adaptability and decreased development time that LabVIEW provided for this system has led to a development effort to replace the software written in ADA with new software written in LabVIEW.

■ **Contact Information**

Michael Fortenberry
NSBF/PSL
P.O. Box 319
Palestine, TX 75802
903-723-8008
forten@psl.nmsu.edu
http://pslwww.nmsu.edu
http://www.wff.nasa.gov/~web/balloons.html

An Interactive Instrumentation System for Wafer Emissivity Estimation

Rus Belikov
Princeton University

Maurizio Fulco
NJIT

Sergey Belikov
CVC, Inc.

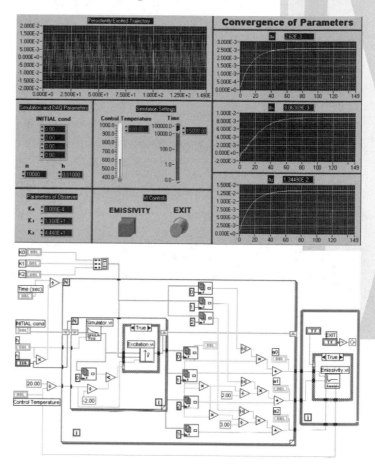

The Challenge Estimating a silicon wafer's emissivity (as a function of wavelength emitted) when the wafer is being processed in an RTP (Rapid Thermal Processing) system.

The Solution Building an interactive LabVIEW-based application (see Figure 12–11 and Figure 12–13) that utilizes various physical models and mathematical methods to estimate the emissivity.

Figure 12–11
Main VI Interface and Block Diagram

Introduction

RTP (Rapid Thermal Processing) is a single-wafer technology to thermally process semiconductor wafers [1]. The basic idea is to heat the wafer inside a special chamber using radiation from an array of lamps. The RTP installation at NJIT, for example, uses 3 concentric rings of tungsten-halogen lamps developed at Texas Instruments, Inc. Radiation would naturally provide a much shorter processing time per wafer in a single-wafer processing environment than conventional furnace methods. However, RTP has some problems that are difficult to overcome. One of the most fundamental ones is temperature control.

This involves measuring the temperature of the wafer being processed as well as making the temperature repeatable and uniform across its surface. This project concerns itself mostly with the first problem. (To achieve temperature uniformity, as well as controlling the temperature when it is measured by thermocouples, there is a model-based adaptive control algorithm with real-time parameter estimation, developed at NJIT [2]. It makes use of the NI LabWINDOWS interface [3].)

Emissivity model

Temperature in RTP environment is usually measured by a pyrometer, since we can't go on and touch the wafer with a thermocouple when it has sensitive devices on top. A pyrometer measures the thermal radiation from the wafer, which then could be used to calculate the temperature using a specific model [4]. However, a pyrometer needs to be calibrated before it could be used. To do this, we go backwards—we calculate what the emissivity is from the temperature of the wafer, which for these purposes is measured by thermocouples. The following model is used [3]:

$$\rho c(T)\dot{T} = 2h^{-1}[E(T) + S + G(U)] \tag{12.4}$$

$$E(T) = \int_0^\infty \varepsilon(\lambda, T)\text{Planck}(\lambda, T)d\lambda \tag{12.5}$$

Without going into details, Equation 12.4 states that the temperature gradient T is linear with respect to the sum of radiation emitted from the wafer E, radiation from the heating lamps

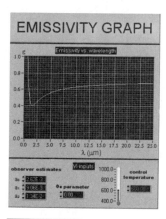

Figure 12–12
Emissivity

$G(U)$, and unmodelled heat flux S. ρ and $c(T)$ are the density and the temperature dependent specific heat of the wafer material, h is the thickness of the wafer. The second equation just states that E is the emission density at a particular wavelength, described by Planck's formula, integrated over the wavelength. As could be seen from the Equation 12.5, the emission at a given wavelength involves the emissivity function $\varepsilon(\lambda, T)$. All other quantities in the equations are either known or estimated.

Parameter Estimation

Therefore, the procedure is as follows. First, we generate temperature variations on the wafer (a persistently excited temperature trajectory) by using the aforementioned adaptive controller developed at NJIT [3]. An example of this trajectory is given in Figure 12–11. Using Equation 12.4, this gives us the total emission $E(T)$ for some range of T. Now, by using Equation 12.5, we have some clue as to what the emissivity is. At this point, a method of trial and error must be used to pick an emissivity function that agrees with the data for $E(T)$ according to Equation 12.5.

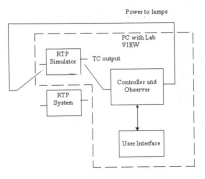

Figure 12–13
Emissivity

This is done by picking some form for the emissivity function with certain adjustable parameters. The following emissivity model seems to work well for certain types of wafers:

$$\varepsilon(\lambda, T) = p^0 + p^1\lambda^{-1} + p^2\lambda^{-2}$$

where λ is the wavelength and p's are the unknown parameters. The parameters are then estimated by using a nonlinear reduced order observer [5]. Figure 12–11 shows the action as the a_i parameters converge. (These are not actually the p's in the emissivity model, but are directly related to them.) The estimated emissivity function can then be viewed by pressing the green "emissivity" button (Figure 12–12).

Conclusions

With everything above in mind, clearly the importance of an interactive interface cannot be overemphasized. The model, initial conditions and other parameters must be easily modifiable; and the application must provide enough flexibility to design a desired real-time experiment and data analysis. LabVIEW provides an ideal environment to build just this interface. In addition, NI's G-Math Toolkit can provide a very convenient feature of real-time formula manipulation, adding one more degree of freedom to estimating the emissivity function.

Acknowledgments

This work was supported by National Science Foundation under the grant ECS-9312451.

References

1. Moslehi, M., Lee, Y. J., Schaper, C., et al. (1996). Single-Wafer Process Integration and Process Control Techniques. In *Advances in Rapid Thermal and Integrated Processes* (ed. F. Roozeboom), Kluwer Acad. Publisher, pp. 163–169.
2. Belikov, S., and Friedland, B. (1995). Closed-Loop Adaptive Control for Rapid Thermal Processing. 34th IEEE Conf. On Decision and Control, New Orleans, LA, pp. 2476–2485.
3. Belikov, S., Hur, D., Friedland, B., et al. (1996). Estimation of Emissivity of a Wafer in an RTP Chamber by a Dynamic Observer, in *Rapid Thermal and Integrated Processing V* (Mater. Res. Sos. Proc. 429, Pittsburgh, PA), pp. 335–340.
4. Wood, S., Apte, P., King, T-J., et al. (1991). Pyrometer Modeling in Rapid Thermal Processing, in *Rapid Thermal and Related Processing Techniques*, R. Singh, M. Moslehi, Editors, Proc. SPIE 1393, pp. 337–348.
5. Friedland, B. (1996). *Advanced Control System Design*, Prentice Hall, Englewood Cliffs, NJ.

■ Contact Information

Sergey Belikov
650- 813-0181
sbelikov@cvc.com

Appendix A
Mathematical Background

Overview

In this appendix, we will discuss some of the mathematical techniques that are commonly used in this book. This background material will help you review these techniques and will make reading the material in the book much easier.

A-1 Laplace Transforms

The Fourier transform is a domain transformation technique for analyzing continuous-time signals and systems. It is a method of representing a time signal $f(t)$ as a continuous summation of exponential functions of the form $e^{j\omega t}$. The Fourier transform exists only for those signals that are absolutely integrable. For example, the Fourier transform cannot handle infinitely rising exponential functions. To extend the transform technique to such signals, a convergence factor needs to be introduced in the transformation definition. In this definition, the signal is expressed as a continuous sum of exponentials of the type e^{st}, where $s = \sigma + j\omega$. The figure below shows a sinusoidal wave on the $j\omega$ axis and how it will appear in the right-half and the left-half of the complex s-plane.

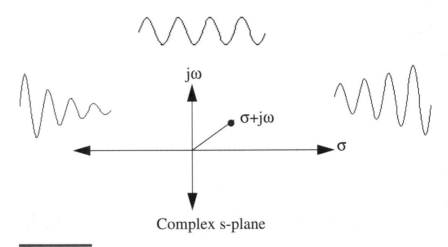

Sinusoidal wave behavior in the complex s-plane.

By transforming the signal to the complex s-plane, we introduce a convergence factor $e^{-\sigma t}$ in the transform definition. Such a transform is known as the Laplace transform. While the Fourier transform is most useful in dealing with frequency domain concepts such as the frequency response of filtering systems, the Laplace transform is very powerful in obtaining a closed-form solution of linear integro-differential equations of a system which may or may not be stable. Equations A.1 and A.2 define a Laplace transform pair:

$$F(s) = \int_{-\infty}^{\infty} f(t)e^{-st}dt \tag{A.1}$$

$$f(t) = \frac{1}{2\pi j}\int_{(\sigma - j\infty)}^{(\sigma + j\infty)} F(s)e^{st}ds \tag{A.2}$$

They are also referred to as the two-sided Laplace transform pair because they cover the time range from negative infinity to positive infinity. However, in a majority of applications of system dynamics, the time functions are causal, i.e., $f(t) = 0$ for $t < 0$. For such functions, the transform pair of Equations A.1 and A.2 reduces to

$$F(s) = \int_0^\infty f(t)e^{-st}dt \tag{A.3}$$

$$f(t) = \frac{1}{2\pi j}\int_{(\sigma - j\infty)}^{(\sigma + j\infty)} F(s)e^{st}ds \tag{A.4}$$

and is referred to as the one-sided Laplace transform. The Laplace transform is unique, and its inverse transform can be found by using a transform table along with certain transform properties. Table A-1 shows some useful one-sided Laplace transforms.

Table A-1   *Some useful one-sided Laplace transforms*

Function	Laplace Transform
$\delta(t)$	1
$u(t)$	$\dfrac{1}{s}$
$tu(t)$	$\dfrac{1}{s^2}$
e^{-at}	$\dfrac{1}{s + a}$
te^{-at}	$\dfrac{1}{(s + a)^2}$

When performing system analysis using the Laplace transform, the following steps have to be carried out.

1. Step 1: Formulate a time-domain differential equation that describes the input-output relationship of a physical system.
2. Step 2: Compute the Laplace transform of the equation. This step incorporates the effects of the inputs (excitations) as well as initial conditions of the system.
3. Step 3: Solve for the output variable in the s-domain by algebraic manipulation of the result in Step 2.
4. Step 4: Obtain the solution of the output variable in the time-domain by inverse Laplace transformation.

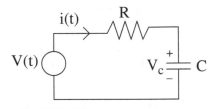

A simple RC circuit

Consider a circuit consisting of a voltage source V, a resistor R, and a capacitor C, as shown in the figure above. We can find the output voltage across the capacitor using the four steps outlined above.

Step 1: Apply Kirchoff's Voltage Law to formulate the time domain differential equation for the system

$$V(t) - Ri(t) - \frac{1}{C}\int i(t)dt = 0 \tag{A.5}$$

Step 2: Compute the Laplace transform of the equation, assuming zero initial conditions

$$V(s) - Ri(s) - \frac{1}{Cs}i(s) = 0 \tag{A.6}$$

Step 3: First solve for the current flowing through the circuit by algebraic manipulation of Equation A.6.

$$i(s) = \frac{sC \times V(s)}{1 + RCs}$$

Then solve for the output voltage (voltage across the capacitor).

$$V_c(s) = \frac{1}{Cs}i(s) = \frac{V(s)}{1 + RCs} = \frac{1}{RC} \times \frac{V(s)}{s + 1/(RC)}$$

Step 4: Use the Inverse Laplace transform to obtain the solution in the time-domain.

$$V_c(t) = \frac{V(t)}{RC}e^{-\frac{1}{RC}t}$$

The Laplace transform thus serves as a useful tool to analyze systems that are described in the time-domain by mathematical descriptions using integration and differentiation.

Appendix B
Error Codes

■ Analysis Error Codes

Table B-1   *Analysis error codes*

Code	Name	Description
0	NoErr	No error; the call was successful.
−20001	OutOfMemErr	There is not enough memory left to perform the specified routine.
−20002	EqSamplesErr	The input sequences must be the same size.
−20003	SamplesGTZeroErr	The number of samples must be greater than zero.
−20004	SamplesGEZeroErr	The number of samples must be greater than or equal to zero.
−20005	SamplesGEOneErr	The number of samples must be greater than or equal to one.

(Continued)

Table B-1 *Analysis error codes*

Code	Name	Description
–20006	SamplesGETwoErr	The number of samples must be greater than or equal to two.
–20007	SamplesGEThreeErr	The number of samples must be greater than or equal to three.
–20008	ArraySizeErr	The input arrays do not contain the correct number of data values for this VI.
–20009	PowerOfTwoErr	The size of the input array must be a power of two: size $= 2^m$, $0 < m < 23$
–20010	MaxXformSizeErr	The maximum transform size has been exceeded.
–20011	DutyCycleErr	The duty cycle must meet the condition: $0 \leq$ duty cycle ≤ 100
–20012	CyclesErr	The number of cycles must be greater than zero and less than or equal to the number of samples.
–20013	WidthLTSamplesErr	The width must meet the condition: $0 <$ width $<$ samples
–20014	DelayWidthErr	The delay must meet the condition: $0 \leq$ (delay $+$ width) $<$ samples
–20015	DtGEZeroErr	dt must be greater than or equal to zero.
–20016	DtGTZeroErr	dt must be greater than zero.
–20017	IndexLTSamplesErr	The index must meet the condition: $0 \leq$ index $<$ samples
–20018	IndexLengthErr	The index must meet the condition: $0 \leq$ (index $+$ length) $<$ samples
–20019	UpperGELowerErr	The upper value must be greater than or equal to the lower value.
–20020	NyquistErr	The cutoff frequency, f_c, must meet the condition: $$0 \leq f_c \leq \frac{f_s}{2}$$
–20021	OrderGTZeroErr	The order must be greater than zero.

(Continued)

Table B-1   *Analysis error codes*

Code	Name	Description
−20022	DecFactErr	The decimating factor must meet the condition: $0 < \text{decimating} \leq \text{samples}$
−20023	BandSpecErr	The band specifications must meet the condition: $0 \leq f_{\text{low}} \leq f_{\text{high}} \leq \dfrac{f_s}{2}$
−20024	RippleGTZeroErr	The ripple amplitude must be greater than zero.
−20025	AttenGTZeroErr	The attenuation must be greater than zero.
−20026	WidthGTZeroErr	The width must be greater than zero.
−20027	FinalGTZeroErr	The final value must be greater than zero.
−20028	AttenGTRippleErr	The attenuation must be greater than the ripple amplitude.
−20029	StepSizeErr	The step-size, μ, must meet the condition: $0 \leq \mu \leq 0.1$
−20030	LeakErr	The leakage coefficient must meet the condition: $0 \leq \text{leak} \leq \mu$
−20031	EqRplDesignErr	The filter cannot be designed with the specified input values.
−20032	RankErr	The rank of the filter must meet the condition: $1 \leq (2\,\text{rank} + 1) \leq \text{size}$
−20033	EvenSizeErr	The number of coefficients must be odd for this filter.
−20034	OddSizeErr	The number of coefficients must be even for this filter.
−20035	StdDevErr	The standard deviation must be greater than zero for normalization.
−20036	MixedSignErr	The elements of the **Y Values** array must be nonzero and either all positive or all negative.
−20037	SizeGTOrderErr	The number of data points in the **Y Values** array must be greater than two.

(Continued)

Table B-1 *Analysis error codes*

Code	Name	Description
–20038	IntervalsErr	The number of intervals must be greater than zero.
–20039	MatrixMulErr	The number of columns in the first matrix is not equal to the number of rows in the second matrix or vector.
–20040	SquareMatrixErr	The input matrix must be a square matrix.
–20041	SingularMatrixErr	The system of equations cannot be solved because the input matrix is singular.
–20042	LevelsErr	The number of levels is out of range.
–20043	FactorErr	The level of factors is out of range for some data.
–20044	ObservationsErr	Zero observations were made at some level of a factor.
–20045	DataErr	The total number of data points must be equal to the product of the levels for each factor and the observations per cell.
–20046	OverflowErr	There is an overflow in the calculated F-value.
–20047	BalanceErr	The data is unbalanced. All cells must contain the same number of observations.
–20048	ModelErr	The random effect model was requested when the fixed effect model was required.
–20049	DistinctErr	The x values must be distinct.
–20050	PoleErr	The interpolating function has a pole at the requested value.
–20051	ColumnErr	All values in the first column in the **X** matrix must be one.
–20052	FreedomErr	The degrees of freedom must be one or more.
–20053	ProbabilityErr	The probability must be between zero and one.
–20054	InvProbErr	The probability must be greater than or equal to zero and less than one.
–20055	CategoryErr	The number of categories or samples must be greater than one.

(Continued)

Table B-1 *Analysis error codes*

Code	Name	Description
−20056	TableErr	The contingency table must not contain a negative number.
−20061	InvSelectionErr	One of the input selections is invalid.
−20062	MaxIterErr	The maximum iterations have been exceeded.
−20063	PolyErr	The polynomial coefficients are invalid.
−20064	InitStateErr	This VI has not been initialized correctly.
−20065	ZeroVectorErr	The vector cannot be zero.

■ G Math Toolkit Error Codes

Table B-2 *G Math Toolkit error codes*

Error Code Number	Error Code Description
0	No error
−23001	Syntax error of parser
−23002	Discrepancy between function, variables, and coordinates
−23003	Number of contours out of range
−23004	Number of color palettes out of range
−23005	Negative distance
−23006	Not a valid path
−23007	Not a graphs file
−23008	Wrong input, Euler method
−23009	Wrong input, Runge Kutta method
−23010	Wrong input, Cash Karp method
−23011	Nonpositive step rate
−23012	Nonpositive accuracy
−23013	Matrix vector conflict

(Continued)

Table B-2 G Math Toolkit error codes

Error Code Number	Error Code Description
−23014	A and X_0 have different dimensions
−23015	Empty X_0
−23016	Singular eigenvector matrix
−23017	Multiple roots
−23018	Left point is a root
−23019	Right point is a root
−23020	Left point greater than right point
−23021	Both function values have the same sign
−23022	Nonpositive accuracy or nonpositive delta $x(h)$
−23023	Wrong dimension of start
−23024	No root found
−23025	Nonvalid triplet (a,b,c)
−23026	No optimum found
−23027	Not exactly one variable
−23028	Wrong model equation
−23029	Levenberg Marquardt has failed
−23030	$m \geq n \geq 0$ is violated or the matrix of derivatives has the wrong dimension
−23031	No valid point
−23032	Maximum does not exist
−23033	Vectors have different dimensions or empty vectors
−23034	Ill conditioned system
−23035	Nonpositive number
−23036	Different parameters
−23037	Not exactly two functions

(Continued)

Table B-2   *G Math Toolkit error codes*

Error Code Number	Error Code Description
–23038	No variables in expression
–23039	Parameter problem
–23040	Derivative out of range
–23041	Not exactly two variables
–23042	Negative argument
–23043	Argument out of range (0,1]
–23044	Argument out of range [0,1]
–23045	$n < k$
–23046	Empty array
–23047	Argument out of range [0,100]
–23048	Invalid time increment
–23049	Invalid window length
–23050	Signal length not a multiple of number
–23051	Signal length not a power of two
–23052	Signal length not a prime and ≥ 5
–23053	Signal length not a power of two and ≥ 4
–23054	Nonunique variables

■ Functions for Use with G Math Toolkit Parser VIs

Table B-3   *Functions for use with G Math Toolkit Parser VIs*

GMath Function Name	Corresponding Function	Description
abs(x)	Absolute Value	Returns the absolute value of x.
acos(x)	Inverse Cosine	Computes the inverse cosine of x.

(Continued)

Table B-3 *Functions for use with G Math Toolkit Parser VIs*

GMath Function Name	Corresponding Function	Description
acosh(x)	Inverse Hyperbolic Cosine	Computes the inverse hyperbolic cosine of x in radians.
asin(x)	Inverse Sine	Computes the inverse sine of x in radians.
asinh(x)	Inverse Hyperbolic Sine	Computes the inverse hyperbolic sine of x in radians.
atan(x)	Inverse Tangent	Computes the inverse tangent of x in radians.
atanh(x)	Inverse Hyperbolic Tangent	Computes the inverse hyperbolic tangent of x in radians.
ci(x)	Cosine Integral	Computes the cosine integral of x where x is any real number.
ceil(x)	Round to +Infinity	Rounds x to the next higher integer (smallest integer $\geq x$).
cos(x)	Cosine	Computes the cosine of x in radians.
cosh(x)	Hyperbolic Cosine	Computes the hyperbolic cosine of x in radians.
cot(x)	Cotangent	Computes the cotangent of x in radians $(1/\tan(x))$.
csc(x)	Cosecant	Computes the cosecant of x in radians $(1/\sin(x))$.
exp(x)	Exponential	Computes the value of e raised to the power x.
expm1(x)	Exponential(Arg)–1	Computes the value of e raised to the power of $x - 1$ $(e^x - 1)$
floor(x)	Round to –Infinity	Truncates x to the next lower integer (largest integer $\leq x$)
gamma(x)	Gamma Function	$\Gamma(n + 1) = n!$ for all natural numbers n.
getexp(x)	Mantissa and exponent	Returns the exponent of x.

(Continued)

Table B-3   *Functions for use with G Math Toolkit Parser VIs*

GMath Function Name	Corresponding Function	Description
getman(x)	Mantissa and exponent	Returns the mantissa of x.
int(x)	Round to nearest integer	Rounds its argument to the nearest even integer.
intrz	Round toward zero	Rounds x to the nearest integer between x and zero.
ln(x)	Natural Logarithm	Computes the natural logarithm of x (to the base e).
lnpl(x)	Natural Logarithm (Arg + 1)	Computes the natural logarithm of $(x + 1)$.
log(x)	Logarithm Base 10	Computes the logarithm of x (to the base 10).
log2(x)	Logarithm Base 2	Computes the logarithm of x (to the base 2).
pi(x)	Represents the value $\pi = 3.14159\ldots$	$pi(x) = x^*\pi$ $pi(1) = \pi$ $pi(2.4) = 2.4^*\pi$
rand( )	Random Number (0–1)	Produces a floating-point number between 0 and 1.
sec(x)	Secant	Computes the secant of x $(1/\cos(x))$.
si(x)	Sine Integral	Computes the sine integral of x where x is any real number.
sign(x)	Sign	Returns 1 if x is greater than 0. Returns 0 if x is equal to 0. Returns -1 if x is less than 0.
sin(x)	Sine	Computes the sine of x in radians.
sinc(x)	Sinc	Computes the sine of x divided by x in radians $(\sin(x)/x)$.
sinh(x)	Hyperbolic Sine	Computes the hyperbolic sine of x in radians.

(Continued)

Table B-3   *Functions for use with G Math Toolkit Parser VIs*

GMath Function Name	Corresponding Function	Description
spike(x)	Spike function	spike(x) returns: 1 if $0 \leq x \leq 1$ 0 for any other value of x.
sqrt(x)	Square Root	Computes the square root of x.
square(x)	Square function	Square (x) returns 1 if $2n \leq x \leq (2n+1)$ 0 if $2n+1 \leq x \leq (2n+2)$ where x is any real number and n is any integer.
step(x)	Step function	Step(x) returns 0 if $x < 0$ 1 if any other condition obtains.
tan(x)	Tangent	Computes the tangent of x in radians.
tanh(x)	Hyperbolic Tangent	Computes the hyperbolic tangent of x in radians.

■ G Math Toolkit Parser Error Codes

Table B-4   *G Math Toolkit Parser error codes*

Error Code Number	Error Code Description	Error Example
0	No error	sin(x)
1	Bracket problem at the beginning	1+x)
2	Incomplete function expression	sin(x)+
3	Bracket problem	()
4	Bracket problem at the end	(1+x
5	Wrong decimal point	1,2 (U.S.)
6	Wrong number format	1e–3 instead of 1E–3

(Continued)

Table B-4   *G Math Toolkit Parser error codes*

Error Code Number	Error Code Description	Error Example
7	Wrong function call	sin()
8	Not a valid function	sins(x)
9	Incomplete expression	$x+$
10	Wrong variable name	a11
11	Wrong letter	sin(X)
12	Too many decimal points	1.23.45
21	Contains more than one variable	$1+x+y4$
22	Inconsistency in variables or numbers	Depends on application
23	Contains variables	Depends on application
24	Variables output problem	Depends on application

Appendix C
Frequently Asked Questions

■ Background

1. **Can the antialiasing filter be a digital filter?**

    In general, when going from a domain P to a domain Q, the filter should be in domain P to avoid the problem of aliasing in the domain Q. Thus, when sampling to convert an analog signal to a digital signal, the antialiasing filter should be an analog filter. However, when performing decimation to reduce the sampling rate, you are going from the digital domain (with a higher sampling rate) to the digital domain (with a lower sampling rate). In this case, the filtering before the decimation is performed in the digital domain by a digital filter.

■ Signal Generation

1. **Why are there Pattern VIs, when the Wave VIs will do all that the Pattern VIs can and more?**

    The historical reason is that Pattern VIs were created first. Also, the Pattern VIs are meant to be used to generate a block of data at a time, whereas the Wave VIs can be called iteratively in a loop.

2. **Why do I need normalized frequency? Why not just use the common frequency unit of Hertz in the input controls and have the VI automatically convert it internally into normalized frequency if necessary?**

Because some people think in terms of Hertz (cycles per second) and some think in terms of cycles. Normalized frequency handles both.

3. **What does it mean that the Wave VIs are "reentrant"?**

You can have several of them on the same block diagram, and each one will have its own internal memory space. They will all work independently of each other. This allows each VI to maintain its own internal phase.

■ Signal Processing

1. **What are the effects zero padding?**

Zero padding in the time domain performs interpolation in the frequency domain. In addition to improving (that is, lowering) the frequency resolution, faster algorithms (using the FFT instead of the DFT) are made possible.

2. **What is the physical meaning of the cross power spectrum?**

It is a measure of the similarity of two signals in the frequency domain. It is the frequency-domain equivalent of the cross-correlation function. It shows the joint presence of energy in the two signals.

■ Windows

1. **Why do different VIs and toolkits provide only some specific window choices? Why not all window choices?**

One reason is that the windows chosen are the most common windows for that particular application. The other is that several of the windows give very similar results, so there is no point in including all of them.

■ Digital Filters

1. **What is meant by linear phase?**

    Linear phase in digital filters means that the phase distortion is nothing more than a digital delay. All input samples will be shifted by some constant number of samples, so this phase change can be easily compensated and/or modeled.

    Nonlinear phase means that the individual sine waves at different frequencies that make up the input signal get shifted in time by different amounts. This sort of phase change is very difficult to work around. Some signals (like the square wave) are very sensitive to this sort of phase distortion.

2. **What does it mean when the IIR filters "execute in place"?**

    In place means that the input array space (memory locations) are being reused as the output array space. In place usually implies lower demands on memory.

3. **Do the FIR filters have only zeros?**

    FIR filters do have zeros, but they also have poles at the origin of the z-plane. If you take the Z-transform of an FIR filter and rewrite it as the product of factored terms, you will find that in addition to zeros, they also have poles at the origin.

■ Curve Fitting

1. **What is the limit in LabVIEW and BridgeVIEW on the number of parameters used for curve fitting?**

    There is no limit. However, keep in mind that the memory requirements, the time required to find the solution, and fitting accuracy are all affected by increasing the number of parameters.

2. **Some curve fitting algorithms allow you to give weight to certain data points. Can I do that in the VIs from the advanced analysis library of LabVIEW and BridgeVIEW?**

No, it is not possible at the time of this writing.

3. **In performing a fit, is there any rule of thumb as to how many data points to use?**

   Normally, you need at least one more than the number of parameters for which you are trying to solve. But there is no such rule as to whether you should use at least five times more, 10 times more, and so on. As an example, in performing a polynomial fit, the number of data points to be used to obtain a "good" fit may be correlated to how close the data is to the underlying polynomial (that is, how much noise).

4. **In the Levenberg-Marquardt VI, why is the derivative information needed?**

   The information is needed to calculate the Jacobian, which is needed in the algorithm to solve for the coefficients that you are trying to determine. See the LabVIEW Online Help for details.

5. **What is the default tolerance value in the Non Linear Least Square Fit VI in the Analysis Library?**

   The default tolerance is 1e-3 (0.001)

6. **How is the algorithm for polynomial interpolation implemented in the Analysis Library?**

   Polynomial interpolation is implemented using the type of basis functions known as Lagrange basis functions. The straightforward way to implement these basis functions is not very efficient. So we implement them using Neville's algorithm. Using this algorithm, it is very easy to build the interpolating function, and it requires less work to actually evaluate the interpolant.

7. **Why are rational functions used only for interpolation and not for curve fitting?**

   By using rational functions, you can only find approximate solutions. This is okay for interpolation applications where you are only interested in the value at one specific point. In the case of curve fitting, the function that describes the data has to fit the given data points exactly. Rational functions do not guarantee this aspect.

Appendix D
References

The following documents contain more detailed information about the topics discussed in this book.

■ National Instruments Manuals

- *Control and Simulation Toolkit for G Reference Manual*
- *Digital Filter Design Toolkit Reference Manual*
- *G Math Toolkit Reference Manual*
- *LabVIEW Analysis VI Reference Manual*
- *LabVIEW Data Acquisition Course Manual*

■ National Instruments Application Notes

- Application Note #023—*Digital Signal Processing Fundamentals*

■ Application Note #041—*The Fundamentals of FFT-Based Signal Analysis and Measurement in LabVIEW and LabWindows*

■ Application Note #040—*Fast Fourier Transforms and Power Spectra in LabVIEW*

■ Application Note #091—*G Math—A New Paradigm for Mathematics*

■ Application Note #097—*Designing Filters Using the Digital Filter Design Toolkit*

■ Other Documents

■ *Advanced Engineering Mathematics*, Erwin Kreyszig, 5th ed., Wiley Eastern Limited, 1983

■ *ANSI S1.11-1986 Specification for Octave-Band and Fractional-Octave-Band Analog and Digital Filters*

■ *Designing Digital Filters*, Charles S. Williams, Prentice Hall, 1986

■ *Digital Filters*, R.W. Hamming, 3rd ed., Prentice Hall, 1989

■ *Discrete-Time Signal Processing*, Alan V. Oppenheim and Ronald W. Schafer, Prentice Hall, 1989

■ *The FFT: Fundamentals and Concepts*, Robert W. Ramirez, Prentice Hall 1985

■ *Joint Time-Frequency Analysis: Methods and Applications*, Shie Qian and Dapang Chen, Prentice Hall, 1996

■ *Multirate Systems and Filter Banks*, P.P. Vaidyanathan, Prentice Hall, 1993

■ *Numerical Recipes, The Art of Scientific Computing*, William H. Press, Brian P. Flannery, Saul A. Teukolsky, William T. Vetterling, Cambridge University Press, 1986

■ "On the Use of Windows for Harmonic Analysis with the Discrete Fourier Transform," Frederic J. Harris,

Proceedings of the IEEE, Vol. 676, No. 1, January 1978, pp. 51-84

■ "Some Windows with Very Good Sidelobe Behaviour," Albert H. Nuttall, *IEEE Transactions on Acoustics, Speech, and Signal Processing*, Vol. ASSP-29, No. 1, February 1981, pp. 84-91

■ "Windows to FFT Analysis," Svend Gade and Henrik Herlufsen, *Sound and Vibration*, March 1988, pp. 14-22

■ *LabVIEW for Everyone, Graphical Programming Made Even Easier,* Lisa K. Wells and Jeffrey Travis, Prentice Hall, 1997

■ *A Digital Signal Processing Primer, with Applications to Digital Audio and Computer Music,* Ken Steiglitz, Addison-Wesley, 1996

■ *Digital Filter Design,* T.W. Parks and C.S. Burrus, John Wiley & Sons, 1987

■ *Theory and Application of Digital Signal Processing,* L.R. Rabiner and B. Gold, Prentice Hall, 1975

■ *DFT/FFT and Convolution Algorithms, Theory and Implementation,* C.S. Burrus and T.W. Parks, John Wiley & Sons, 1987

■ *Who Is Fourier? A Mathematical Adventure,* Transnational College of LEX, Language Research Foundation, 1997

■ *Understanding Digital Signal Processing,* Richard G. Lyons, Addison-Wesley, 1997

■ *Linear System Theory and Design,* Chi-Tsong Chen, Harcourt Brace College Publishers, Austin, 1984.

■ *Modern Control Systems,* R. C. Dorf and R. H. Bishop, Addison-Wesley, California, 1995

■ *Scientific Computing, An Introduction Survey,* M.T. Heath, McGraw-Hill, New York, 1997

■ *Matrix Computations,* G. H. Golub and C. F. VanLoan, The Johns Hopkins University Press, Baltimore, 1989

- *Theory and Problems of Probability and Statistics*, Murray Spiegel, Schaum's Outline Series, McGraw Hill, New York, 1975

- *Mathematics for Business, Life Sciences, and Social Sciences*, Abe Mizrahi and Michael Sullivan, 5th Edition, John Wiley & Sons, Inc. New York, 1993

- *Elementary Differential Equations and Boundary Value Problems*, William E. Boyce and Richard C. DiPrima, 5th ed., John Wiley & Sons, USA, 1992

- *LabVIEW Applications*, Rahman Jamal and Herbert Pichlik, Prentice Hall, 1999

Appendix E
National Instruments Contact Information, Resources, and Toolkits

How to Contact National Instruments

Corporate Office
National Instruments
6504 Bridge Point Parkway
Austin, TX 78730-5039 USA
Tel. (512) 794-0100
Fax. (512) 794-8411
E-mail: info@natinst.com
Web Address: http://www.natinst.com

Australia
National Instruments Australia
P.O. Box 466
Ringwood, Victoria 3134
Australia
Tel. (03) 9879 5166
Fax. (03) 9879 6277
E-mail: info.australia@natinst.com
Web Address: http://www.natinst.com/australia

Austria
National Instruments Ges.m.b.H.
Plainbachstr. 12
5101 Salzburg-Bergheim
Austria
Tel. 43 662 457990-0
Fax. 43 662 457990-19
E-mail: natinst-austria@magnet.at
Web Address:
http://www.natinst.com/austria

Belgium
National Instruments Belgium nv/sa
Leuvensesteenweg 613
B-1930 Zaventem
Belgium
Tel. (352) 405 120 (Luxemburg)
Tel. 32 2 757 00 20
Fax. 32 2 757 03 11
E-mail: info.belgium@natinst.com
Web Address:
http://www.natinst.com/belgium

Brazil
National Instruments Brazil
Avenida Paulista, 509-cj301-3
01311-910
São Paulo-SP-Brazil
Tel. 55 11 288 3336
Fax. 55 11 288 8528
E-mail: ni.brazil@natinst.com
Web Address:
http://www.natinst.com/brazil

Canada
National Instruments Canada
15-6400 Millcreek Drive
Suite #424
Mississauga, ON
Canada, L5N 3E7
Tel. (905) 785-0085
Fax. (905) 785-0086
E-mail: info@natinst.com
Web Address:
http://www.natinst.com/canada

Denmark
National Instruments Danmark
Christianshusvej 189
2970 Hørsholm
Danmark
Tel. 45 45 76 26 00
Fax. 45 45 76 26 02
E-mail: ni.denmark@natinst.com
Web Address:
http://www.natinst.com/denmark

Finland
National Instruments Finland Oy
PL 2, Sinikalliontie 9
02631 Espoo, Finland
Tel. 358-9-725-725-11
Fax. 358-9-725-725-55
E-mail: ni.finland@natinst.com
Web Address:
http://www.natinst.com/finland

France
National Instruments France
Centre d'Affaires Paris-Nord
Immeuble Le Continental - BP 217
93 153 Le Blanc-Mesnil CEDEX
France
Tel. 33 01 48 14 24 24
Fax. 33 01 48 14 24 14
Web Address:
http://www.natinst.com/france

Germany
National Instruments Germany GmbH
Konrad-Celtis Str. 79
81369 München
Germany
Tel. 49 89 74 13 13 0
Fax. 49 89 7 14 60 35
E-mail: nig.cs@natinst.com
Web Address:
http://www.natinst.com/germany

Hong Kong
National Instruments Hong Kong Limited
Suite 210
New Commerce Center
19 On Sum Street
Shantin, N.T.
Hong Kong
Tel. 852 2645 3186
Fax. 852 2686 8505
E-mail: general@nihk.com.hk
Web Address:
http://www.natinst.com/hongkong

Israel
National Instruments Israel, LTD.
10 Sokolov Street, 3rd Floor
Ramat-Gan 52571
Israel
Tel. 972 3 6120092
Fax. 972 3 6120095
E-mail: ni.Israel@natinst.com
Web Address:
http://www.natinst.com/israel

Italy
National Instruments Italy S.r.l.
Via Anna Kuliscioff, 22
20152 Milan, Italy
Tel. 39 2 413 091
Fax. 39 2 4130 9215
E-mail: ni.italy@natinst.com
Web Address:
http://www.natinst.com/italy

Japan
Nihon National Instruments K.K.
Shuwa Shiba Park Bldg. B-5F
Shibakoen 2-4-1, Minato-ku
Tokyo, Japan 105-0011
Tel. 81 3 5472 2970
Fax. 81 3 5472 2977
Web Address:
http://www.natinst.com/nni

Korea
National Instruments Korea Limited
#203 Poongsung B/D
51-12 Banpo4-dong, Seocho-ku
Seoul, Korea 137-044
Tel. 82 2 596-7456
Fax. 82 2 596 7455
Web Address:
http://www.natinst.com/korea

Mexico
National Instruments Mexico
Galileo, 31B
Suite 570
Col. Polanco
11560 Mexico, D.F.
Tel. 525 520 2635
Fax. 525 520 3282
E-mail: info.mexico@natinst.com
Web Address:
http://www.natinst.com/mexico

Netherlands
National Instruments Netherlands BV
Vijzelmolenlaan 8A
3447 GX Woerden
Netherlands
Tel. 31 348 433466
Fax. 31 348 430673
Email: info.netherlands@natinst.com
Web Address:
http://www.natinst.com/netherlands

Norway
National Instruments Norge
Industrigt. 15
Postboks 592
3412 Lierstranda
Norway
Tel. 47 32 84 84 00
Fax. 47 32 84 86 00
E-mail: ni.norway@natinst.com
Web Address:
http://www.natinst.com/norway

Singapore
National Instruments Singapore Pte Ltd
8 Jalan Kilang Timor
#03-03B Kewalram House
Singapore 159305
Tel. 65 226 5886
Fax. 65 226 5887
Email: natinst@singnet.com.sg
Web Address:
http://www.natinst.com/singapore

Spain
National Instruments Spain, S.L.
Europa Empresarial
C/ Rozabella N°. 6
Edf. París, 2° Planta, Oficina N° 8
28230 - Las Rozas, Madrid, Spain
Tel. (34) 1 640 0085
Fax. (34) 1 640 0533
E-mail: ni.spain@natinst.com
Web Address:
http://www.natinst.com/spain

Sweden
National Instruments Sweden AB
Box 2004
Råsundavägen 166
169 02 Solna
Sweden
Tel. 46 8 7304970
Fax. 46 8 7304370
E-mail: ni.sweden@natinst.com
Web Address:
http://www.natinst.com/sweden

Switzerland
National Instruments Switzerland
Sonnenbergstr. 53
CH-5408 Ennetbaden
Switzerland
Tel. 41 56 200 5151
Fax. 41 56 200 5155
E-mail: ni.switzerland@natinst.com
Web Address:
http://www.natinst.com/switzerl

Taiwan
National Instruments Taiwan
7th Fl, No. 135
Keelung Road, Section 2
Taipei, Taiwan
R.O.C.
Tel. 8862 2377 1200
Fax. 8862 2737 4644
Web Address:
http://www.natinst.com/taiwan

United Kingdom
National Instruments Corporation (UK)
Ltd.
21 Kingfisher Court
Hambridge Road
Newbury, Berkshire RG14 5SJ
United Kingdom
Tel. 44 1635 523 545
Fax. 44 1635 523 154
E-mail: info.uk@natinst.com
Web Address:
http://www.natinst.com/uk

Resources to Help You

You have many options for getting help and finding additional information about LabVIEW and virtual instrumentation. Read on to discover the vast array of help and information available.

■ LabVIEW Documentation and Online Help

The LabVIEW manuals provide excellent tutorials to get you up to speed with LabVIEW. They also contain plenty of reference information. You can find the answers to almost all of your questions in these manuals. In addition, LabVIEW has extensive online help to assist you as you build your application.

■ EXAMPLES Directory

The LabVIEW EXAMPLES directory, which is indexed by the Search Examples Help option and ships with the full version of LabVIEW, contains hundreds of LabVIEW programs that you can run as is or modify to suit your application. You can learn a lot about LabVIEW programming just by browsing through these examples and looking at the techniques used. You can get a description of each example by opening it and then selecting **Show VI Info...** from the **Windows** menu.

■ Info-LabVIEW

Info-LabVIEW is a user-sponsored Internet mailing list that you can use to communicate with other LabVIEW users. You can post messages containing questions, answers, and discussions about LabVIEW on this mailing list. These messages will be sent to LabVIEW users worldwide.

If you want to subscribe to Info-LabVIEW, send an e-mail message to info-labview-request@pica.army.mil requesting that your e-mail address be added to the subscription list.

All messages posted to Info-LabVIEW will then be forwarded to your e-mail address. You can cancel your subscription by sending a message to the above address requesting that your e-mail address be removed from the list.

If you want to post a message, send e-mail to `info-labview@pica.army.mil`. Although Info-LabVIEW is user-sponsored, many National Instruments engineering, marketing, and sales personnel read the postings. Your discussions, questions, and product requests will be heard by National Instruments.

■ Fax-on-Demand Information Retrieval System

You can have product data sheets, answers to common questions, technical and application notes, user solutions, and other information sent to you from National Instruments' automatic fax-on-demand information retrieval system. Fax-on-demand is available 24 hours a day, seven days a week, from a touch-tone telephone. Simply call (800) 329-7177 or (512) 418-1111 and follow the instructions provided. You can request indexes of product information, technical support information, and User Solutions; these indexes will list the documents available to you on the automated fax system.

■ World Wide Web

You can access the National Instruments World Wide Web site at `http://www.natinst.com`. The Instrumentation Web site contains up-to-date online information about National Instruments products, services, developer programs, and many other topics.

■ LabVIEW Internet FTP Site

You can connect to the LabVIEW Internet FTP site to receive the following services:

- ■ Software and technical support
- ■ Electronic correspondence
- ■ Common questions
- ■ Technical publications, including application notes
- ■ Software updates and the latest hardware drivers
- ■ Utility VIs for specialized applications
- ■ Free instrument drivers
- ■ Educational VI Exchange

Internet FTP Site	
Address	ftp.natinst.com
Login	anonymous
Password	your e-mail address

■ Educational VI Exchange

The Educational VI Exchange (EVE) gives you access to Lab-VIEW VIs and VI libraries that are used in colleges and universities around the world. The Exchange is an excellent way for users to see new ways to apply LabVIEW in their fields. You can download these VIs to see educational uses of LabVIEW in many disciplines and then run them in your own applications. You can also upload your programs to share with others who can benefit from your ideas.

The Exchange is located on the National Instruments FTP site in the `contrib\labview\education` directory. You can also access it from the worldwide web. Inside the education directory, you will find three folders: `windows`, `mac`, and `all_platforms`. Most VIs should be placed in the `all_platforms` directory. If your VIs contain code interface nodes (CINs) or platform specific code such as Apple Event or dynamic data exchange (DDE) calls, then place the VIs in the appropriate Mac or Windows directory. Files compressed with ZIP and self-extracting archives should also be placed in the appropriate platform-specific directory. Virtual instruments and VI libraries should be transferred in binary mode.

National Instruments encourages you to put VIs written in both the professional version and student edition on the Exchange. They request that you provide a detailed description in the **Show VI Info...** section of your top-level VIs so that others can easily understand what you are doing. You should include the following information:

- Description of the VI
- Course(s) or setting in which you use the VI
- Hardware and other equipment you use
- Whether the VI was written in the professional version or student edition
- Any special instructions or recommendations
- Your name and institution
- Your e-mail address, regular mail address, and phone number (optional)

Also, please upload an accompanying "readme" file containing the same information so that users can browse through your descriptions. The "readme" file should have the same name as your VI or VI library file, with a `.txt` extension.

■ Application and Technical Notes

National Instruments has many application and technical notes available to you at no cost. These notes are available from our Internet site and fax-on-demand system.

■ Technical Support

National Instruments provides top-quality technical support to LabVIEW users. You can get help using e-mail, fax, and of course, the telephone. To make your life easier, be sure to compose a detailed description of the problem before you contact them.

In the U.S. and Canada	
Fax	(512) 794-5678
E-mail	support@natinst.com
Telephone	(512) 795-8248

International
Contact your local branch office

■ National Instruments Customer Education

National Instruments offers hands-on customer education classes worldwide on LabVIEW, DAQ, Signal Processing, Image Processing, GPIB, VXI, and LabWindows/CVI. These courses help you get up to speed quickly. For more information, call National Instruments at (512) 794-0100 in the United States and Canada or contact the local branch office in your country.

■ LabVIEW Basics—Interactive: A Multimedia Training CD-ROM

LabVIEW Basics—Interactive is a self-paced multimedia course on CD-ROM that teaches LabVIEW concepts through an interactive user interface. Experience high quality video and audio effects as a National Instruments applications engineer builds sample VIs on screen. Based on material from the highly acclaimed LabVIEW Basics course, LabVIEW Basics—Interactive progresses from LabVIEW fundamentals and the construction of simple VIs to developing data acquisition and instrument control applications.

If you are new to LabVIEW, you can take advantage of the step-by-step instructions included with each hands-on exercise. If you are already familiar with certain LabVIEW programming concepts, you may choose to advance at a quicker pace. You can pause, review, or skip lessons to suit your level of expertise and learning style. You can also add bookmarks for noteworthy material, so that you (and other users) can exit the course and pick up where you left off. An index feature provides direct access to content using key-word searches.

LabVIEW Basics—Interactive ships on a single CD-ROM that will run on PCs under Windows 95, Windows 3.1 or greater, and Macintosh/Power Macintosh running System 7.1 or greater. LabVIEW version 4.0 or greater must be installed separately.

■ Training Videos

You can learn LabVIEW at your own pace using an instructional video available from National Instruments. This two-hour video teaches engineers and scientists how to use LabVIEW to develop data acquisition and control applications. For more information, call National Instruments at (512) 794-0100 in the United States and Canada or contact your local branch office.

■ National Instruments Catalogue

The National Instruments *Instrumentation Catalogue* contains detailed information about all National Instruments hardware and software products.

■ LabVIEW Graphical Programming

LabVIEW Graphical Programming, Second Edition, is a book about LabVIEW written by an avid and versatile LabVIEW user of many years, Gary Johnson. It contains valuable information on LabVIEW, data acquisition, instrument control, and specific application configurations, presented in a user-friendly manner. To order LabVIEW Graphical Programming, contact McGraw-Hill at (800) 822-8158 and reference ISBN# 0-07-032915-X. Outside the United States, call (717) 794-2191.

■ LabVIEW for Everyone

LabVIEW for Everyone: Graphical Programming Made Even Easier by Lisa Wells of National Instruments and Jeffrey Travis of VI Technology (Austin, Texas) is a text that helps beginning to experienced LabVIEW users learn the skills needed to implement fast and efficient graphical programs in a wide variety of applications. The book contains clear definitions, numerous photos and illustrations, and step-by-step tutorials to give readers a feel for what LabVIEW can do. It includes a complete introduction to the basic structures of LabVIEW programs, as well as more advanced topics such as programming techniques, data acquisition theory and programming, and instrument control. To order *LabVIEW for Everyone: Graphical Programming Made Even Easier,* contact Prentice Hall at (800) 947-7700 and reference ISBN# 013-268194-3. Readers can also access the Prentice Hall web site at www.prenhall.com

■ LabVIEW Power Programming

LabVIEW Power Programming, a new text by *LabVIEW Graphical Programming* author, Gary Johnson, is a compilation of advanced graphical programming techniques used by leading programmers in industry and education. The text details the power and flexibility of graphical programming as embodied in LabVIEW, and offers its readers insight into new and innovative ways of leveraging this powerful software. Scientists and engineers will enjoy reading about LabVIEW applications in space, uses of LabVIEW in computer science, LabVIEW-based cryptography, and more. To order *LabVIEW Power Programming*, contact McGraw-Hill at (800) 722-4726 and reference ISBN# 0-07-913666-4. Readers can also access the McGraw-Hill web site at `www.ee.mcgraw-hill.com`

■ Seminars

National Instruments regularly holds seminars around the world. The seminars are designed to keep you abreast of the latest technologies and trends in measurement and instrumentation. The material is usually presented in a hands-on format, with equipment donated by leading instrumentation and computer manufacturers.

■ Instrupedia™ 97

Instrupedia™ 97, the encyclopedia of instrumentation, is an interactive and searchable reference CD-ROM that you can request from National Instruments for no charge. It includes a glossary of common instrumentation terms, a summary of books on instrumentation and measurement, collections of actual LabVIEW user applications, and application notes covering specific topics in GPIB, VXI, DAQ, data analysis, and software. Instrupedia 97 also features National Instruments' online catalog, a technical support reference library, demonstration programs, and configuration-advising programs that help you design your system. It's available for Windows platforms only.

Instrupedia™ 97

■ Software Showcase

The National Instruments Software Showcase is a CD-ROM that provides one-stop shopping for anyone interested in demos of their software. The Software Showcase will teach users about the concept of virtual instrumentation and will allow them to investigate the virtual instrumentation software tools of their choice. Available for Windows 3.1, Windows 95, WindowsNT, Macintosh, and Power Macintosh, the Software Showcase will also discuss existing user applications, common technical uses, and available complementary products. To request one, contact National Instruments.

■ The Instrumentation Newsletter

The *Instrumentation Newsletter*, produced quarterly by National Instruments, features User Solutions along with updates on the latest products for instrumentation. Academic, Radio Frequency/Microwave, Automotive, and international versions are available. You can request this free newsletter by contacting National Instruments.

■ LabVIEW Technical Resource

LabVIEW Technical Resource is a quarterly newsletter that provides technical information for LabVIEW systems developers. It offers solutions to common problems, programming tips, tools, and techniques, and each issue comes with a disk containing LabVIEW programs. For more information about LabVIEW Technical Resource, call (214) 827-9931 or write to LTR Publishing, 5614 Anita, Dallas, TX 75206.

■ National Instruments Alliance Members

National Instruments' Alliance Program is comprised of several hundred third-party companies that provide consulting services, turnkey solutions, and valuable add-on products. Located all over the world, Alliance members can help you design a system or write a specific application to meet your needs. For a complete listing of Alliance member companies and their services, refer to our web site at http://www.natinst.com or request a *Solutions* guide from National Instruments.

LabVIEW Add-On Toolkits

You can purchase the following special add-on toolkits from National Instruments to increase LabVIEW's functionality. In addition, new toolkits are created frequently, so if you have a particular goal, it's worthwhile to check and see if a toolkit already exists to accomplish it. If one doesn't exist, National Instruments will appreciate your suggestions on what to develop in the future!

Many third-party developers, often Alliance Members, also make add-ons to LabVIEW to do all sorts of things. We suggest posting to the info-labview user forum (described in the previous section of this appendix) if you have a specific task and you want to know if someone's already done it.

■ Application Builder

The LabVIEW Application Builder is an add-on package for creating stand-alone, executable applications. When accompanied by the Application Builder, a LabVIEW system can create VIs that operate as stand-alone applications. You can run the executable file but cannot edit it. Stand-alone applications minimize RAM and disk requirements by saving only those resources needed for execution. The application contains the same execution resources as those in the LabVIEW Full Development System, so VIs will execute at the same high-performance rates.

■ JTFA Toolkit

The Joint Time-Frequency Analysis (JTFA) Toolkit is a LabVIEW add-on library and stand-alone executable that enhances the signal processing capabilities of LabVIEW software on nonstationary signals. It is useful for precise signal analysis of data whose frequency content changes with time. The toolkit features the award-winning and patented Gabor Spectrogram, which is ideal in applications of speech processing, sound analysis, sonar, radar, vibration analysis, and dynamic signal monitoring.

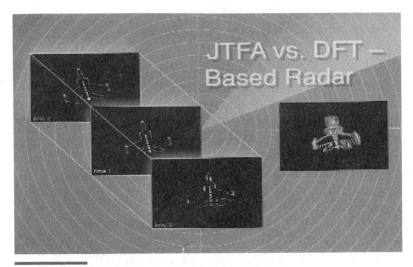

The Joint Time-Frequency Analysis (JTFA) Toolkit

■ Digital Filter Design Toolkit

The Digital Filter Design Toolkit is a ready-to-run virtual instrument for interactively designing Finite Impulse Response (FIR) filters and Infinite Impulse Response (IIR) filters. Outputs include pole zero plots, magnitude and phase plots, impulse response, and step response.

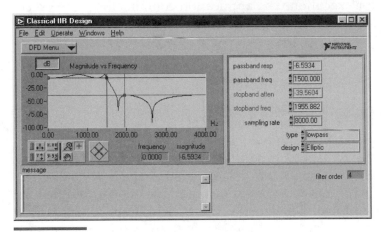

The Digital Filter Design Toolkit

■ Third Octave Analyzer Toolkit

The Third Octave Analyzer Toolkit, combined with National Instruments DAQ hardware, is a ready-to-run virtual instrument. The Third Octave Analyzer provides a standard instrument interface used widely in acoustics and vibration analysis. The toolkit includes LabVIEW source code so that you can make modifications for custom applications.

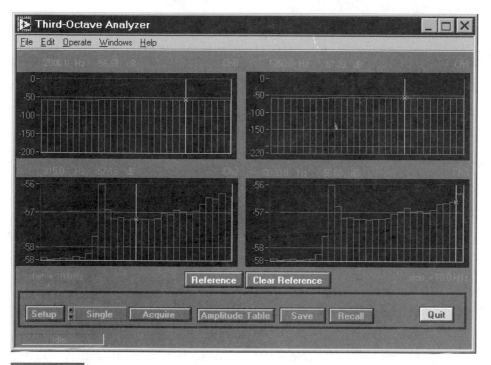

The Third Octave Analyzer Toolkit

■ Wavelet and Filter Bank Toolkit

The Wavelet and Filter Bank Toolkit is a multipurpose LabVIEW and BridgeVIEW add-on package for wavelet and filter bank design and analysis. It also includes a ready-to-run application for designing wavelet and filter bank coefficients that can be used with other software applications, such as LabWindows/CVI, C, and Visual Basic.

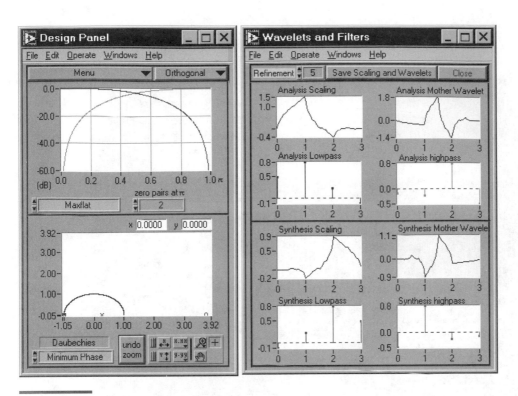

The Wavelet and Filter Bank Toolkit

■ Virtual Bench/DSA Toolkit

The Virtual Bench DSA measures total harmonic distortion (THD), harmonic content, frequency response, impulse response, power spectrum, amplitude spectrum, coherence, and cross power spectrum. Built-in windowing functions used to reduce spectral leakage include Hanning, Flat top, Hamming, Blackman-Harris, and Exact Blackman.

■ Signal Processing Suite

The Signal Processing Suite gives users ready-to-run signal processing capabilities. Scientists and engineers can use the JTFA Toolkit, the Digital Filter Design Toolkit, the Virtual Bench/DSA

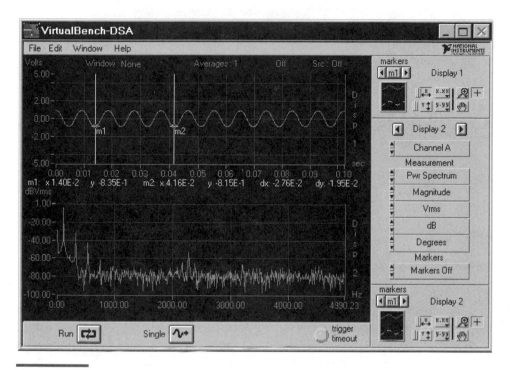

The Virtual Bench Dynamic Signal Analyzer

dynamic signal analyzer, the Third Octave Analyzer Toolkit, and the Wavelet and Filter Bank Toolkit in applications such as acoustics, analog telephony, radar, seismology, remote sensing, and vibration analysis.

■ PID Control Toolkit

The PID Control Toolkit adds sophisticated control algorithms to LabVIEW. With this package, you can quickly build data acquisition and control systems for your own control application. By combining the PID Control Toolkit with the math and logic functions in LabVIEW, you can quickly develop programs for control.

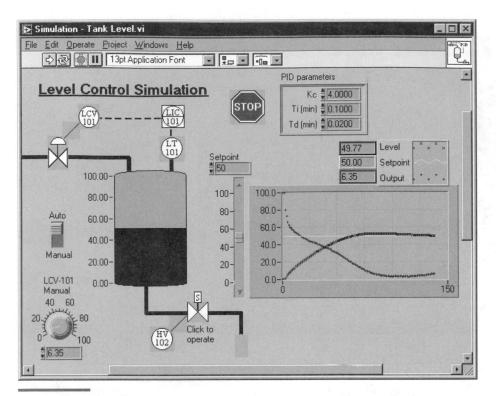

The PID Control Toolkit

■ Control and Simulation Toolkit

The Control and Simulation Toolkit is LabVIEW/BridgeVIEW add-on softwae for simulating, designing, analyzing, and optimizing linear and nonlinear control systems. You can use this toolkit to accelerate your system design by using VIs modeled after symbols and blocks used in control engineering. Plus you can immediately integrate National Instruments DAQ hardware into your control system after you have designed and simulated your system.

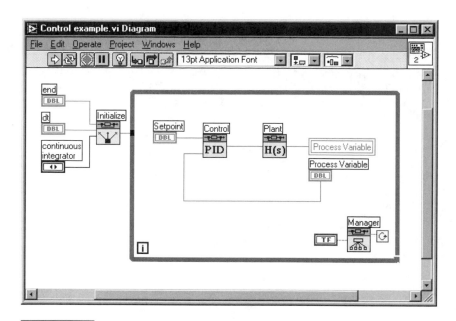

The Control and Simulation Toolkit

■ Fuzzy Logic Control Toolkit

The Fuzzy Logic Control Toolkit is application software for designing fuzzy logic control systems for LabVIEW or BridgeVIEW. In addition to helping you design your control system, the point-and-click software includes functions for implementing your fuzzy control system in either LabVIEW or BridgeVIEW. Fuzzy Logic can be combined with NI-DAQ, PID Control Toolkit, and the Control and Simulation Toolkit for advanced control applications.

■ G Math Toolkit

The G Math Toolkit is a multipurpose LabVIEW and Bridge-VIEW add-on package for math, data analysis, and data visualization. The math toolkit consists of numerical recipes in G which let you solve advanced mathematical problems using the power and speed of graphical programming. This toolkit includes more than 100 high-level math VIs for ordinary differential equations (ODEs), root solving, optimization, integration, differentiation, transforms, and function evaluation.

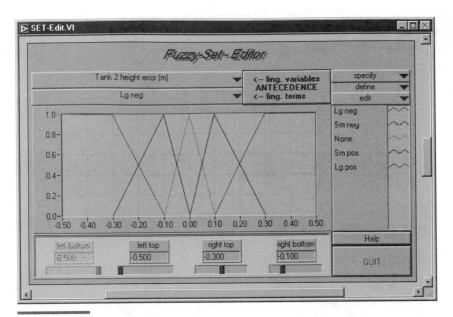

The Fuzzy Logic Toolkit

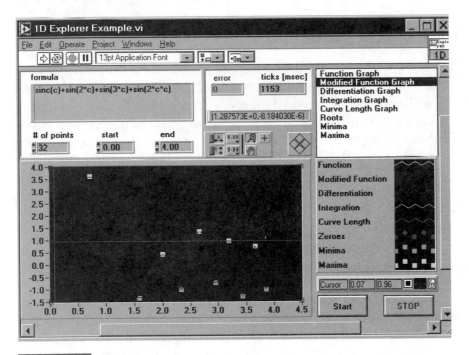

The G Math Toolkit

Glossary

Prefix	Meaning	Value
M-	mega-	10^6
K-	kilo-	10^3
m-	milli-	10^{-3}
μ-	micro-	10^{-6}
n-	nano-	10^{-9}

Numbers/Symbols

1D One-dimensional.

2D Two-dimensional.

∞ Infinity.

π Pi.

Δ Delta. Difference. Δx denotes the value by which x changes from one index to the next.

AC   Alternating current.

ADC   Analog-to-digital converter, the hardware that converts an analog signal into a digital signal.

analog signal   A signal whose values are continuous and are defined at each and every instant of time.

ANSI   American National Standards Institute.

antialiasing filter   An analog lowpass filter used to limit the frequency content of the analog signal, before sampling, to less than the Nyquist frequency (half the sampling rate). This prevents aliasing in the digital domain.

array   Ordered, indexed set of data elements of the same type.

asymptotic stability   A system whose response to an arbitrary set of initial conditions with zero input tends towards zero.

bandpass filter   A filter that passes signals within a certain band of frequencies.

bandstop filter   A filter that attenuates signals within a certain band of frequencies.

BPF   Bandpass filter.

BridgeVIEW   An application software package for developing human machine interface (HMI) and supervisory control and data acquisition (SCADA) solutions for manufacturing and process control applications. It leverages National Instruments' patented graphical programming language G.

BSF   Bandstop filter.

Butterworth filter   A filter characterized by a smooth response at all frequencies, and the absence of ripple in both the passband and the stopband. Due to the absence of ripple it is also known as a maximally flat filter.

Chebyshev filter   A filter characterized by ripple in the passband, but a smooth response in the stopband.

Chebyshev II filter   *See* inverse Chebyshev filter.

coherent gain (CG)   The coherent gain is the zero frequency gain (or DC gain) of a window.

complex conjugate transpose   A matrix operation consisting of taking the complex conjugate of each element of the matrix and then transposing the resulting matrix.

complex matrix   A matrix with at least one element that is a complex number.

condition number   A measure of how close the matrix is to being singular. The condition number of a square nonsingular matrix is defined as

$$\text{cond}(\mathbf{A}) = \|\mathbf{A}\|_p \cdot \|\mathbf{A}^{-1}\|_p$$

where p corresponds to the pth norm of the matrix. The condition number can vary between 1 and infinity. A matrix with a large condition number is nearly singular, while a matrix with a condition number close to 1 is far from being singular.

continuous system   A system whose behavior is defined at every instant of time.

controllability   A system property that guarantees that any unconstrained control input can transfer the state of that system from any initial state to any other state.

curve fitting   Technique for extracting a set of curve parameters or coefficients from a data set to obtain a functional description of the data set.

DAQ   Data acquisition.

data acquisition   Process of acquiring data, typically from A/D or digital input plug-in boards.

dB   Decibels.

DC   Direct current.

decibel   A logarithmic scale used to compress large amplitudes and expand small amplitudes. It is given by

$$\text{one dB} = 10 \log_{10} (\text{Power Ratio}) = 20 \log_{10} (\text{Voltage Ratio})$$

DFT   Discrete Fourier transform, the algorithm used to transform samples of the data from the discrete time domain into the discrete frequency domain.

difference equations   Equations that describe the operation of a system (for example, a filter) in the discrete time domain.

differential equations   Equations which involve derivatives of the variables and describe the operation of a system in the continuous time domain.

digital signal   A signal that can take on only specific amplitude values that are defined at discrete points in time.

discrete-time signal   A signal whose values are continuous in amplitude, but which is defined only at discrete points in time.

DSP   Digital signal processing.

elliptic filter   A filter characterized by ripple in both the passband and the stopband.

equivalent noise bandwidth ENBW   The equivalent noise bandwidth of a window is the width of an ideal rectangular response that will pass the same amount of noise power as the frequency response of the window.

FFT   Fast Fourier transform. A fast method for calculating the discrete Fourier transform. It is used when the number of samples is a power of two.

filter bank   A set of filters connected in parallel. Each filter may be tuned to a different frequency range. The filters in the filter bank may or may not have the same bandwidth.

FIR filter   Finite impulse response filter. A type of filter whose output depends only on the current and past inputs. It is also known as a nonrecursive filter.

forward coefficients   Forward coefficients are those that multiply inputs.

frequency response   A measure of how a signal or system varies with frequency. It consists of two parts, the magnitude response and the phase response. The magnitude response is the magnitude of the frequency response at different frequencies, whereas the phase response is the phase of the frequency response at different frequencies. (*See* impulse response.)

G   National Instruments' patented graphical programming language

Gaussian probability density function   A density function that is completely characterized by its mean and standard deviation and is given by

$$f(x) = \frac{1}{\sigma\sqrt{2\pi}}\exp\left[-\frac{1}{2}\left(\frac{x-\mu}{\sigma}\right)^2\right]$$

where μ is the mean and σ is the standard deviation. It is also known as the Normal probability density function.

harmonic distortion   The distortion inherent in a nonlinear system that results in generation of frequencies at its output that are harmonics of the input frequency. The more the degree of nonlinearity of the system, the higher the frequencies of the harmonics.

Hermitian matrix   A complex matrix whose complex conjugate transpose is equal to the matrix itself. (*See also* complex conjugate transpose.)

highpass filter   A filter that passes frequencies above a certain cutoff frequency. It passes high frequencies but attenuates low frequencies.

HiQ   An interactive problem-solving environment using a notebook methodology that combines interactive analysis, data visualization, and report generation in a single, intuitive environment.

HPF   Highpass filter.

Hz   Hertz, or cycles per second.

IEEE   Institute of Electrical and Electronics Engineers.

IIR filter   Infinite impulse response filter. A type of filter whose output depends not only on the current and past inputs, but also on past outputs. It is also known as a recursive filter.

impulse   A signal that has a value of one at a particular time instant and zero everywhere else.

impulse response   The time domain response of a system with an input that is an impulse.

inf   Digital display value for a floating-point representation of infinity.

infinitesimal magnitude   Very very small magnitude.

in place   In place means that the input array space (memory locations) are being reused as the output array space. In place usually implies lower demands on memory.

inverse Chebyshev filter   A filter characterized by a smooth response in the passband, but with ripple in the stopband.

I/O   Input/output. The transfer of data to or from a computer system involving communications channels, operator input devices, and/or data acquisition and control interfaces.

joint time-frequency analysis (JTFA)   A method of analysis that simultaneously provides both time and frequency information. It shows how the frequency spectrum of a signal varies with time.

LabWindows/CVI   An integrated ANSI C environment designed for engineers and scientists creating virtual instrument applications.

Levenberg-Marquardt   A general curve fitting algorithm used to estimate the coefficients of a curve to fit a set of samples. It can be used for both linear and nonlinear relationships but is almost always used to fit a nonlinear curve.

linear phase   Linear phase in digital filters means that the phase distortion is nothing more than a digital delay. All input samples will be shifted by some constant number of samples, so this phase change can be easily compensated and/or modeled. Nonlinear phase means that the individual sine waves at different frequencies that make up the input signal get shifted in time by different amounts. This sort of phase change is very difficult to work around. Some signals (like the square wave) are very sensitive to this sort of phase distortion.

logarithmic plot   A plot in which one of the axes is represented in the log scale.

lower triangular matrix   A matrix whose elements above the main diagonal are all zero.

lowpass filter   A filter that passes frequencies below a certain cutoff frequency. It passes low frequencies but attenuates high frequencies.

LPF   Lowpass filter.

LU decomposition   A method that factors a matrix as a product of an upper and a lower triangular matrix.

matrix   Two-dimensional array.

MB   Megabytes of memory.

MSE   Mean squared error. The MSE is a relative measure of the residuals between the expected curve values and the actual observed values.

multivariable system   A system with many inputs and many outputs.

NaN   Digital display value for a floating-point representation of *not a number*, typically the result of an undefined operation, such as $\log(-1)$.

norm   The norm of a vector or matrix is a measure of the magnitude of the vector or matrix. There are different ways to compute the norm of a matrix. These include the *2-norm* (Euclidean norm), the *1-norm*, the *Frobenius norm* (F-norm), and the *Infinity norm*(inf-norm). Each norm has its own physical interpretation. The LabVIEW/BridgeVIEW **Matrix Norm** VI can be used to compute the norm of a matrix.

Normal probability density function   *See* Gaussian probability density function.

nonlinear phase   *See* linear phase.

nonrecursive filter   *See* FIR filter.

nonsingular matrix   Matrix in which no row or column is a linear combination of any other row or column, respectively.

normalized frequency   A frequency that is specified as a ratio of frequency (in Hertz) with respect to the sampling frequency (in samples/second). Its units are in cycles/sample. However, if the frequency is given in terms of cycles, then the normalized frequency is given by the ratio of frequency (in cycles) to the number of samples.

Nyquist frequency (f_N)   Half the sampling frequency, $f_N = f_s/2$, where f_s is the sampling frequency.

Nyquist stability criteria   A criteria to determine the stability of a system.

Nyquist theorem   A theorem stating that to recover an analog signal from its samples, the sampling frequency should be at least twice the highest frequency in the signal.

observability   A system property that guarantees that the output of the system possesses a component due to each state variable.

observation matrix (H)   A matrix used as an input to the **General LS Linear Fit** VI. If there are N data points, and k coefficients (a_0, a_1, ..., a_{k-1}) for which to solve, H will be an $N\,k$ matrix with N rows and k columns. Thus, the number of rows of H is equal to the number of data points, whereas the number of columns of H is equal to the number of coefficients for which we are trying to solve.

octave   A doubling in frequency.

ODE   Ordinary Differential Equation.

one-sided transform   A representation consisting of only the positive frequency (and DC) components.

passband   The range of frequencies that are passed by a filter with a gain of almost one (0 dB).

passband ripple   The amount of variation of the passband gain from unity (0 dB). It is usually specified in dB. It is given by

$$\text{ripple (dB)} = 20\log_{10}\frac{A_o(f)}{A_i(f)}$$

where $A_i(f)$ and $A_o(f)$ are the amplitudes of a particular frequency before and after the filtering, respectively.

pattern VIs   These are signal generation VIs that do not keep track of the phase of the signal that they generate each time they are called.

poles   Values for which the transfer function equals infinity.

pole-zero plot   A plot showing the positions of the poles and zeros of a system. The pole-zero plot is useful in determining the stability of the system.

power spectrum   The power spectrum of a signal gives you the power in each of its frequency components. It can be calculated by squaring the magnitude of the Fourier transform (DFT or FFT) of the signal.

quality factor (Q)   A measure of how selective a bandpass filter is in passing frequencies around the center frequency and attenuating unwanted frequencies. It is defined as the ratio of the center frequency of the filter to its bandwidth.

$$Q = f_m/B_m$$

where f_m is the center frequency, and B_m is the bandwidth of the filter. Thus, for a fixed center frequency, the larger the bandwidth the smaller the quality factor, and vice versa.

rank   The *rank* of a matrix **A**, denoted by $\rho(\mathbf{A})$, is the maximum number of linearly independent columns in **A**. The number of linearly independent columns of a matrix is equal to the number of independent rows. So, the rank can never be greater than the smaller dimension of the matrix. Consequently, if **A** is an n m matrix, then

$$\rho(\mathbf{A}) \le \min(n, m)$$

where min denotes the minimum of the two numbers. The rank of a square matrix pertains to the highest order nonsingular matrix that can be formed from it. So, the rank pertains to the highest order matrix that we can obtain whose determinant is not zero. A square matrix is said to have full rank if and only if its determinant is different from zero.

recursive filter   *See* IIR filter.

regression analysis   *See* curve fitting.

reverse coefficients   The reverse coefficients are those that multiply the outputs.

RHS   Right-hand side.

ripple   A measure of the deviation of a filter from the ideal filter specifications.

RMS   Root mean square.

sampling frequency   The number of samples acquired per second. Its units are samples/second.

short-time Fourier transform (STFT)   The term for taking a Fourier transform of shorter time intervals of samples of a signal, rather than on the entire set of samples. Also known as the windowed Fourier transform.

single-variable system   A system with one input and one output.

singular value decomposition (SVD)   A method that decomposes a matrix into the product of three matrices $A = USV^T$ where U and V are orthogonal matrices, and S is a diagonal matrix. SVD is useful for solving analysis problems such as computing the rank, norm, condition number, and pseudoinverse of matrices.

spectral leakage   A phenomenon where it appears as if energy has leaked out from one frequency into another. It occurs because of the discontinuities introduced when the sampled waveform does not contain an integral number of cycles of the original signal. The larger the discontinuity, the more the leakage. Leakage can be reduced by reducing the amplitude of the discontinuities. The reduction is achieved by multiplying the time-domain waveform by a window function. Note that if there are an integer number of cycles in the sampled waveform, there is no leakage.

state of a system   A set of numbers such that the knowledge of these numbers and the input function will, with the equations describing the dynamics, provide the future behavior of the system.

state-space representation   A representation which depicts the internal dynamics of the system.

stability   A property which guarantees that the system will operate correctly under favorable conditions.

step response   The response of a system to a step input.

stopband   The range of frequencies that are attenuated by the filter.

stopband attenuation   The amount of attenuation in the stopband of a filter, usually specified in dB. It is given by

$$A(\mathrm{dB}) = 20\log_{10}\frac{A_o(f)}{A_i(f)}$$

where $A_i(f)$ and $A_o(f)$ are the amplitudes of a particular stopband frequency before and after the filtering, respectively.

SVD   Singular value decomposition.

symmetric matrix   A matrix whose transpose is equal to the matrix itself.

time-invariant system   A system whose input-output response characteristics do not vary with time.

total harmonic distortion   A relative measure of the amplitudes of the fundamental to the amplitudes of the harmonics. If the amplitude of the fundamental is A_1, and the amplitudes of the harmonics are A_2 (2nd harmonic), A_3 (3rd harmonic), A_4 (4th harmonic), ... A_N (Nth harmonic), then the Total Harmonic Distortion (THD) is given by

$$\text{THD} = \frac{\sqrt{A_2^2 + A_3^2 + ... + A_N^2}}{A_1}$$

When the THD is expressed as a percentage, it is known as the Percentage Total Harmonic Distortion (%THD) and is given by

$$\% \, \text{THD} = \frac{100 \times \sqrt{A_2^2 + A_3^2 + ... + A_N^2}}{A_1}$$

transfer function   A mathematical description relating the input and output of the system.

transient response   The transient that initially appears at the output of a filter when the filter is run for the first time.

transition region   The region between the passband and the stopband where the gain of the filter varies from one (0 dB) or almost one (in the passband) to a very small value (in the stopband).

transpose   A matrix operation that consists of interchanging the rows and columns of a matrix.

two-sided transform   A representation consisting of both the positive and negative (and DC) frequency components.

upper triangular matrix   A matrix whose elements below the main diagonal are all zero.

vector   One-dimensional array.

wave VIs   Signal generation VIs that keep track of the phase of the signal that they generate each time they are called.

window   A smoothing function applied to a time-domain waveform, before it is transformed into the frequency domain, so as to minimize spectral leakage.

windowed Fourier transform   *See* short-time Fourier transform (STFT).

zeros   Values at which the transfer function equals zero.

zero padding   Addition of zeros to the end of a sequence so that the total number of samples is equal to the next higher power of two. When zero padding is applied to a set of samples in the time domain, faster computation is possible by using the FFT instead of the DFT. In addition, the frequency resolution (Δf) is improved (made smaller) because $\Delta f = f_s/N$, where f_s is the sampling frequency and N is the total number of samples.

zero-pole plot   See pole-zero plot.

Index

E

T

U

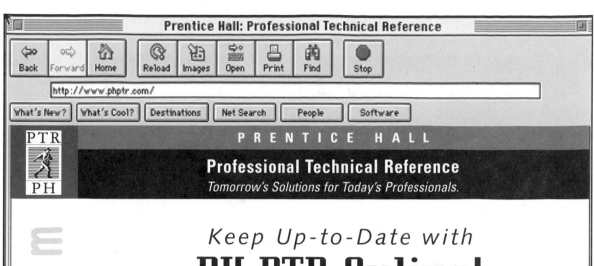

National Instruments Corporation
6504 Bridge Point Parkway
Austin, TX 78730-5039 USA
Tel: (512) 794-0100
E-mail: info@natinst.com
FTP Site: ftp.natinst.com
Web Address: http://www.natinst.com

Telephone Numbers of International Offices:

Australia	03 9 879 9422	Italy	02 413091
Austria	0662 45 79 90 0	Japan	03 5472 2970
Belgium	02 757 00 20	Korea	02 596 7456
Canada (Ontario)	905 785 0085	Mexico	95 800 010 0793
Canada (Quebec)	514 694 8521	Netherlands	0348 433466
Denmark	45 76 26 00	Norway	32 84 84 00
Finland	90 527 2321	Singapore	2265886
France	1 48 14 24 24	Spain	91 640 0085
Germany	089 741 31 30	Sweden	08 730 49 70
Hong Kong	2645 3186	Switzerland	056 200 51 51
Israel	03 573 4815	Taiwan	02 377 1200
		U.K.	01635 523545

8. **LIMITED WARRANTY AND DISCLAIMER OF WARRANTY:** The Company warrants that the SOFTWARE, when properly used in accordance with the Documentation, will operate in substantial conformity with the description of the SOFTWARE set forth in the Documentation. The Company does not warrant that the SOFTWARE will meet your requirements or that the operation of the SOFTWARE will be uninterrupted or error-free. The Company warrants that the media on which the SOFTWARE is delivered shall be free from defects in materials and workmanship under normal use for a period of thirty (30) days from the date of your purchase. Your only remedy and the Company's only obligation under these limited warranties is, at the Company's option, return of the warranted item for a refund of any amounts paid by you or replacement of the item. Any replacement of SOFTWARE or media under the warranties shall not extend the original warranty period. The limited warranty set forth above shall not apply to any SOFTWARE which the Company determines in good faith has been subject to misuse, neglect, improper installation, repair, alteration, or damage by you. EXCEPT FOR THE EXPRESSED WARRANTIES SET FORTH ABOVE, THE COMPANY DISCLAIMS ALL WARRANTIES, EXPRESS OR IMPLIED, INCLUDING WITHOUT LIMITATION, THE IMPLIED WARRANTIES OF MERCHANTABILITY AND FITNESS FOR A PARTICULAR PURPOSE. EXCEPT FOR THE EXPRESS WARRANTY SET FORTH ABOVE, THE COMPANY DOES NOT WARRANT, GUARANTEE, OR MAKE ANY REPRESENTATION REGARDING THE USE OR THE RESULTS OF THE USE OF THE SOFTWARE IN TERMS OF ITS CORRECTNESS, ACCURACY, RELIABILITY, CURRENTNESS, OR OTHERWISE.

IN NO EVENT, SHALL THE COMPANY OR ITS EMPLOYEES, AGENTS, SUPPLIERS, OR CONTRACTORS BE LIABLE FOR ANY INCIDENTAL, INDIRECT, SPECIAL, OR CONSEQUENTIAL DAMAGES ARISING OUT OF OR IN CONNECTION WITH THE LICENSE GRANTED UNDER THIS AGREEMENT, OR FOR LOSS OF USE, LOSS OF DATA, LOSS OF INCOME OR PROFIT, OR OTHER LOSSES, SUSTAINED AS A RESULT OF INJURY TO ANY PERSON, OR LOSS OF OR DAMAGE TO PROPERTY, OR CLAIMS OF THIRD PARTIES, EVEN IF THE COMPANY OR AN AUTHORIZED REPRESENTATIVE OF THE COMPANY HAS BEEN ADVISED OF THE POSSIBILITY OF SUCH DAMAGES. IN NO EVENT SHALL LIABILITY OF THE COMPANY FOR DAMAGES WITH RESPECT TO THE SOFTWARE EXCEED THE AMOUNTS ACTUALLY PAID BY YOU, IF ANY, FOR THE SOFTWARE.
SOME JURISDICTIONS DO NOT ALLOW THE LIMITATION OF IMPLIED WARRANTIES OR LIABILITY FOR INCIDENTAL, INDIRECT, SPECIAL, OR CONSEQUENTIAL DAMAGES, SO THE ABOVE LIMITATIONS MAY NOT ALWAYS APPLY. THE WARRANTIES IN THIS AGREEMENT GIVE YOU SPECIFIC LEGAL RIGHTS AND YOU MAY ALSO HAVE OTHER RIGHTS WHICH VARY IN ACCORDANCE WITH LOCAL LAW.

ACKNOWLEDGMENT
YOU ACKNOWLEDGE THAT YOU HAVE READ THIS AGREEMENT, UNDERSTAND IT, AND AGREE TO BE BOUND BY ITS TERMS AND CONDITIONS. YOU ALSO AGREE THAT THIS AGREEMENT IS THE COMPLETE AND EXCLUSIVE STATEMENT OF THE AGREEMENT BETWEEN YOU AND THE COMPANY AND SUPERSEDES ALL PROPOSALS OR PRIOR AGREEMENTS, ORAL, OR WRITTEN, AND ANY OTHER COMMUNICATIONS BETWEEN YOU AND THE COMPANY OR ANY REPRESENTATIVE OF THE COMPANY RELATING TO THE SUBJECT MATTER OF THIS AGREEMENT.

Should you have any questions concerning this Agreement or if you wish to contact the Company for any reason, please contact the publisher, in writing at the address below.

Robin Short
Prentice Hall PTR
One Lake Street
Upper Saddle River, New Jersey 07458

About the CD

In this CD-ROM You Will Find

The CD-ROM included with this book contains an evaluation version of LabVIEW, additional software for the toolkits discussed in this book, activities, and solutions to the activities. The CD-ROM also contains software associated with the real-world applications in Chapters 5 (Measurement), 6 (Digital Filtering), 8 (Linear Algebra), and 10 (Control Systems). If you do not have the full version of LabVIEW already installed on your computer, you can use the evaluation version to work through all the activities in Chapters 1-9. The CD-ROM only contains a portion of the Control and Simulation Software for LabVIEW, the Digital Filter Design Toolkit, and the G Math Toolkit needed to work through the activities in Chapters 10, 11, and 12 respectively.

The folder structure on the CD-ROM is as follows:

```
Activities and Solutions
      Activities.llb
      Solutions.llb

Additional Software
      Digital Filter Design Toolkit
      Controls.llb
      GMath.llb

LabVIEW

Real-World Applications
```

How to Install the Software

1. Install the LabVIEW evaluation software on your computer. To install this software, run the `setup.exe` program from the `LabVIEW` folder on the CD-ROM. Follow the instructions on the screen. If you already have the full version of LabVIEW installed, you do not need to install the evaluation version.

2. Create an `Addons` folder in your LabVIEW `Vi.lib` directory, if it does not already exist.

3. Copy `Controls.llb` and `GMath.llb` from the `Additional Software` folder on the CD-ROM to the `Addons` directory.

4. Copy the `Digital Filter Design Toolkit` folder from the `Additional Software` folder on the CD-ROM to your hard drive.

5. Copy the `Real-World Applications` folder from the CD-ROM to your `LabVIEW` folder. This folder contains software corresponding to the real-world applications in Chapters 5, 6, 8, and 10.

6. Copy the contents of the `Activities and Solutions` folder from the CD-ROM to your `LabVIEW` folder. This folder contains the libraries `Activities.llb` and `Solutions.llb`. While performing the activities in this book, you will open and/or save VIs in `Activities.llb`. The solutions to the activities can be found in `Solutions.llb`. You will have to change the access permissions for both these libraries. To do so, follow the instructions below:
   a) Right click on either `Activities.llb` or `Solutions.llb`.
   b) Select Properties, then select General.
   c) Deselect the Read-only attribute.
7. Repeat Steps 6a, 6b, and 6c for the libraries `Controls.llb` and `GMath.llb` in the `Vi.lib >> Addons` folder.

Read the contents of the **readme.txt** file on the CD-ROM extremely carefully.

Restrictions of the LabVIEW Evaluation Version

To run the evaluation version, launch the `Labview.exe` program from the folder in which you installed LabVIEW. Select the `Exit to LabVIEW` button in the lower right corner. This opens a window which gives you general information about LabVIEW and National Instruments. After reading this information, click on the OK button. This will then open the LabVIEW window. You can access all the features of the full version in the evaluation version with some restrictions. See the `readme.txt` file on the CD-ROM for information about these restrictions and minimum hardware and memory requirements.

Performing the Activities

1. While performing the activities in this book, you will open and/or save VIs in the library `Activities.llb`. The solutions to the activities can be found in the library `Solutions.llb`. After completing the installation instructions as outlined above, both these libraries can be found in your `LabVIEW` folder.

2. As shown in the example below,

 Sine Wave VI (Analysis >> Signal Generation subpalette)

in an activity, when you come across a VI icon followed by the name of the VI (**Sine Wave VI**) and its location (**Analysis >> Signal Generation** subpalette) in the **Functions** palette, choose the particular VI from the specified subpalette and drop it on the block diagram. Then follow the instructions in the activity to complete the block diagram.